AF545060

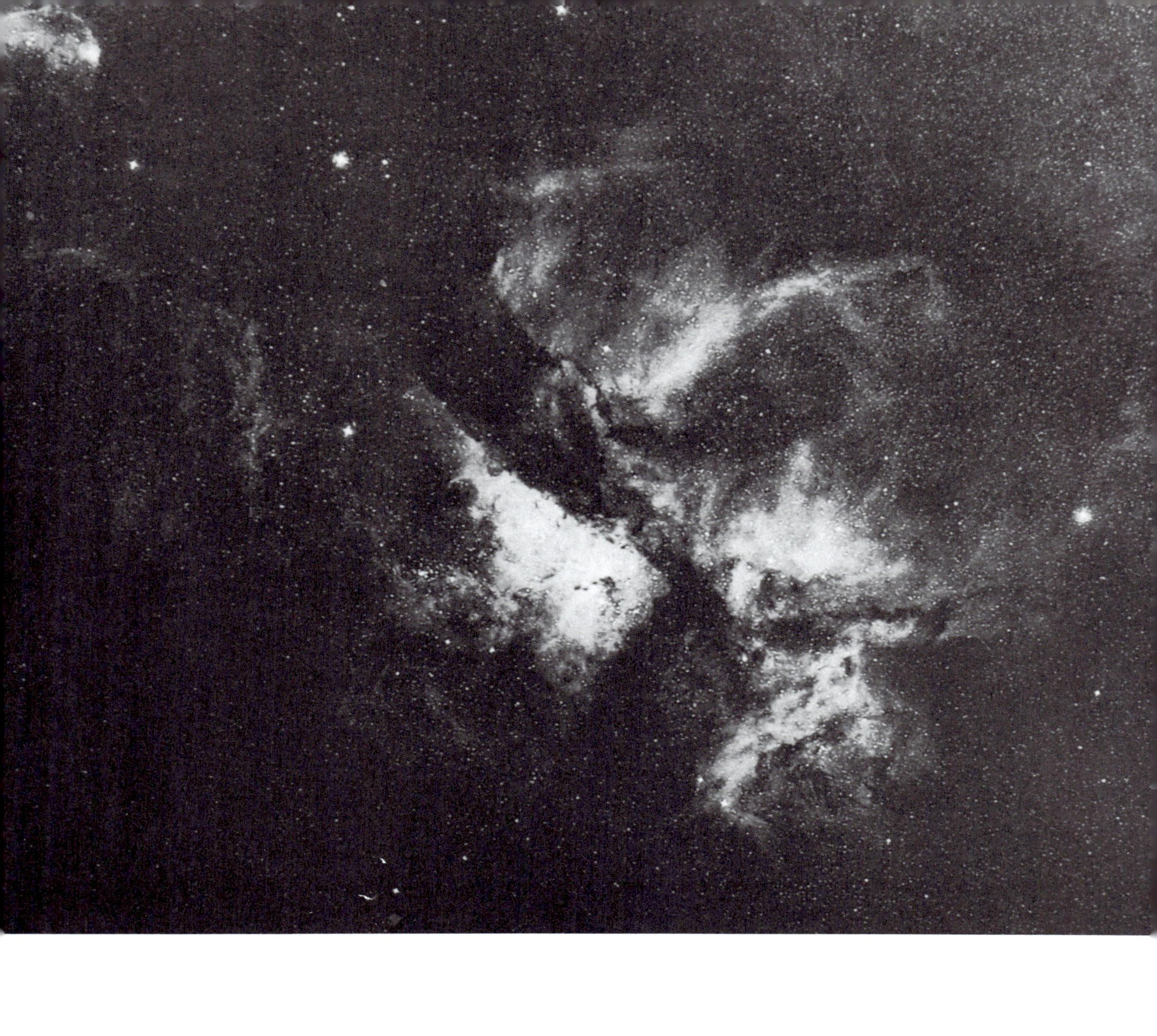

ORIGIN AND EVOLUTION OF THE UNIVERSE

Joy and amazement at the beauty and grandeur of this world of which man can just form a faint notion.

Albert Einstein
(1879–1955)

ORIGIN AND EVOLUTION OF THE UNIVERSE

G. Siegfried Kutter
The Evergreen State College

Jones and Bartlett Publishers
Boston/Portola Valley

Publisher's Note: *Origin and Evolution of the Universe* was published originally as Part One of a book written by the same author entitled *The Universe and Life*, published by Jones and Bartlett Publishers, Inc., 1987.

Editorial, Sales and Customer Service Offices:
Jones and Bartlett Publishers
20 Park Plaza
Boston, MA 02116

Manuscript editor: Elizabeth Sorenson
Illustrations: John G. Hamwey
Photo research: Traci A. Sobocinski
Text and cover design: Rafael Millán
Typesetting: Publication Services, Inc.

Printed in the United States of America
10 9 8 7 6 5 4 3 2 1

Library of Congress Cataloging-in-Publication Data
Kutter, G. Siegfried.
Origin and evolution of the universe.

Rev. ed. of pt. 1 of: The universe and life. c1987
Bibliography: p.
Includes index.
1. Cosmology. I. Kutter, G. Siegfried, Universe and life . II. Title.
QB981.K924 1989 523.1 88-32858

ISBN: 0-86720-121-5

Cover photograph: Supernova 1987A (lower right) and 30 Doradus, also known as the Tarantula Nebula, in the Large Magellanic Cloud at a distance of 160,000 light-years. The supernova was first sighted shortly after midnight on February 23, 1987. (Courtesy M. Bass, Cerro Tololo Inter-American Observatory.)

Frontispiece: The Eta Carinae Nebula ("Keyhole Nebula").
This enormous cloud of luminous interstellar gas and dark dust lanes, located some 9,000 light-years from Earth, is one of many regions in the Milky Way in which stars have recently formed. Cerro Tololo Inter-American Observatory, Curtis Schmidt telescope photograph (Courtesy National Optical Astronomy Observatories).

To my wife and children,
in appreciation of their love and support

Preface

THIS TEXT IS AN INTRODUCTION TO ASTRONOMY for liberal arts majors at the college and university level. It begins with our current theories of the origin of the Universe and then describes chronologically the evolution that produced the galaxies, stars, and the Solar System. This approach differs from that of most astronomy texts, which follow, more or less, the history of astronomical discovery, starting with the Solar System.

This evolutionary approach has two advantages, particularly for courses aimed at liberal arts majors. Its theme is clear: It starts from the beginning and works its way forward in time. And it focuses on the age-old question: "What gave rise to the Earth, the Sun, and the rest of the Solar System, our birthplace in the Universe?" Today we know that it was not a single event, but a long sequence of events that began with the Big Bang some 10 to 15 billion years ago. A description of this sequence, even though incomplete and imperfect, lends meaning and content to our lives, and it offers motivation and purpose to the student.

Chapter 1 of this text deals with the Big Bang, during which conditions were far from simple and as different from our experiences as can be imagined. The same is true of the origin and evolution of the galaxies and stars, the topics of chapters 2 and 3. In chapters 4 and 5 the Earth and the rest of the Solar System are introduced.

Each chapter contains numerous comments and short essays that are boxed to set them off from the main text. They offer additional information that, if included in the text, would interrupt the flow of the presentation. For those who wish to explore a given topic in greater depth, each chapter contains an extensive, annotated bibliography.

My original thought for writing a text that follows the evolutionary theme dates back to the mid-1970s and included both physical *and* biological evolution. That thought came to fruition in 1987, when I published with Jones and Bartlett the text *The Universe and Life: Origins and Evolution*. This text starts with the Big Bang and ends with *Homo sapiens*, according roughly equal space and emphasis to physical and to biological evolution. It concludes with an epilogue in which I offer my estimate of how frequently extraterrestrial life, including beings capable of communicating across interstellar distances, might have arisen on other planets circling other stars.

The present text is a new edition of the physical evolution part of the 1987 text, revised and updated to the middle of 1988. It retains the cross references to the biological chapters for those who wish to pursue the biological applications of some of the topics.* Furthermore, it retains the epilogue and several of the tables and charts of the biological part that summarize the major stages of terrestrial biological evolution. Supernova 1987A, which many scientists judge to be the dominant astronomical event of the decade and which occurred shortly after the 1987 text had gone to press, is described in the following Note to the Student.

In conclusion, I'd like to thank the instructors who have selected this text, and I wish them and their students much success in using it.

Olympia, Washington
November 1988

G. Siegfried Kutter

*Table of Contents of the biological part of *The Universe and Life: Origins and Evolution:*
Introduction to Biological Evolution
Chapter 6. Origin of Life on Earth
Chapter 7. Evolution of Early Life
Chapter 8. Origin and Evolution of the Major Animal Phyla
Chapter 9. Evolution of the Vertebrates
Chapter 10. Origin and Evolution of the Primates

Note to the Student*

SHORTLY AFTER MIDNIGHT ON FEBRUARY 23, 1987, Oscar Duhalde, a native Chilean and telescope operator at Las Campanas Observatory in the Andes of Northern Chile, took a break from guiding the observatory's one-meter telescope and stepped briefly outside. Glancing up at the Large Magellanic Cloud, a prominent galaxy in the southern sky, he noticed a bright object near the Tarantula nebula that he had never seen before. Duhalde was seeing the explosion of a massive star—a supernova—the first such event in our cosmic neighborhood in nearly 400 years. Ian Shelton of the University of Toronto, who was also working that night at Las Campanas Observatory, and Albert Jones, an amateur astronomer in New Zealand, sighted the explosion at nearly the same time.

Supernova (SN) 1987A,† as the event has been named, resulted from the explosion of a blue supergiant, Sanduleak (SK) −69°202. It created an enormous amount of excitement and activity among astronomers, confirming some of their theories of stellar evolution and raising questions about others. Here are some of the major observational and theoretical results of SN 1987A:

1. Sanduleak −69°202 was born as a 20 $M_\odot$ star, but lost roughly 4 $M_\odot$ of its envelope in the course of its evolution, most of it probably during a red supergiant phase. That left the star with about 16 $M_\odot$ at the time of the explosion. (The symbol $M_\odot$ means *a mass equal to that of the Sun*.)

2. At the time of the explosion, SK −69°202 was a blue supergiant, not a red one as astronomers would have predicted. It was blue probably because of mass loss during an earlier red supergiant phase and because of the relatively low abundance of elements heavier than helium in its envelope.

3. The first signal on Earth of the explosion was neutrinos from the collapse of SK −69°202's core. The neutrinos were detected in Japan and the U.S. at universal time 7:35:42 on February 23, nearly 22 hours before the supernova was visually sighted. (Note that the neutrinos and the light from the outburst had been travelling for 160,000 years through space, for the Large Magellanic Cloud is 160,000 light-years from us.)

4. The collapse of the core of SK −69°202 created a shock wave that travelled towards the star's surface in just under 22 hours and accelerated the surface

*You may not be familiar with some of the terms and concepts used in this note. Don't be discouraged. You will learn about them as you read this text.

†See cover photo, figure 2.16, and pp. 150–163.

layers to 30,000 km/sec, as deduced from the blue-shift of absorption lines in the first spectra taken of the supernova.

5. The luminosity of the supernova peaked 85 days after the initial outburst and has been fading since. The rate of fading since June 1987 indicates that the light energy comes primarily from the radioactive decay of cobalt-56 into iron-56 (half-life 77 days). The presence of the short-lived element cobalt-56 confirms the astronomers' basic theory of stellar nucleosynthesis.
6. Ultraviolet emission lines in light from interstellar matter ½ light-year from the supernova reveals that a pre-existing low-velocity shell of gas surrounds the supernova. Quite likely that shell is the matter lost by SK $-69°202$ during its red supergiant phase.
7. One month after the explosion, a second object appeared roughly 18 light-days to the south of the supernova. Two months later, the "mystery object" had disappeared. All attempts to explain this transient phenomenon have so far failed.
8. Since the fall of 1987, gamma rays have been arriving from the supernova. The gamma rays come from the radioactive decay of cobalt-56 into iron-56 (see no.5 above). The cobalt itself was produced by the radioactive decay of nickel-56 (half-life 6.1 days). The nickel was synthesized when the shock wave of the explosion heated the precursor star's envelope and induced nuclear burning. Approximately 0.07 $M_\odot$ of iron are being synthesized by the decay of cobalt-56.
9. Since early in 1988, optical and infrared "echoes" from the supernova have been detected (see photo on opposite page). The echoes are reflections by interstellar dust of radiation emitted by the supernova and, therefore, they advance radially outward with the speed of light. They have the appearance of faint arcs or rings.

Of the numerous questions raised by SN 1987A, the following are the most compelling ones:

1. What is the full range of elements synthesized during the supernova explosion?
2. Is the ejected matter spherically symmetric or not? Is the matter homogeneous or clumped? Are there jets? What was the "mystery object?"
3. Is the collapsed core of SK $-69°202$ a neutron star or a black hole? What is its mass and its size? Will it pulsate as a pulsar once the ejecta thins and it can be seen?

Supernova 1987A is a striking example of the dynamic and exciting nature of astronomy. There are countless further examples in our Galaxy and beyond capable of stirring our intellect and imagination. Many are just now being discovered. In fact, astronomy is presently in a renaissance of discovery, with the development of ever more sensitive observing tools both on the ground and in space. But as our exploration of the Universe continues, Einstein's words still ring true: "Joy and amazement at the beauty and grandeur of this world of which man can just form a faint notion." I hope you will feel some of this joy and amazement when reading this book.

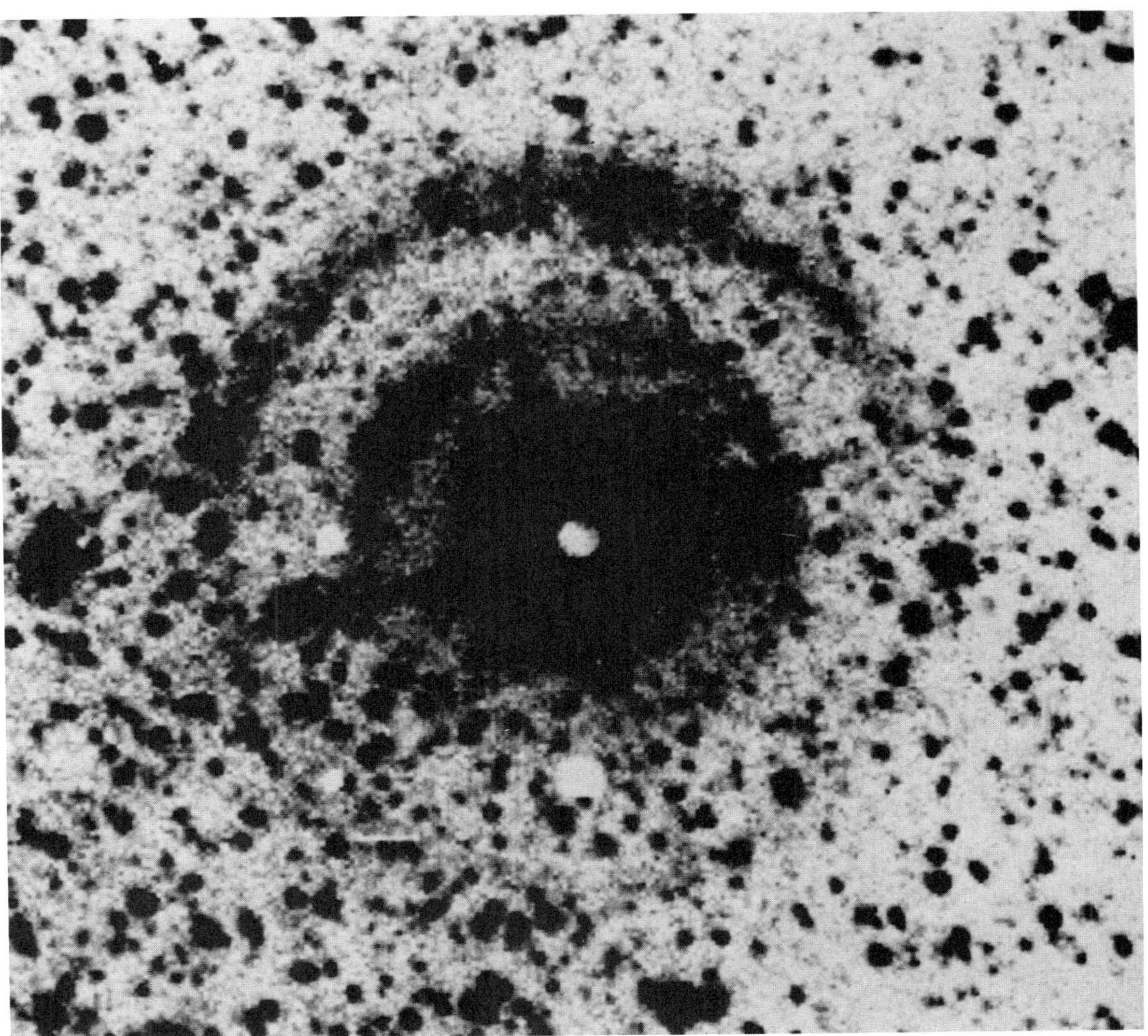

Light Echoes from Supernova 1987A in the Large Magellanic Cloud (LMC), observed with the 3.6 meter telescope of the European Southern Observatory, La Silla, Chile. The two circular echoes are reflections by interstellar clouds in the LMC of light emitted by the supernova. This computer-enhanced negative print was taken on February 13, 1988, nearly a year after the supernova was first observed. In order to see the faint echoes, the image of the supernova was dimmed by a small obscuring disk, here seen as a lighter spot at the center. (Copyright European Southern Observatory.)

Foreword

PROFESSOR KUTTER has actually accomplished what most scientist-scholars experience only in their wildest dreams: a readable treatise on cosmic evolution. Are we the only intelligent (interstellar-communicating) life in the universe? The calculation of the probability of the existence of the book you hold in your hands follows a logic similar to that Kutter leads us through in his fascinating Epilogue on the probability of other intelligent life in the universe. I base my rough estimate that Kutter's *The Universe and Life: Origins and Evolution* could ever exist at all on the following probabilities.

1. The probability of the rise of modern astronomy which employs the properties of light (this includes of course the construction and use of telescopes, spectroscopes, satellites, and the rest.)

2. The probability of the entire geological (geophysical and oceanographic) revolution. I refer to the radical shift of our thinking about the surface of the Earth that followed from the plate tectonics explanation of Wegener's concept of continental drift.

3. The probability of emergence of our awareness of the "biosphere-Gaia" phenomenon. What I refer to here is the recognition of life as a planetary surface phenomenon, one that modulates its own environment. Specifically I am thinking of V. I. Vernadsky's understanding of life as a geological force coupled with J. E. Lovelock's Gaia hypothesis, according to which life produces and regulates the chemistry, the salt concentration, the temperature and other aspects of its immediate environment. These local phenomena have led to a global system making the Earth different from its planetary neighbors, Mars and Venus. Life has lead to Earth surface regulation over immense periods of time.

4. In figuring out the probability of the existence of this book we must factor in the chances that the Darwinian revolution, including its most recent manifestations, would ever have occurred. I would include here the new recognition of the antiquity of Archean and Proterozoic life, the glimpse of understanding of symbiosis, karyotypic fissioning, and other kinds of "macromutation" as important mechanisms in the Darwinian evolutionary process. Here too is the recent success in the paleontological reconstruction of primate history and the "DNA revolution." We now understand far more clearly the cell and molecular basis of evolution through the techniques of biochemistry, molecular genetics, electron microscopy, and the like. Even

techniques developed by the petroleum geochemical industry have been integrated into our investigations of the evolution of early life.

5. We must include in our analysis of the chance of the existence of this book the probability that curious scientists, each limited to a tiny aspect of these grand themes, would be supported in their investigations to the extent that their work reaches a stage of publication. No one, after all, even the most dedicated philosopher is paid to contemplate cosmic evolution on the scale of the Kutter treatise. And Kutter, in all his wisdom, could never have woven together these scientific threads without the lost weekends and all-night labors in every corner of the world of thousands of obsessive and compulsively communicative scientists.

6. Given that the scientific information required to produce this long essay on evolution of the universe even exists anywhere, we must include in our calculation the probability that a G. Siegfried Kutter also can exist. That is, we must first estimate the chances that an academic institution encourages a research astronomer to read the disparate, contradictory, sometimes boring and inaccurate literature of organic chemistry, biology, paleontology and anthropology. Then we must ask what the probability is that such an institution—The Evergreen State College in Kutter's case—fosters intercommunication among scientists and students in entirely different specialties and departments. Remembering that science is paid for by those who want to see the development of new drugs, weapons, communication systems and other marketable commodities, we must decrease the estimate of our probability of the existence of this book. Finally, we must ask what are the chances that such a single G. S. Kutter, who can assimilate, evaluate, integrate and formulate the slippery information into a scientific story line can make his tale compelling enough that it is readable with interest by the rest of us. This is not all. In our calculation of probabilities we must inquire: even if our able, articulate, hard-working scientist-scholar-writer exists, what are the chances that his personal situation will permit the time and effort needed to actually create a book? Even if we assume that such a G. S. Kutter can produce a scientifically accurate yet fascinating manuscript, we have to ask another probability question. What are the odds that some publisher (knowing that few other scientist-scholars are educated well enough even to *teach* from such a manuscript) will undertake the financial risk in bringing out a large textbook for a course that does not exist?

Like Kutter, although I believe that quantification is essential to science, I want to limit the amount of arithmetic forced upon the reader. In any case it is clear that the probability that the book you now hold in your hands can exist at all is vanishingly small. A conservative scientist would tend to dismiss the entire problem: estimating the probability to be so close to zero he would assume that no such book could ever have been published. Yet here it is.

Will Kutter's text be too broad, too intellectual and too difficult for the undergraduate and other readers for which it was written? Will its erudition and sense-of-the-whole frighten teachers of undergraduate general science courses who are unaccustomed to instruct in a mode more familiar to literature and humanities experts? We hope not. Indeed we expect an alternative scenario. Broadly-based university and college courses on science that, from lack of an adequate text, could never be taught in the past, fertilized by Kutter, should spring up like mushrooms in September after rain. Science, for the first time since 19th century academic programs in Natural Theology, will be integrated. Science will begin to be recognized as the humanity it is. On this basis, not of preconceived religious views, but of hard-earned answers to questions asked directly to nature, science and scientific teaching will be ushered through a renaissance. Science will become a reliable way of knowing about the whole. Science is not just a method for the calculation of the targets of missiles, or a means for developing flexible plastics and cures for infectious diseases. Science can become the way of knowing about the eternal questions of the philosophers.

I cannot help but notice that Kutter as a biologist still has some learning to do. I am sure that any anthropologist, vertebrate zoologist, geophysicist, organic chemist, enzymologist or sedimentologist will see small errors and differences in interpretation. My response to this inevitability is that all of you specialists, students and teachers using this tome celebrate the magnitude of Kutter's effort by bringing to his attention your criticisms. The first edition of the book is a unique event in scientific publishing; a later edition, responding to comments and corrections by readers, might provide a product of concern to everyone in the academic world. From bureaucrat to parent, dean and faculty member . . . everyone lamenting our scientific illiteracy and ignorance will have cause to celebrate. The accomplishment is within view and one expects many imitators. The definitive, scientifically accurate, integrated one-year course of study at the college level revealing (in T. H. Huxley's terms) "man's place in nature" can now be brought into the curriculum with an appropriate course number, to be taught at an appropriate level. Why now? Because such a course, formerly a product only of a scientist's wildest dream, now has a textbook.

Lynn Margulis
University Professor
Boston University

Table of Contents

Expanded Table of Contents

Chapter 3: Stars and Heavy Element Synthesis 117

Photograph Credits

Part One: Introduction to Physical Evolution

Page 3 National Aeronautics and Space Administration
4 Fig. I.1—National Aeronautics and Space Administration
5 Fig. I.2—National Aeronautics and Space Administration
5 Fig. I.3—National Aeronautics and Space Administration
5 Fig. I.4—National Aeronautics and Space Administration
8 Fig. I.7—G.J. Hills, John Innes Institute/Science Photo Library, Science Source/Photo Researchers, Inc.
13 Fig. I.11—United States Department of Energy
30 Fig. I.20—Mount Wilson and Las Campanas Observatories (Hale Observatories)

Chapter 1: Cosmology

Page 35 Fig. 1.1—National Optical Astronomy Observatories
37 Fig. 1.2—Lick Observatory
38 Fig. 1.3—© 1960 National Geographic Society—Palomar Sky Survey. Reproduced by permission of the California Institute of Technology
38 Fig. 1.4—Copyright © 1983 Royal Observatory, Edinburgh
40 Fig. 1.5—Harvard College Observatory
41 Fig. 1.6—Reprinted, by permission, from *The Hubble Atlas of Galaxies* (*Plate 3, lower left*), Carnegie Institution of Washington, (1961) and Allan Sandage
42 Fig. 1.8—Palomar Observatory 48-inch Photograph, Copyright, California Institute of Technology
43 Fig. 1.9—Reprinted, by permission, from *The Hubble Atlas of Galaxies*, (*Plate 1, lower right*), Carnegie Institution of Washington, (1961) and Allan Sandage
44 Fig. 1.10—Mount Wilson and Las Campanas Observatories (Hale Observatories)
45 Fig. 1.11—National Optical Astronomy Observatories
45 Fig. 1.12—Reprinted, by permission, from *The Hubble Atlas of Galaxies* (*Plate 2*), Carnegie Institution of Washington, (1961) and Allan Sandage
46 Fig. 1.13—National Optical Astronomy Observatories
48 Fig. 1.14—M. Seldner, B. Siebers, E. J. Groth, and P. J. Peebles, "New Reduction of the Lick Catalog of Galaxies." *Astronomical Journal 82*, 249 (1977)
49 Fig. 1.15—Mount Wilson and Las Campanas Observatories (Hale Observatories)
52 Box 1.1, Fig. C—© 1960 National Geographic Society—Palomar Sky Survey. Reproduced by permission of the California Institute of Technology
61 Fig. 1.19—A. Stockton, Institute for Astronomy, University of Hawaii, Honolulu

Chapter 2: Galaxies

Page 81 Fig. 2.1—Reprinted, by permission, from *The Hubble Atlas of Galaxies* (*Plate 26*), Carnegie Institution of Washington (1961) and Allan Sandage
84 Fig. 2.3—Carlos Frenk (University of Durham), Simon White (University of Arizona), and Marc Davis (University of California at Berkeley)
88 Fig. 2.4—Joseph Silk *The Big Bang* (San Francisco: W.H. Freeman & Co., 1980), p. 171
88 Fig. 2.5—Donald Schneider and James Gunn
89 Fig. 2.6—Drs. Alar and Juri Toomre, Massachusetts Institute of Technology and the University of Colorado
89 Fig. 2.7— Copyright © 1980 Royal Observatory, Edinburgh
90 Fig. 2.8—National Optical Astronomy Observatories
91 Fig. 2.9—Adapted from Valerie de Lapparent, Margaret J. Geller, and John P. Huchra "A Slice of the Universe" *The Astrophysical Journal*, 302, L1–L5, © 1986, March 1, The American Astronomical Society
93 fig. 2.11—Reprinted, by permission, from *The Hubble Atlas of Galaxies* (*Plate 1, upper right*), Carnegie Institution of Washington (1961) and Allan Sandage
95 Fig. 2.13—Reprinted, by permission, from *The Hubble Atlas of Galaxies* (*Plate 25*), Carnegie Institution of Washington (1961) and Allan Sandage
96 Fig. 2.14—Reprinted, by permission, from *The Hubble Atlas of Galaxies* (*Plate 19*), Carnegie Institution of Washington (1961) and Allan Sandage
97 Fig. 2.15—Reprinted, by permission, from *The Hubble Atlas of Galaxies* (*Plate 45*), Carnegie Institution of Washington (1961) and Allan Sandage
98 Fig. 2.16—(*Left*) Copyright European Southern Observatory. (*Right*) Reprinted, by permission, from *The Hubble Atlas of Galaxies* (*Plate 38*), Carnegie Institution of Washington (1961) and Allan Sandage
99 Fig. 2.17—Reprinted, by permission, from *The Hubble Atlas of Galaxies* (*Plate 41*), Carnegie Institution of Washington (1961) and Allan Sandage
99 Fig. 2.18—Max-Planck-Institut für Radioastronomie, F.R.G.
100 Fig. 2.19—National Optical Astronomy Observatories
101 Fig. 2.20—The Yerkes Observatory
102 Fig. 2.21—National Optical Astronomy Observatories
103 Fig. 2.22—Dr. Hyron Spinrad and *The Astrophysical Journal*
108 Fig. 2.24—Lund Observatory
110 Fig. 2.26—Lick Observatory
111 Fig. 2.27—Illustration: U.K. Schmidt/E. Becklin

Chapter 3: Stars and Heavy Element Synthesis

Page 117 Fig. 3.1—Mount Wilson and Las Campanas Observatories
124 Fig. 3.3—Harvard College Observatory
130 Fig. 3.6—National Optical Astronomy Observatories
132 Fig. 3.7—Mount Wilson and Las Campanas Observatories
137 Fig. 3.9—Mount Wilson and Las Campanas Observatories
138 Fig. 3.10—Mount Wilson and Las Campanas Observatories
138 Fig. 3.11—Mount Wilson and Las Campanas Observatories
139 Fig. 3.12—Mount Wilson and Las Campanas Observatories
144 Fig. 3.15—Lick Observatory
147 Fig. 3.17—Lick Observatory
148 Fig. 3.18—*Photos a* and *b*, Lick Observatory Photographs; *Photo c*, H. Ford
149 Fig. 3.21—National Optical Astronomy Observatories
159 Fig. 3.23—Lick Observatory
161 Fig. 3.24—Lick Observatory
162 Fig. 3.25—Lick Observatory

Chapter 4: The Solar System: Orbital Layout

Page 167 Fig. 4.1—National Aeronautics and Space Administration
170 Fig. 4.2—Reproduced, by permission, from J. Kepler, *Mysterium Cosmographicum* (p. 86, pp. 227–228), Abaris Books, (1981)
177 Fig. 4.3—William M. Sinton, University of Hawaii at Manoa
178 Fig. 4.4—Yerkes Observatory
179 Fig. 4.5—Official U.S. Naval Observatory Photograph
186 Fig. 4.8—National Aeronautics and Space Administration
189 Box 4.4 Fig. A—Max-Planck-Institut für Astronomie, Heidelberg, F.R.G.

Chapter 5: The Solar System: Structure of Planetary Bodies and Origin

Page 195 National Aeronautics and Space Administration
196 Fig. 5.1—National Aeronautics and Space Administration
197 Fig. 5.2, and Fig. 5.3—National Aeronautics and Space Administration
202 Fig. 5.6—United States Geological Survey
202 Fig. 5.7—TASS from SovFoto
205 Fig. 5.9—John A. Wood, Harvard-Smithsonian Center for Astrophysics
207 Fig. 5.10—National Aeronautics and Space Administration
214 Fig. 5.12—National Aeronautics and Space Administration
216 Fig. 5.13—National Aeronautics and Space Administration
217 Fig. 5.14—National Aeronautics and Space Administration
217 Fig. 5.15—National Aeronautics and Space Administration
218 Fig. 5.16—National Aeronautics and Space Administration
220 Fig. 5.18—National Aeronautics and Space Administration
221 Fig. 5.19—National Aeronautics and Space Administration
222 Fig. 5.20—National Aeronautics and Space Administration
223 Fig. 5.21—TASS from SovFoto
224 Fig. 5.22—U.S. Geological Survey, Flagstaff, AR
225 Fig. 5.23—National Aeronautics and Space Administration
226 Fig. 5.24—United States Geological Survey #439
227 Fig. 5.25—National Aeronautics and Space Administration
228 Fig. 5.26—National Aeronautics and Space Administration
229 Fig. 5.27—National Aeronautics and Space Administration
230 Fig. 5.28—National Aeronautics and Space Administration
232 Fig. 5.29—National Aeronautics and Space Administration
233 Fig. 5.30—National Aeronautics and Space Administration
235 Fig. 5.31—National Aeronautics and Space Administration
237 Fig. 5.33—National Aeronautics and Space Administration
245–247 Fig. 5.35 (a–c)—From "Formation of the Earth from Planetesimals," George W. Wetherill. Copyright © June, 1981 by Scientific American, Inc. All rights reserved.

ORIGIN AND EVOLUTION OF THE UNIVERSE

PART ONE

PHYSICAL EVOLUTION

In the beginning, this world was nothing at all,
heaven was not, nor earth, nor space.
Because it was not, it bethought itself:
I will be. It emitted heat.
Ancient Egyptian text

There is nothing permanent except change.
Heraclitus of Ephesus (ca. 540–480 B.C.)

If simple perfect laws uniquely rule the universe, should not pure thought be capable of uncovering this perfect set of laws without having to lean on the crutches of tediously assembled observations? True, the laws to be discovered may be perfect, but the human brain is not. Left on its own, it is prone to stray, as many past examples sadly prove. In fact, we have missed few chances to err until new data freshly gleaned from nature set us right again for the next steps. Thus pillars rather than crutches are the observations on which we base our theories . . .
Martin Schwarzschild (Professor Emeritus, Princeton University)

Earth as viewed from the *Apollo 13* spacecraft on its return from the Moon (April 1970). Views such as this one prompted Apollo astronaut James Lovell to declare: "We do not realize what we have on Earth until we leave it."

I

Introduction to Physical Evolution

WHEN THE APOLLO astronauts stood on the Moon more than a decade ago, the most beautiful object they saw anywhere in the sky was Earth. Satellite photographs of our planet confirm this for all of us (figures I.1–I.4): its spherical symmetry, its cover of clouds and oceans, the outlines of the continents, the meandering of the rivers, the green of the forests, and the yellow and purple of the deserts. Earth is our home—our refuge—in the vastness of space. Its solid crust provides a stable platform upon which we and the rest of life have come into existence and are evolving, its atmosphere and magnetic field offer protection from the hostile environment of outer space, and its distance from the Sun is such that it absorbs just enough radiant energy for warmth and comfort. How could such a planet, with conditions so well suited for life, have come to be? Or, reversing the perspective, how could life that is so well suited for this planet have come to be? Answering the first of these two questions will be the main goal of the first part of this book. Answering the second question will be the main goal of the second part.

The Earth, the Sun, and the rest of the Solar System are not isolated in space and they were not created in isolation. They originated as byproducts of the general physical evolution of the Universe. Thus, in order to understand our origin, we need to look beyond our immediate surroundings and learn about the origin and evolution of the Universe and its major components—the stars, galaxies, and clusters of galaxies. In our thinking we must switch from the familiar scales of inches, feet, and miles to scales that are measured in light-years. At the same time, we must also consider the microscopic world of molecules, atoms, and nuclei of atoms, for many of the large-scale features of the Universe are rooted in processes occurring on the smallest scale. For instance, all the matter in the Universe today is the result of countless microscopic interactions that occurred under conditions of immense temperatures and pressures during the first few seconds of the Universe's existence more than 10 billion years ago. And nearly all the light emitted by stars and galaxies originates

Figure I.1. Earth as seen from the *Apollo 8* Command Module, 110 km above the Moon's surface (December 1968). *Apollo 8* was the first manned spacecraft to orbit the Moon.

Figure I.2. The eastern Mediterranean, showing the Nile Delta, Gulf of Suez (bottom right), Sinai Peninsula, Israel, Dead Sea, Lebanon, Syria, and Cyprus (top middle). The photograph was taken from *Gemini 7* (December 1965), at an altitude of 300 km.

Figure I.3. The "top of the world," as seen from an altitude of 250 km by the *Apollo 7* astronauts. The Himalayan Mountains bisect this photograph diagonally, with the Ganges River of India to the left and Tibet to the right. The circle indicates Mt. Everest, Earth's highest peak (8848 m).

Figure I.4. Hurricane Gladys off the west coast of Florida, as photographed from *Apollo 7* on October 17, 1968. The clouds are aligned along an enormous spiral pattern, extending over hundreds of square kilometers and focused on the eye of the hurricane. The eye itself is hidden under a layer of clouds that is pushed by a vigorous updraft against the cold, stable air of the tropopause (at 16,500 m).

from the energy liberated by the conversion of atomic nuclei of one kind into those of another kind deep in the interior of stars.

Despite the psychological obstacles we may face in thinking simultaneously of the very large and the very small, our task is made easier by the fact that the fundamental laws of nature are few in number and they apply on all scales, large and small. Furthermore, there are only five stable elementary particles that make up the known content of the Universe: *electrons, protons, neutrons, neutrinos*, and *photons* (but see box I.1). The first three—electrons, protons, and neutrons—are the basic building blocks of the material content of the Universe, from galaxies and stars down to molecules and atoms. Neutrinos are very elusive particles created in certain nuclear reactions, and photons are quanta of light. Both neutrinos and photons play important roles in nature, as I shall explain in the following chapter. Mostly, however, this book deals with just electrons, protons, and neutrons, and the various structures created from them. This introduction describes some of the properties of these five elementary particles and the laws by which they interact. The second part of this book—Biological Evolution—shows that living organisms are also made of electrons, protons, and neutrons and hence are subject to the same natural laws as inanimate matter. Thus, the laws of nature and the existence of a few elementary particles allow us to view the origin and evolution of the Universe—including life on Earth—in a unified way.

BOX I.1.
Quarks and Gravitons

SINCE the 1930s physicists have been discovering a host of "elementary" particles other than the five mentioned above. These particles were first detected among the cosmic rays and then produced with particle accelerators. They are all extremely short-lived and decay into stable elementary particles fractions of seconds after they are produced.

Recent theoretical work suggests that protons and neutrons are perhaps not really elementary, but consist of still more fundamental particles—the *quarks*. The name *quark* was given to these particles by the American physicist Murray Gell-Mann, after the phrase "three quarks for Muster Mark" in James Joyce's *Finnegan's Wake*. Six quarks are believed to exist—*up, down, charm, strange, top*, and *bottom* quarks. A triad of up and down quarks makes up protons and neutrons. Much effort is currently devoted to the detection of quarks in laboratories around the world.

If quarks do exist, are they the ultimate building blocks of matter? This is one of the most profound questions confronting science today and there is at present no answer.

There may exist a sixth stable elementary particle—the *graviton*. Gravitons have never been detected. They have merely been predicted by theory. However, the theoretical arguments for their existence are very compelling. They would be the gravitational analogue of electromagnetic radiation (photons), be massless, and travel with the speed of light. They are believed to be created whenever mass is violently accelerated, as during the gravitational collapse of a star, the falling of matter into a black hole, or the Big Bang origin of the Universe.

Because the gravitational force is the weakest of all known forces in nature, gravitons—if they exist—would interact only exceedingly weakly with the other particles. That may be the reason why they have not yet been detected, despite concerted efforts to do so.

THE FUNDAMENTAL BUILDING BLOCKS OF MATTER

To begin the discussion of elementary particles, consider a familiar object, a snowflake (see figure I.5). It measures about one centimeter across (see box I.2 for comparison of the metric and English systems of units). A look through a high-power microscope reveals its very beautiful and intricate structure, one that a master architect might have put together. Myriads of particles, called *water molecules*, are joined together in very precise ways so as to give the snowflake its characteristic macroscopic appearance. Each water molecule is about 100 million times smaller than the snowflake. The water molecules are composed of still smaller particles called *atoms* (from the Greek, meaning "not cut" [a-tom] or "indivisible"). Two atoms of hydrogen and one of oxygen, bonded together as shown in figure I.5, make up one water molecule. Finally, the atoms are made of electrons, protons, and neutrons. Protons and neutrons are 100,000 times smaller than the atoms, and electrons are smaller still. These particles are so small that even the most powerful electron microscopes cannot reveal their existence. Exotic instruments such as X-ray diffraction machines or particle accelerators are required to "see" them. In figure I.6 the sizes of the elementary particles, atoms, and molecules are compared with those of some of the larger objects far and near in the Universe.

If, instead of a snowflake, we examine the internal architecture of a water droplet, a lunar rock, a piece of gold (see figure I.7), tissues from our own bodies, or any other kind of matter, we find that each of them, too, is constructed of molecules and atoms. These, in turn, are always made of the same fundamental building blocks—electrons, protons, and neutrons. Thus, all matter is made of the same stuff. The only difference between one kind of matter and another lies in the ways in which the building blocks are assembled.

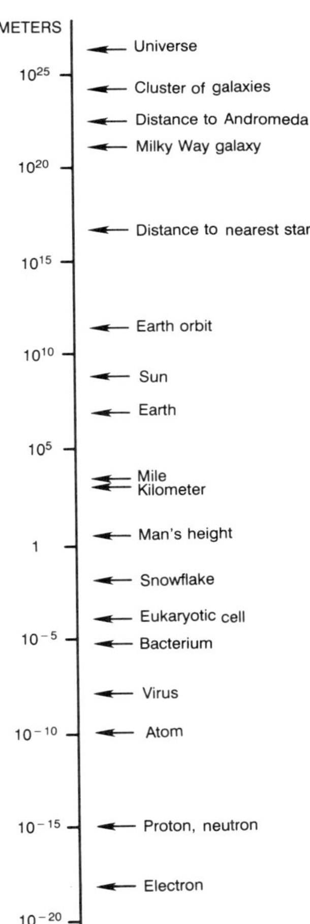

Figure I.6. Range of sizes of some of the objects dealt with in this book.

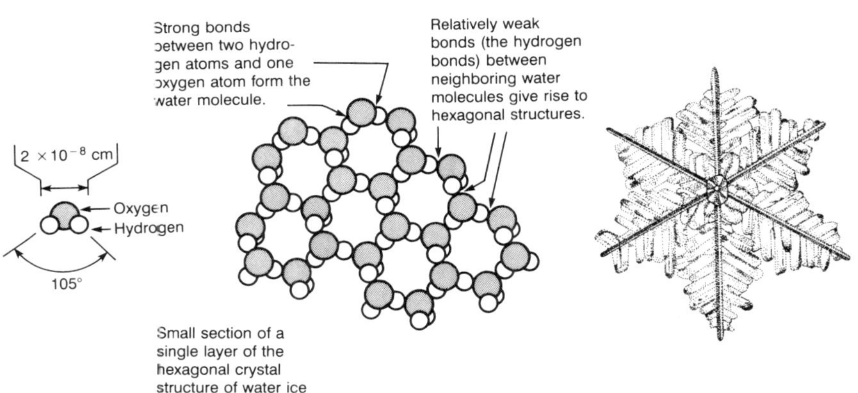

Figure I.5. Snowflakes, liquid water, and steam are composed of the same building blocks—water molecules. Each water molecule consists of one oxygen atom and two hydrogen atoms, firmly bonded together by the electric force. In addition, weak attractive forces (also electric) exist between the oxygen atom of one water molecule and the hydrogen atoms of neighboring water molecules. At temperatures below freezing, these weak attractions allow water molecules to arrange themselves spontaneously into hexagonal crystal structures. Snowflakes and ice crystals are evidence on the macroscopic scale of the precision of structure existing on the molecular level.

BOX I.2.

Units of Measurement Used in This Book

THE natural sciences are experimental and observational in nature. That means scientific theories are not purely intellectual constructs, but are based on experiments and observations. In the natural sciences, nature is the ultimate judge.

Scientific experimentation and observations require the carrying out of measurements: measuring distance or a time interval, reading the temperature or pressure on an instrument dial, determining the mass or electric charge of a particle, comparing the intensity of a light signal against some standard signal, counting objects or events, and so forth. In order to carry out such measurements, the scales or units to be used must be defined carefully. Are the units for distance in inches, meters, or light-years? Is temperature expressed in degrees Kelvin, Celsius, or Fahrenheit? Is the pressure scale in units of atmospheres, centimeters of mercury, or dynes per square centimeters? It must be clear from the outset what units are being used in order to avoid confusion.

In this text I shall follow the practice of the scientific community. Lengths, masses, and time intervals will be expressed in units of centimeters, grams, and seconds or in convenient multiples of these units. Temperature will be referred to in degrees Kelvin or Celsius. Energy and pressure will be described by making comparisons with familiar standards, such as the energy output of the Sun and the terrestrial atmospheric pressure at sea level.

The unit of length—the *centimeter* (cm)—is described by the markings on the following line:

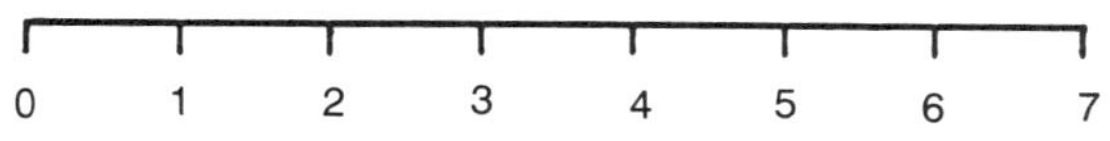

The multiples of the centimeter used in the text are:

1/10,000 cm	=	1 micrometer (abbreviated as μm)
1/10 cm	=	1 millimeter (mm)
100 cm	=	1 meter (m)
100,000 cm	=	1 kilometer (km)
6.38×10^8 cm	=	1 earth radius (equatorial)
7.0×10^{10} cm	=	1 solar radius($R_\odot$)
1.5×10^{13} cm	=	1 astronomical unit (AU), the average Sun–Earth distance
9.5×10^{17} cm	=	1 light-year (LY)

The unit of mass—the *gram* (g)—is the mass (or weight, if the weighing is done at sea level on Earth) of one cubic centimeter of water. The multiples of the gram used are:

1000 g	=	1 kilogram (kg)
6.0×10^{27} g	=	1 earth mass
2.0×10^{33} g	=	1 solar mass ($M_\odot$)

The metric system of length and mass compares with the English system as follows:

1 cm	= 0.3937	inch	1 inch	=	2.540 cm
1 m	= 3.281	feet	1 foot	=	30.48 cm
1 km	= 0.6214	mile	1 mile	=	1.609 km
1 g	= 0.03527	ounce	1 ounce	=	28.35 g
1 kg	= 2.205	pound	1 pound	=	453.6 g

The unit of time—the *second* (sec)—is the same in the metric and English systems of units. One *year* contains 3.16×10^7 seconds.

Figure I.7. A thin layer of gold, as seen through an electron microscope, magnified 40 million times. The regularly spaced bright dots are individual gold atoms. The distance between adjacent gold atoms is about 0.0002 μm.

The universality of electrons, protons, and neutrons raises the question: How can just three kinds of particles produce the staggering variety of objects found in the Universe? The answer is that each of the three elementary particles possesses unique properties that determine how the particles interact with each other and how they combine into larger objects. The number of properties of the particles and the number of kinds of interactions are small, but the number of ways in which the properties and interactions allow the particles to combine is virtually infinite. That is why the material objects all around us have such a great variety of shapes, forms, and sizes.

There are four known ways in which the elementary particles interact with each other. They are the gravitational, electromagnetic, nuclear, and weak interactions or forces.* We are all familiar with the first two of them because we are continuously exposed to them. The nuclear and weak interactions are somewhat more remote because generally they remain hidden deep within the atoms. They do affect us, however, for they are responsible for such phenomena as the energy production of the Sun, the explosive power of hydrogen bombs, and the generation of energy in nuclear power plants. The main features of the four interactions are described in the next section.

THE FOUR FORCES OF NATURE

Gravitational Force

Gravity is the force that holds us to the surface of the Earth. Without it we, the atmosphere, and everything else on our planet would fly apart and off into space. Gravity also holds the Sun, the stars, and the galaxies together. It keeps the Moon in orbit around the Earth and the Earth in orbit around the Sun. In short, *gravity is an attractive force that acts on all matter in the Universe*. It arises because all matter has mass. (Gravity also acts on photons, even though they carry no mass, as explained in chapter 1.)

The mass of an object is the property that we refer to colloquially as weight and that we measure in such units as pounds or grams (see box I.2). The greater the masses of particles or objects, the greater is the gravitational attraction between them. The greater their separation, the weaker is the gravitational force. More precisely, the gravitational force between two particles or objects (such as two elementary particles, the Earth and the Moon, the Sun and the Galaxy) is directly proportional to the product of the masses (M_1 and M_2) of the particles or objects and inversely proportional to the square of their separation (R):

$$F_{gravity} \propto \frac{M_1 M_2}{R^2}.$$

However, the gravitational force never becomes zero, no matter how great the separation between the two objects. For instance, the Solar System is approximately 30,000 light-years from the center of the Milky Way galaxy. Yet the

*The words *interaction* and *force* have the same meaning. They refer to such actions as pulling, pushing, pressing, and lifting. For instance, a person pulling a cart or lifting a box does so by exerting a force.

gravitational force exerted by the concentration of stars at the Galaxy's center prevents the Sun and the planets from flying off into intergalactic space despite their distance from the center. Similarly, galaxies are separated by millions and billions of light-years, but they still influence each other through gravity. The gravitational force, even though it weakens with distance, has an infinite range.

Compared to the other forces, gravity is extremely weak. We feel it strongly on Earth only because the mass of the Earth is large and we stand right on the planet's surface. Similarly, the gravitational force exerted by the center of our Galaxy is able to keep the Solar System in orbit because the concentration of mass at the Galaxy's center is so great. And galaxies affect each other gravitationally over the immense distance separating them because each contains a mass of millions to many billions of suns. However, the gravitational attraction between two elementary particles is virtually negligible, even at distances comparable to their sizes, for their masses are so small.

Electromagnetic Force

The second force that we are continuously exposed to is the *electromagnetic force*. It is responsible for the properties of all the objects in our immediate, macroscopic environment. It makes rocks hard, water liquid, and air gaseous. It participates in such processes as the dissolving of a sugar cube in a cup of tea, the burning of wood, and the digestion of food. Quite generally, the *electromagnetic force is responsible for all chemistry, including the chemistry of life*. In addition, it makes electrical appliances work, from TVs to electric motors and pocket calculators. Finally, it is responsible for light and radio waves. It is a very versatile kind of force.

As the name implies, the electromagnetic force consists of two parts—the electric force and the magnetic force. The electric force exists because two of the three elementary particles have electric charges and, hence, are surrounded by electric fields: *The proton is positively charged and the electron is negatively charged*. Because their charges are opposite, protons and electrons attract each other. Electrons repel each other and so do protons, because their charges are alike. The electric force does not affect the neutron because it has no electric charge. It is electrically neutral, as its name implies.

The mathematical expression for the strength of the electric force between two charged particles is similar to that of the gravitational force, except that the electric charge takes the place of mass. It is directly proportional to the product of the charges of the particles (Q_1, Q_2) and inversely proportional to the square of their separation (R):

$$F_{electric} \propto \frac{Q_1 Q_2}{R^2}.$$

And, like the gravitational force, the electric force has infinite range even though it weakens with distance.

The electric force comes into play in a flashlight (see figure I.8). Pressure on the button closes the flashlight's circuit and an electric current flows from the negative terminal of the battery through a wire or the metal casing to the bulb and on to the positive terminal. This current is just a flow of electrons. They flow because the first terminal of the battery is negatively charged and repels the electrons, while the second terminal is positively charged and attracts them. In the process the electrons are forced through the thin, high-resistance filament of the bulb and heat it to incandescence by friction.

Another, more spectacular demonstration of the electric force is lightning (see figure I.9). When the high winds of a thunderstorm sweep past each other, the resulting friction gives the lower regions of the rain clouds a negative charge (excess of electrons) and their upper regions, as well as the ground right below the clouds, a positive charge (excess of protons). As the charges build up, the attractive electric force between the negatively and positively charged regions increases until suddenly a surge of electrons neutralizes them, which we observe as lightning. This discharge of electricity between the upper and lower regions of the clouds, or between the clouds and the ground, is accompanied by the release of a great deal of energy in the form of light, thunder, and localized heat.

Whenever charged particles are in motion they create a magnetic field, and objects possessing magnetic fields interact by the magnetic force. This force is evident when we play a radio near a power line. The electric current in the power line creates a magnetic field that interferes with the radio reception and

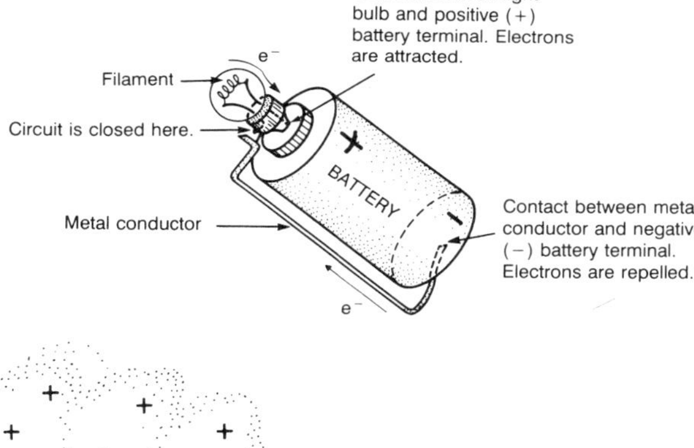

Figure I.8. A flashlight demonstrating the existence of electric charges as well as the attractive force between opposite charges and the repulsive force between like charges.

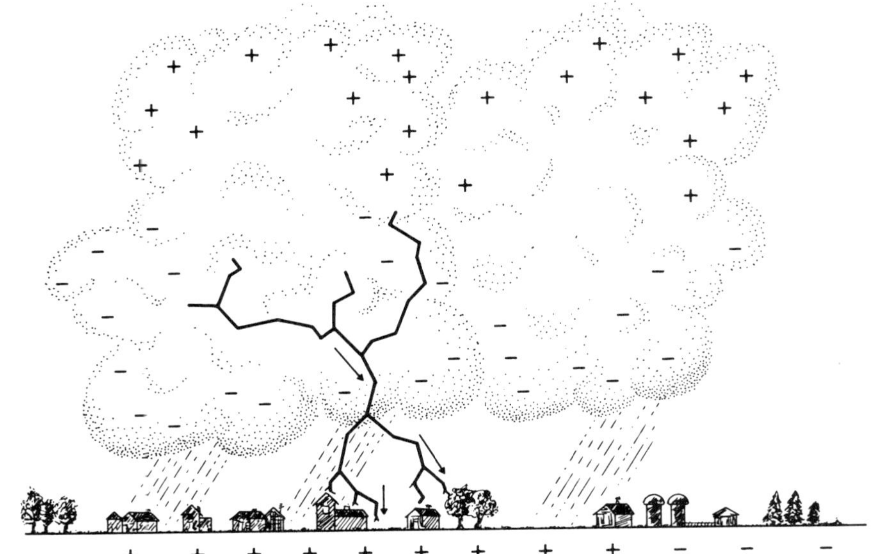

Figure I.9. Electron distribution during thunderstorms. Arrows in the diagram indicate the flow of electrons during a cloud–ground discharge.

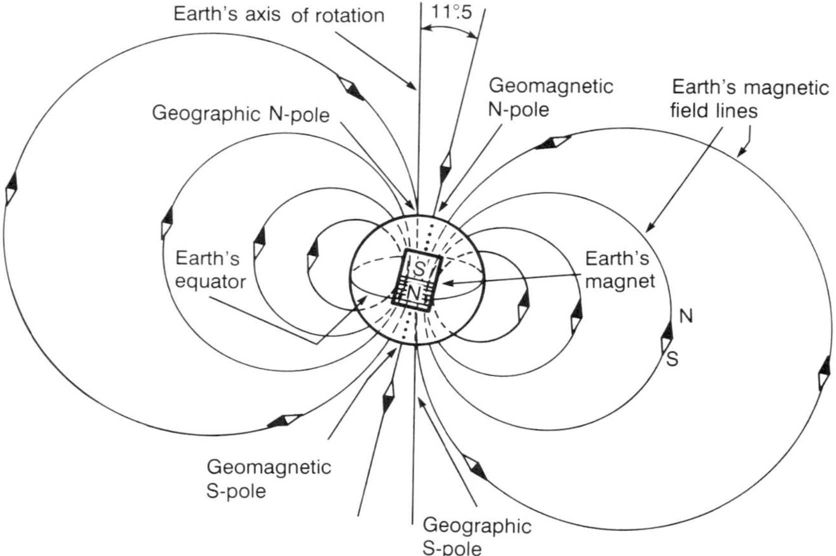

Figure I.10. The Earth's magnetic field. The configuration of the field is as if a short, strong bar magnet were buried at its center. The S-pole of the magnet points north and its N-pole points south. The magnet's axis is inclined by about 11.5° to the Earth's axis of rotation. In reality, the magnet is the result of motions of electrically conducting matter deep in the Earth's interior. Compass needles align themselves along the Earth's magnetic field lines such that their N-poles point toward the geomagnetic N-pole and their S-poles point toward the geomagnetic S-pole.

causes static. Two bar magnets provide a more direct experience of the magnetic force. When the magnets are close together, their north poles (and their south poles) repel each other, but the N-pole of one bar magnet attracts the S-pole of the other. A third example of the action of the magnetic force is the alignment of a compass needle along the magnetic field of the Earth (see figure I.10). The Earth's magnetic field is caused by the motion of electrically conducting matter deep in its interior. The compass needle also possesses a magnetic field, caused by microscopic motions of electrons in the atoms of the needle. The two magnetic fields interact via the magnetic force such that the needle's N-pole points toward the geomagnetic N-pole and its S-pole points toward the geomagnetic S-pole. Other examples of the presence of the magnetic force are the auroras known as the northern and southern lights. Their shimmering patterns occur when electrons and protons arriving from the Sun (the solar wind) are funneled by the Earth's magnetic field into the upper atmosphere near the polar regions and interact with nitrogen and oxygen atoms.

Moving electric charges give rise to magnetic fields and magnetic forces. Similarly, moving magnetic poles give rise to electric fields and electric forces. Thus, the electric and magnetic forces are inextricably linked and that is why collectively they are called the electromagnetic force. Later in this introduction I shall discuss one more important application of this linkage, namely light and radio waves.

The electromagnetic force between an electron and a proton is far stronger than the gravitational force between them. Yet, large aggregates of matter, like rocks or our bodies, interact very little with each other by the electromagnetic force. The reason is that they contain roughly equal numbers of electrons and protons and hence their electrical charges and magnetic poles cancel each other. The magnetic field of the Earth and electrically charged clouds in thunderstorms are exceptions.

No such canceling occurs with mass, for mass always has a positive sign. No negative mass exists. That is why the masses of the electrons, protons, and neutrons of our bodies, the Earth, the Sun, and the galaxies add up to large total masses. They create large gravitational effects, even though the contribution of a single elementary particle is very small.

Nuclear Force

Two forces remain to be discussed—the *nuclear* and the *weak forces*. Both participate in nuclear reactions, which are reactions between the nuclei of atoms (more about the structure of atomic nuclei in the following section; for a discussion of nuclear reactions see chapters 1 and 3). The roles the two forces play in these reactions are quite different.

The nuclear force (also referred to as the *strong* or *strong nuclear* force) is responsible for the large amounts of energy released in nuclear reactions, such as the energy generated by nuclear power stations, the tremendous explosive power of atomic and hydrogen bombs (figure I.11), and the more gentle,

Figure I.11. The first hydrogen bomb explosion, at Bikini Atoll in the Pacific Ocean (November 1, 1952). The explosion was so powerful that it obliterated a small island and left a crater more than one mile in diameter. The energy of hydrogen bomb explosions results from the conversion of hydrogen (in particular, the isotopes ^{2}H and ^{3}H) into helium, similar to the energy generation in the core of the Sun (where the nuclear fuel is ^{1}H).

though still potentially harmful, release of energy by radioactive decay (see box 5.1). Most important, it is responsible for the nuclear reactions that take place at the center of the Sun and that are the source of the radiant energy the Sun pours forth so prodigiously year after year. Without this steady supply of energy, life on Earth would quickly come to an end.

Processes involving the nuclear force are very energetic because this force is by far the strongest of the four forces. It is an attractive force that acts between protons and neutrons and holds them together to form the nuclei of atoms. The nuclear force does not affect electrons, neutrinos, or photons.

Protons and neutrons interact by the nuclear force only over distances comparable to their own sizes. At greater distances the nuclear force drops quickly to zero. Thus, unlike the gravitational and electromagnetic forces, the range of the nuclear force is very short. That is most fortunate, for if there were no limit on its range, this force would overpower all other forces and compress matter together into tightly bound lumps. Extended objects like living organisms, planets, stars, and galaxies could not exist.

Weak Force

In contrast to the nuclear force, the weak force (also referred to as the *weak nuclear* force) does not lead to the release of large amounts of energy in nuclear reactions. As its name implies, this is a very weak force with a strength that falls between that of the electromagnetic force and the gravitational force. Its range is even shorter than the range of the nuclear force.

The weak force comes into play only in the presence of a neutrino, the fourth elementary particle introduced above. The name *neutrino* means "little neutron" and was given to the particle by the great Italian physicist Enrico Fermi (1904–1954). It is a most appropriate name because the neutrino is electrically neutral like the neutron, but has a much smaller mass. In fact, up to the mid-1970s it was assumed that the neutrino resembled the photon (see below) by having no mass at all. This is now questioned because of new theoretical results, according to which the neutrino has a tiny mass, corresponding to about 1/100,000 of an electron's mass. This conjecture has not yet been confirmed by measurement.

Neutrinos are restless and elusive particles, and they are not part of ordinary matter. They come spontaneously into existence during the conversion of protons into neutrons or neutrons into protons (more about this in later chapters). As soon as they are created, neutrinos fly off into space with nearly the speed of light. When they come upon a star, a planet, or any other chunk of matter, they pass right through with only a negligible chance of interacting. In fact, neutrinos pass through our bodies all the time, yet we never feel them. It is very difficult to detect these particles.

Neutrinos are so elusive because they interact by the weak force, which has a very short range and is exceedingly weak. They do not interact by the electromagnetic or nuclear forces. Like all particles, they do interact by the gravitational force. However, the gravitational effect of the neutrinos passing through our bodies or the Earth is entirely negligible. Still, as discussed in chapter 1, the total number of neutrinos in the Universe may be so great that

Table I.1. Properties of the Stable Elementary Particles

Particle	Symbol	Electric Charge	Mass	Size
proton	p^+	+	1.7×10^{-27} kg	10^{-15} m
neutron	n	0	1.7×10^{-27} kg	10^{-15} m
electron	e^-	–	9.1×10^{-31} kg	10^{-18} m
neutrino	ν*	0	approx. 10^{-35} kg	less than 10^{-18} m
photon	γ*	0	0	no fixed size, depends on wavelengths

*ν and γ are the Greek letters *nu* and *gamma*, corresponding to Roman *n* and *g*.

their combined gravitational effect could have a decisive influence on the eventual fate of the Universe.

Summary

Table I.1 summarizes the most important properties of the elementary particles discussed here and in the previous section—the proton, neutron, electron, and neutrino. The table also includes the fifth elementary particle—the photon—which will be discussed later in this introduction. The symbols used as abbreviations for the proton, neutron, and electron are self-explanatory. The symbols for the neutrino and photon, ν and γ, are the Greek letters *nu* and *gamma* and correspond to the Roman letters *n* and *g*. The Greek letter γ is used because it was chosen when gamma-radiation was named (see box 5.1).

Three of the elementary particles in the table have zero electric charge—the neutron, neutrino, and photon. The charges of the proton and electron are *exactly* equal but opposite in sign. These charges are very small and it takes a great number of them to produce a noticeable, macroscopic effect. For instance, when we turn on a 150-watt light, the electric current carries 10^{19} electrons* through the filament of the bulb every second.

Of the five elementary particles, one is without mass—the photon—and the others have nearly vanishingly small masses, as indicated by the large negative exponents in the fourth column of table I.1. It may help you comprehend just how small these masses are if you know that with every breath you inhale approximately ¾ of a gram of air and that each ¾ gram consists of more than 10^{23} protons, neutrons, and electrons. Thus the mass of any one of these particles is truly vanishingly small. Also note that the masses of the proton and neutron are nearly equal, and that the mass of the electron is almost 2000 times smaller. The estimated mass of the neutrino is smaller still by roughly another factor of 100,000.

*The expression 10^{19} is a shorthand notation for the number 1 followed by 19 zeros. This notation is used in the sciences whenever very large or very small numbers are encountered. It is explained in box I.3.

Table I.2. The Four Known Fundamental Forces of Nature

Force	Relative Strength At 10^{-15} m	Range	Particles Affected	Applications
Gravitational	10^{-38}	infinite	all particles	holding together cosmic structures like the Earth, Sun, Solar System, Galaxy, Universe
Weak	10^{-13}	less than 10^{-18} m	ν	Beta decay: conversion of protons into neutrons or neutrons into protons
Electromagnetic	10^{-2}	infinite in principle, but limited by the canceling between positive and negative electric charges and between north and south magnetic poles	e^-, p^+	atoms and molecules, chemical reactions, electric appliances, magnetic field of Earth, electromagnetic radiation (e.g., light, radiowaves)
Nuclear	1	10^{-15} m	p^+, n	nuclear reactions (e.g., stellar energy sources, atomic energy, radioactive decay)

Table I.2 summarizes the essential features of the four forces. The gravitational force is the weakest force, followed by the weak, the electromagnetic, and the nuclear forces. Their range varies from very short-range to infinite. Only gravity acts on all particles; the weak force plays a role only when a neutrino is present; the electromagnetic force acts only on matter with electric charge or magnetic fields; and the nuclear force acts only between protons and neutrons. Because of the differences in strength and range, each of the four forces has very specific applications in nature, from holding stars and galaxies together to making chemistry work, producing light and radio waves, and releasing enormous amounts of energy in nuclear reactions.

And now a word of caution. Though at present only four fundamental forces are well established experimentally and theoretically, others may exist—perhaps forces that are very weak or of short range, and thus have escaped detection. In fact, the existence of an additional fundamental force has been suggested repeatedly for the past two decades, most recently by Ephraim Fischbach, Daniel Sudarsky, Aaron Szafer, and Carrick Talmadge of Purdue University and by S. H. Aronson of Brookhaven National Laboratory.

These researchers re-analyzed the data of a famous experiment carried out earlier in this century by Baron von Eötvös (1848–1919), a Hungarian physicist. Eötvös and his coworkers carefully measured the gravitational acceleration—that is, the rate at which bodies fall near the surface of the Earth—of bodies made of different materials and possessing different masses, among them metals, water, wood, tallow, and several kinds of minerals. Their aim was to determine whether gravitational acceleration depends on the composition and mass of falling bodies or not. They reported no such dependences. This result confirmed the traditional view of physicists that a feather and coin dropped from the same height in a vacuum fall at the same rate. This view dates back to

Galileo Galilei (1564–1642), who supposedly experimented with different sized cannon balls by dropping them from the Leaning Tower of Pisa.

When Fischbach and his colleagues re-examined the original data of the Eötvös experiment, they found small unexplained discrepancies. The Hungarian scientists had also been aware of these discrepancies, but thought they were unavoidable experimental errors and ignored them. Upon very careful analysis, the Fischbach team interpreted the discrepancies as having arisen from a fifth fundamental force, which acted in addition to the force of gravity. They suggest that this force is much weaker than gravity, that it is repulsive, and that it has a range of about 200 meters. Furthermore, they suggest that this force depends on a property of matter called *hypercharge*, which is related to the number of protons and neutrons present in atomic nuclei. This latter property means that the hypercharge force is composition-dependent. For example, in a vacuum a feather should fall more rapidly than a coin.

At present it is not known whether the suggestion of a fifth fundamental force is correct. The Eötvös experiment needs to be repeated with greater sensitivity and with a variety of materials of precisely known composition to find out. If it turns out to be real, "the force is not likely to change calculations on the scale of planetary systems. [However,] when you see something as fundamental as a new force, it's likely to change many things. We will have to rethink many views of particle physics and cosmology." (Fischbach, quoted in *The New York Times*, January 8, 1986, 1:2.) But even if the hypercharge force is not confirmed by new experiments, the fact that a team of competent scientists suggested its possible existence indicates that current scientific theories are not necessarily complete and final. Scientists have barely begun to investigate nature and, no doubt, many currently held views of physics and the other sciences will have to be rethought in the future.

THE ARCHITECTURE OF MATTER

Now that we are acquainted with the fundamental building blocks of matter and the forces by which they interact, let us examine the architecture of matter in more detail. In particular, let us look at the structure of atomic nuclei, atoms, and molecules. As noted earlier, these particles are far too small to be seen with the unaided eye or felt by our sense of touch. That is why we have no direct experience with them. In order to acquire some understanding of them, let us for the moment give our imagination free rein and do a *Gedanken* or "thought" experiment. We invent a machine that can magnify objects indefinitely and show all their details. (In this regard, the machine is superior to any electron microscope, X-ray diffraction machine, or particle accelerator.) Then let us magnify a water droplet and "see" what it is made of.

As the droplet becomes larger, there is at first no apparent change in its appearance. Even when it has grown to the size of a football field, it still seems to contain clear, liquid water, except for some giant bacteria, protists, and other impurities, which do not interest us at the moment. However, when the droplet becomes greater than one kilometer across, details suddenly emerge. We no longer see water, but an immense swarm of particles that are in continuous disorganized motion. Let us stop the magnification at 100 million, when the

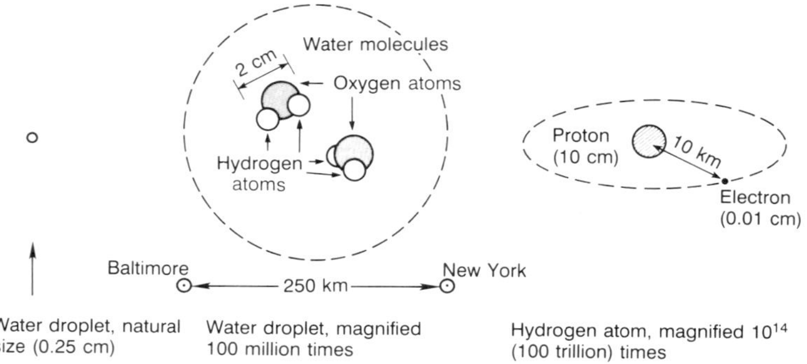

Figure I.12. Structure of a water droplet. At a magnification of 10^{14} the water droplet would extend from the Sun to between the orbits of Venus and Earth.

droplet has grown to a size of 250 km, roughly the distance from New York City to Baltimore (see figure I.12). Each of the particles is now about two centimeters across and they are all alike. If we had sufficient patience to count them, we would find approximately 10^{20} of them in that one water droplet.

Each of the particles making up the water droplet consists of three blurry-looking spheres, which are bonded together by the electric force and are not easily broken up. These particles are water molecules and the spheres are atoms. The central and slightly larger spheres are oxygen atoms and the two peripheral ones are hydrogen atoms. For brevity, hydrogen and oxygen atoms are abbreviated by the capital letters H and O, and water molecules by H_2O.*

At this magnification we cannot recognize any structural details in the atoms. For that we need to enlarge the water droplet further until it is 250,000 km in size and the total magnification is 100 billion. The water droplet is now nearly one-fifth the size of the Sun, the molecules are roughly 20 m across, and the atoms are about 10 m. The atoms turn out to consist mostly of empty space. At each of their centers is a tiny dense particle, just barely visible despite the enormous magnification. They are about 1/100 cm in size and are the nuclei of the atoms. The nuclei of the hydrogen atoms are single protons, while those of the oxygen atoms consist of clusters of eight protons and eight neutrons, held tightly together by the nuclear force.

Circling these nuclei are electrons at distances of up to approximately 10 m, corresponding to the sizes of the atoms. The hydrogen atom has only one electron, but the oxygen atom has eight. The electrons are smaller still than the nuclei, and it would require an additional magnification of one thousand to make them visible.

The electrons are held in their orbits by the electric force, which pulls them (since they are negatively charged) toward the positively charged nuclei. However, the electrons do not fall into the nuclei because of the forward component of their orbital motions, which would carry them off into space if the pull toward the central nucleus were suddenly to disappear. This is quite analogous to the

*The subscript 2 in H_2O means that there are two hydrogen atoms in every water molecule. The absence of a subscript on the O means that only one oxygen atom is present.

BOX I.3.

Exponential Notation

MANY concepts in mathematics and in the sciences are more clearly expressed by numbers or symbols than by words. For instance, "one thousand nine hundred eighty-six" requires considerably more effort to write as well as to read than the numerical notation "1986."

When we deal with very large or very small numbers, even Arabic numerals become cumbersome. For example, "150,000,000,000" and "0.000,000,000,000,000,000,000,000,000,000,91" require that we count zeros before we know the values of these numbers. To ease the task of writing and reading very large and very small numbers, mathematicians have invented *exponential notation*. This notation has the form a^m and is read "*a* to the power *m*," where *a* and *m* stand for numbers and where *a* is called the *base* and *m* the *exponent*. In this book, *a* will always be the number 10 and *m* will always be a positive or negative integer or zero. (In general, however, *a* and *m* may be any number, including fractions and irrational numbers.)

Thus, the exponential notation used in this book is of the form 10^9, 10^8, 10^{-6}, and so forth. This notation is to be interpreted as illustrated by the following examples:

⋮	⋮		⋮
10^9	= 1,000,000,000	one billion	(a one followed by *nine* zeros.)
10^8	= 100,000,000	one hundred million	(— " — *eight* ")
⋮	⋮		⋮
10^3	= 1,000	one thousand	(— " — *three* ")
10^2	= 100	one hundred	(— " — *two* ")
10^1	= 10	ten	(— " — *one* ")
10^0	= 1	one	(— " — *zero* ")
10^{-1}	= 0.1	one-tenth	(a one with the decimal point shifted *one* place to the left.)
10^{-2}	= 0.01	one-hundredth	(*two* places — " —)
⋮	⋮		⋮
10^{-6}	= 0.000,001	one-millionth	(*six* places — " —)
⋮	⋮		⋮

In exponential notation, the two numbers in the second paragraph above can now be written clearly and compactly as 1.5×10^{11} and 9.1×10^{-31}. These two numbers describe the Sun–Earth distance in meters and the mass of the electron in kilograms.

motions of the planets around the Sun or of satellites around the Earth (see box I.4).

It is interesting to carry a bit further the comparison between electrons orbiting around their nuclei and planets circling around the Sun. For instance, electron orbits are approximately 50,000 times larger than the sizes of the atomic nuclei. In contrast, the Earth's orbit is only about 200 times larger than

BOX I.4.
Centripetal and Centrifugal Forces

THE orbital motions of electrons around the central nucleus or of planets around the Sun are sometimes explained as being due to a balance between two forces: the *centripetal (center-seeking) force* that pulls the electrons or planets toward the centers of their orbits and the *centrifugal (center-fleeing) force* that pulls them away from the centers of their orbits. Strictly speaking, this is not a correct way of describing orbital motions because only the centripetal forces (that is, the electric and gravitational forces in these two examples) are real. The centrifugal forces are not. There are no forces that push the electrons or planets away from the centers of their orbits. The centrifugal force merely constitutes a particle's resistance to having the direction of its motion changed. It is not a force on par with the four fundamental forces I introduced in the text. Should the centripetal force suddenly disappear, then the centrifugal force would disappear as well and the particle would fly off in a straight line. (This can be tested easily by whirling a rock in a circle at the end of a string and suddenly cutting the string, or, more unpleasantly, by driving a car around a curve and suddenly skidding on an oil slick or patch of ice.)

Despite its fictitious nature, the concept of centrifugal force is a useful one, particularly when analyzing the behavior of rotating gases and fluids. However, the use of the concept of centrifugal force is logically consistent with the rest of physics only in a reference-frame that rotates along with the gases or orbiting objects. A simple example will explain this. Let us assume we step onto the platform of a merry-go-round and sit on one of the horses. As the merry-go-round starts turning, we feel a force that tends to push us off the horses in the direction away from the merry-go-round's center. That force is the centrifugal force. To avoid falling off we need to exert an equally strong force toward the center of the merry-go-round, the centripetal force. We do this by holding on to the horses or by leaning inward. As long as the two forces, the centrifugal and centripetal forces, are in balance, we will not fall off. Of course, the faster the merry-go-round turns or the farther from its center we sit, the stronger we have to hold on or lean to maintain this balance. The reason is that the centrifugal force increases with rotation speed and distance from the rotation axis.

Note that only we who are on the rotating merry-go-round feel the centrifugal force. Spectators on the ground are not aware of it. All they see is that we are holding on to the horse or leaning inward. Therefore, as far as they are concerned, only the centripetal force is real, the centrifugal force is not. Nevertheless, in the context just described (that is, from the point of view of a rotating reference frame), the concept of centrifugal force is a perfectly good one and many research scientists make use of it. This text will too in discussions of the physics of rotating galaxies, interstellar gas clouds, and the origin of the Solar System.

the Sun. Thus, atoms are, in relative terms, more extended by far than the Solar System. Expressed in units of volume, atoms contain roughly 10 million times more empty space than the Solar System.

There are many different kinds of atoms in nature—hydrogen, carbon, nitrogen, oxygen, iron, gold, and uranium atoms, to name just a few. Quite often, scientists are not so much interested in individual atoms, but in collections of them, like the carbon atoms in our bodies or the oxygen atoms in the air. They then speak of the *element* carbon, the *element* oxygen, or more briefly, just of carbon or oxygen. The total number of elements occurring in nature is 92 and about a dozen more (the transuranic elements) have been created in nuclear accelerator experiments. Of the naturally occurring elements, the simplest is hydrogen and the most complex is uranium. Only a handful of them are of real interest to us in this book: hydrogen, helium, carbon, nitrogen, oxygen, phosphorus, and sulfur, for example.

Every element is characterized by unique physical and chemical properties, such as color, degree of hardness, density, ability to conduct heat or electricity, tendency to combine with other elements and so forth. For instance, the oxygen in the present Earth atmosphere exists in molecular form, O_2, and it is gaseous; hydrogen reacts readily with oxygen to form water, H_2O; iron is a solid up to approximately 1800 K,* but at higher temperatures it melts and becomes a liquid; and gold is a very dense, malleable, noncorroding and yellow metal, with high heat and electrical conductivity.

Why do the elements have unique properties that distinguish them from each other? The answer is that the atoms of every element have their own unique structures: each has a unique nucleus that (in the electrically neutral state) is surrounded by a unique number of electrons. For instance, the nuclei of hydrogen, helium, carbon, nitrogen, and oxygen contain, respectively, one, two, six, seven, and eight protons (see figure I.13). In the electrically neutral state, the protons (because of their positive charge) keep an equal number of electrons in orbit; thus, the nucleus of hydrogen is surrounded by one electron, that of helium by two, that of carbon by six, and so forth.

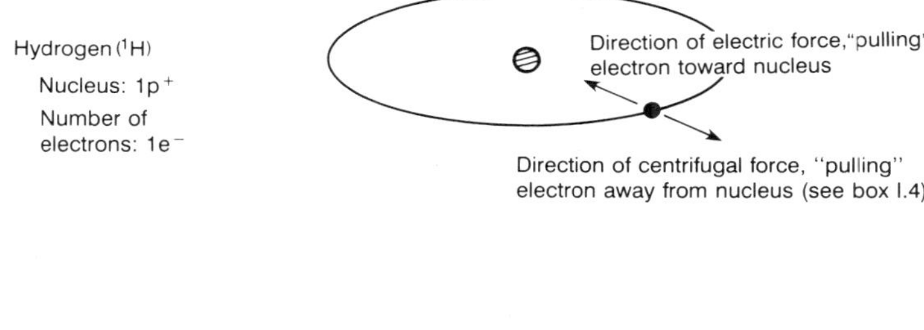

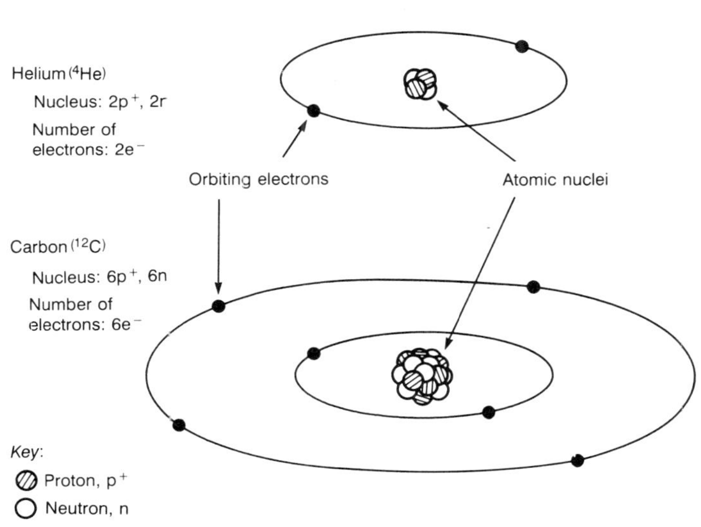

Figure I.13. Diagrams of hydrogen, helium, and carbon atoms (not to scale). The protons and neutrons are roughly the same size. The electrons are approximately 1000 times smaller and their orbits are approximately 100,000 times larger than the protons and neutrons.

*This book uses degrees *Kelvin* or *Celsius* to measure temperature, the temperature scales used in science. Box 1.2 discusses the meaning of temperature and compares degrees Kelvin and Celsius to the Fahrenheit scale used in the United States.

Electrically neutral atoms are the only ones under consideration here. In later chapters I describe how atoms may lose one or more electrons or they may gain an electron. They then acquire a positive or negative charge and are called *ions*. The process by which atoms lose or gain electrons is called *ionization*. It is important to recognize that ionization does not change an atom from one kind to another (such as a nitrogen atom into a carbon or oxygen atom). That can be accomplished only by altering the number of protons in the nucleus.

Remember, *the number of protons present in the nucleus determines the nature of an atom*. Thus, the above description of hydrogen, helium, and carbon and so on may be turned around and stated like this: An atom whose nucleus contains one proton is a hydrogen atom; an atom whose nucleus contains two protons is a helium atom; an atom whose nucleus contains six protons is a carbon atom; and likewise for all the other atoms.

In addition to protons, atomic nuclei usually contain neutrons. The neutrons are necessary to make the nuclei stable, but because they are electrically neutral, neutrons have little effect on the chemical properties of the elements and they are not considered in the naming of the elements.

Most elements come in several versions that differ from each other merely in the number of neutrons present in their nuclei. Such different versions of the same element are called *isotopes*. For instance, there are three isotopes of hydrogen: Their nuclei consist of *one* proton (the most common of the hydrogen isotopes), *one* proton and one neutron, and *one* proton and two neutrons. There are also three isotopes of carbon: Their nuclei consist of *six* protons and six neutrons (the most common of the carbon isotopes), *six* protons and seven neutrons, and *six* protons and eight neutrons. The third carbon isotope is not stable, but decays into nitrogen. (Radioactive decay of unstable isotopes is discussed further in box 5.1.)

There is a shorthand notation for the elements and their isotopes. It consists of one- or two-letter symbols, together with superscripts and subscripts, as shown in table I.3, the Periodic Table of the Elements. The letters are abbreviations of the elements' names; the superscripts refer to the total number of protons *plus* neutrons present in the nuclei; and the subscripts are reminders of the number of protons present in the nuclei. For instance, the notation for the three isotopes of hydrogen is ${}^{1}_{1}H$, ${}^{2}_{1}H$, ${}^{3}_{1}H$, and that for the three isotopes of carbon is ${}^{12}_{6}C$, ${}^{13}_{6}C$, and ${}^{14}_{6}C$.* Only the most common isotopes of the elements are listed in the table.

Generally it is not necessary to distinguish between different isotopes, and in this book most discussions will use the symbols H, C, and so forth without superscripts. Only in discussing nuclear reactions or decays will it be necessary to refer to specific isotopes and to use the superscripts. The subscripts will

*Usually the letter abbreviations are taken from the elements' Latin names. When the English and the Latin names of an element do not coincide, strange abbreviations result. For instance, Na is the abbreviation for sodium and is derived from this element's Latin name *natrium*; Ag is the abbreviation for silver (*argentum*); and Au is the abbreviation for gold (*aurum*).

The isotope ${}^{2}_{1}H$ is called hydrogen-2 or deuterium and the isotope ${}^{3}_{1}H$ is called hydrogen-3 or tritium. The isotope ${}^{12}_{6}C$ of carbon is called carbon-12, ${}^{13}_{6}C$ is called carbon-13, and ${}^{14}_{6}C$ is carbon-14. The isotopes of all the other elements are referred to similarly: helium-4 for ${}^{4}_{2}He$, oxygen-16 for ${}^{16}_{8}O$, and so forth.

Table I.3. Periodic Table of the Elements. Only the most common isotopes of the elements are listed.

1 **H** 1 Hydrogen																	4 **He** 2 Helium
7 **Li** 3 Lithium	9 **Be** 4 Beryllium											11 **B** 5 Boron	12 **C** 6 Carbon	14 **N** 7 Nitrogen	16 **O** 8 Oxygen	19 **F** 9 Fluorine	20 **Ne** 10 Neon
23 **Na** 11 Sodium	24 **Mg** 12 Magnesium											27 **Al** 13 Aluminum	28 **Si** 14 Silicon	31 **P** 15 Phosphorus	32 **S** 16 Sulfur	35 **Cl** 17 Chlorine	40 **Ar** 18 Argon
39 **K** 19 Potassium	40 **Ca** 20 Calcium	45 **Sc** 21 Scandium	48 **Ti** 22 Titanium	51 **V** 23 Vanadium	52 **Cr** 24 Chromium	55 **Mn** 25 Mangenese	56 **Fe** 26 Iron	59 **Co** 27 Cobalt	58 **Ni** 28 Nickel	63 **Cu** 29 Copper	64 **Zn** 30 Zinc	69 **Ga** 31 Gallium	74 **Ge** 32 Germanium	75 **As** 33 Arsenic	80 **Se** 34 Selenium	79 **Br** 35 Bromine	84 **Kr** 36 Krypton
85 **Rb** 37 Rubidium	88 **Sr** 38 Strontium	89 **Y** 39 Yttrium	90 **Zr** 40 Zirconium	93 **Nb** 41 Niobium	98 **Mo** 42 Molybdenum	97 **Tc** 43 Technetium	102 **Ru** 44 Ruthenium	103 **Rh** 45 Rhodium	106 **Pd** 46 Palladium	107 **Ag** 47 Silver	114 **Cd** 48 Cadmium	115 **In** 49 Indium	120 **Sn** 50 Tin	121 **Sb** 51 Antimony	130 **Te** 52 Tellurium	127 **I** 53 Iodine	132 **Xe** 54 Xenon
133 **Cs** 55 Cesium	138 **Ba** 56 Barium	139 **La** 57 Lanthanum	180 **Hf** 72 Hafnium	181 **Ta** 73 Tantalum	184 **W** 74 Tungsten	187 **Re** 75 Rhenium	192 **Os** 76 Osmium	193 **Ir** 77 Iridium	195 **Pt** 78 Platinum	197 **Au** 79 Gold	202 **Hg** 80 Mercury	205 **Tl** 81 Thallium	208 **Pb** 82 Lead	209 **Bi** 83 Bismuth	209 **Po** 84 Polonium	210 **At** 85 Astatine	222 **Rn** 86 Radon
223 **Fr** 87 Francium	226 **Ra** 88 Radium	227 **Ac** 89 Actinium															
				140 **Ce** 58 Cerium	141 **Pr** 59 Praseodymium	142 **Nd** 60 Neodymium	145 **Pm** 61 Promethium	152 **Sm** 62 Samarium	153 **Eu** 63 Europium	158 **Gd** 64 Gadolinium	159 **Tb** 65 Terbium	164 **Dy** 66 Dysprosium	165 **Ho** 67 Holmium	166 **Er** 68 Erbium	169 **Tm** 69 Thulium	174 **Yb** 70 Ytterbium	175 **Lu** 71 Lutetium
				232 **Th** 90 Thorium	231 **Pa** 91 Protactinium	238 **U** 92 Uranium											

always be left off because they merely spell out information about protons that is implied by the letter symbols and given explicitly, for reference purposes, in the table.

ELECTROMAGNETIC RADIATION

The light from the Sun and the stars, radio waves emitted by broadcasting stations, radar used in air traffic control and by the police, and X rays used in medicine are all examples of the same phenomenon—electromagnetic radiation. Clearly, the various kinds of electromagnetic radiation are not all alike. What is the difference?

Electromagnetic radiation consists of waves of oscillating electric and magnetic fields traveling with the speed of light. In order to understand this statement and to answer the above question, let us first consider more familiar waves, namely water waves on a lake or ocean.

When water waves travel across the surface of a lake or the ocean, water does not flow horizontally along with the wave. Instead, the water's motion is a rhythmic up and down oscillation. We can observe this by watching a piece of cork bobbing up and down as a wave passes it or, better yet, by going into the water and enjoying the sensation of being lifted way up and then brought way down by a big wave. The up and down oscillations, which constitute a water wave, vary according to a regular pattern. One swimmer may just reach the crest of the wave, while a little farther away another swimmer is still on the way up and a third is already being carried down. With time, the rhythmic up and down motions of the swimmers repeat as wave after wave passes by (see figure I.14).

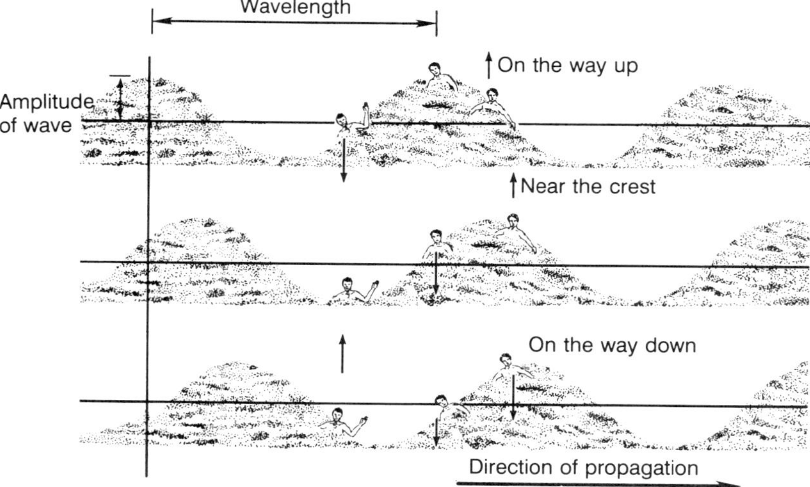

Figure I.14. A water wave. The rhythmic up and down oscillations of the surface of the water are illustrated by three swimmers. (Top) The swimmer on the right is being carried up by the wave; (center) the right-hand swimmer has almost reached the crest of the wave; and (bottom) the crest of the wave has passed and the swimmer is on the way down. Between the top and bottom diagrams, the wave travels one-third of a wavelength to the right.

Water waves are not all alike. Not only do they have different amplitudes, they also differ in wavelength, oscillation frequency, and the speed with which they progress. The wavelength is the distance from crest to crest of the wave and the oscillation frequency is the rate at which the water moves up and down at a given location (expressed in cycles per second). The speed of the wave is equal to the product of wavelength and oscillation frequency. Large ocean waves have wavelengths exceeding 100 meters; their oscillation frequencies are roughly one cycle every 10 seconds; and their speeds are about 30 to 50 kilometers per hour. On a lake, the wavelength may be as short as a fraction of a meter, the oscillation frequency of the order of one cycle per second, and the speed a few kilometers per hour.

The various kinds of electromagnetic radiation resemble water waves in that they, too, consist of rhythmic back and forth motions, and they are characterized by wavelength, oscillation frequency, and the speed of propagation. The main difference is that in the case of electromagnetic radiation the back and forth motions do not involve water or any other material substance. They are oscillations of electric and magnetic fields. The two fields oscillate in unison and they are aligned at right angles to each other and are perpendicular to the direction of wave propagation, as shown in figure I.15. As in the case of water

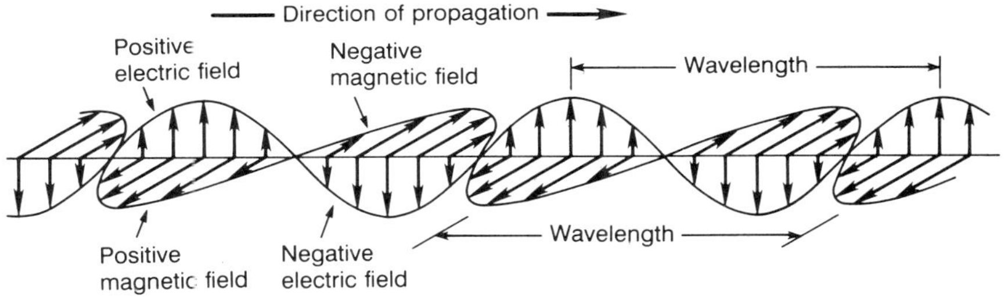

Figure I.15. Schematic diagram of electromagnetic radiation.

waves, the product of wavelength and oscillation frequency of electromagnetic radiation gives the speed of propagation—*300,000 km/sec*. However, unlike the speed of water waves, this speed is the same for all electromagnetic waves. It is the *speed of light*. More precisely, it is the speed of light in vacuum. When electromagnetic radiation passes through a material medium, such as air or glass, the speed of propagation is slightly reduced.

The only difference between the various kinds of electromagnetic radiation is in wavelength and oscillation frequency, as shown in figure I.16. The wavelengths range from 10^{-12} m for gamma rays to several kilometers for the longest radio waves. The rest—namely X rays, ultraviolet (UV) rays, visible light, infrared (IR) radiation, and the various kinds of radio waves (microwaves, radar, and FM, TV, and AM broadcasting waves)—lie between these two extremes.

Visible light has wavelengths between 4×10^{-7} and 7×10^{-7} m, which is a relatively narrow range compared with the full range of electromagnetic radiation (see figure I.17). Yet this range contains all the color components to which human eyes are sensitive. These components become apparent when white light, like that from the Sun, is sent through a prism, which splits it into the familiar spectrum of rainbow colors from violet to blue, green, yellow, orange, and red. The same splitting occurs in a rainbow, where raindrops act as

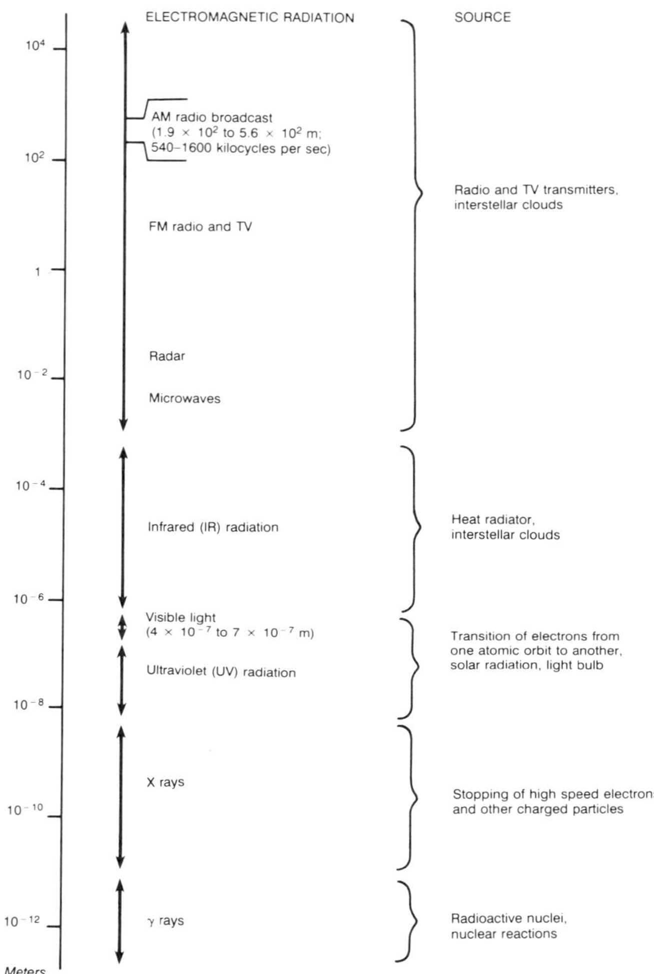

Figure I.16. The wavelength range of electromagnetic radiation.

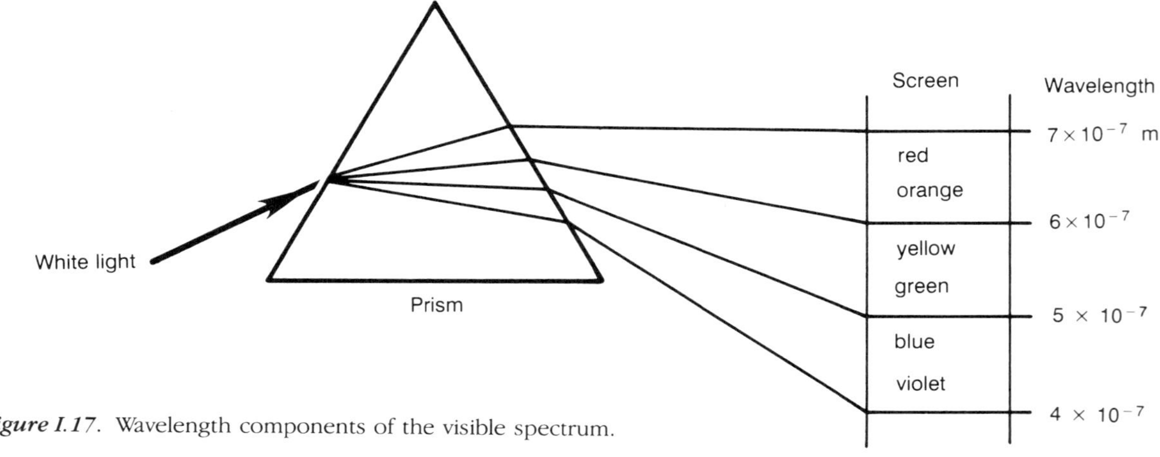

Figure I.17. Wavelength components of the visible spectrum.

prisms. Each of the color components of visible light has its own wavelength, with violet light having the shortest and red light the longest.

In an earlier discussion I referred to electromagnetic radiation as being an elementary particle — the photon. Yet it is also a wave, consisting of oscillating electric and magnetic fields, propagating with the speed of light. How can the same entity be both a particle and a wave?

To answer this question it is important to realize that when scientists speak of elementary particles, of atoms and molecules, and of other natural phenomena, they are constructing models that they hope will correspond to nature. The models are human creations, not reality. Any model that is consistent with observations or experiments is a good model. Photons and electromagnetic waves are two different models — a particle model and a wave model — of the same phenomenon. Both models are equally consistent with what is observed in nature. Hence, both are good models. It turns out that the particle model is more useful when considering the emission or absorption of electromagnetic radiation, while the wave model is more useful when considering the propagation of electromagnetic radiation. Thus scientists need both models and later chapters will refer to both. This is because the emission and absorption as well as the propagation of electromagnetic radiation are important processes in nature and affect the structure and evolution of matter.

ELECTROMAGNETIC SPECTRA

All material bodies — solids, liquids, and gases — emit electromagnetic radiation. The emission is usually the result of very complex processes, involving the random motions of a body's constituent atoms, molecules, and other particles. Generally, if a body is opaque it will emit radiation in all wavelengths. Furthermore, if it is very hot, such as the Sun, burning wood, or a stove-top burner turned on high, most of the wavelengths tend to be short (which is equivalent to saying that individual photons carry a lot of energy) and can be seen by the human eye or by UV and X-ray detectors. If a body is cool, such as our bodies, the ground, or ice, most of the wavelengths tend to be long (that is, the individual photons carry little energy) and can be detected only by IR sensors or radio antennae.

The intensity of the radiation emitted by an opaque body is not the same in all wavelengths but shows a distinct wavelength dependence. A plot representing this dependence is said to be a body's *continuum spectrum*. In figure I.18, continuum spectra are shown for several bodies, each at a different temperature. The curves indicate that the radiation from a hot body is more intense than that from a cool body.

Two features of the curves deserve special notice. First, the peaks of the curves shift to shorter wavelengths with increasing temperature. We all are familiar with this: A piece of iron withdrawn from a very hot fire at first emits whitish-blue radiation (short wavelengths), but as it cools its radiation turns orange and then deep red (long wavelengths). Second, the areas under the curves increase with temperature, meaning that the total energy emitted per unit area every second by the bodies increases as their temperatures increase. We are familiar with this as well: The radiation from the Sun is more intense

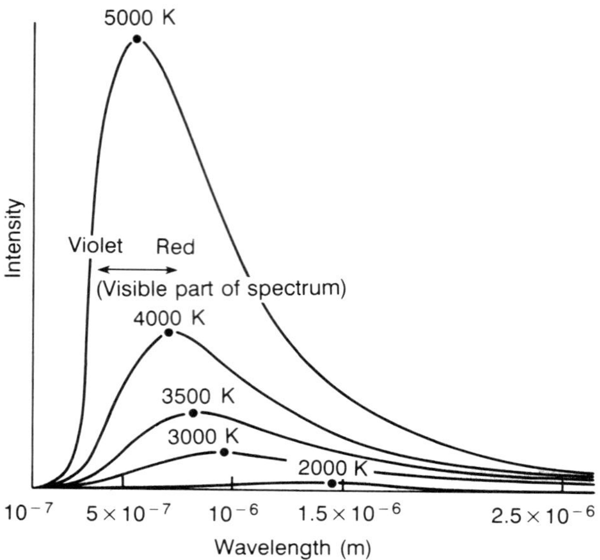

Figure I.18. Continuum spectra of opaque bodies at five different temperatures. (The spectral curves shown are idealized. Those from real bodies have emission and absorption lines superimposed.)

than that from cooler bodies such as burning coals or the ground. (*Intensity* is defined as the energy in a beam of electromagnetic radiation passing through a unit normal area per unit time.)

Exactly how is electromagnetic radiation emitted or absorbed by atoms? First we need to recognize that unlike planets or satellites, which may assume orbits of any size, electrons circling a nucleus can occupy only distinct, or quantized, orbits. If an electron is in the smallest allowed orbit, it is said to be in the *ground state*. If it is in one of the higher orbits, it is said to be in an *excited state*. Normally, the electrons are in the ground state. However, collisions resulting from the random motions of a body's constituent atoms and molecules tend to kick the outermost electrons of atoms into higher orbits (this may also happen by the absorption of photons, see below). When an electron has been raised into a higher orbit, it is not long before it jumps back to lower orbits. Sometimes it jumps back to the ground state in one step. At other times it returns to the ground state in several steps by first jumping to intermediate orbits. As it does so, electromagnetic radiation of specific wavelengths is emitted, corresponding to the energy differences between the higher and lower orbits (see figure I.19).

When the radiation from a large collection of identical atoms, whose electrons are in excited states, is sent through a prism and projected onto a screen or photographic plate, a distinct wavelength pattern, called an *emission spectrum*, can be seen. Emission spectra are unique for each element (because the allowed electron orbits of each element are unique) and can be used to deduce which elements are present in a radiating body, their relative abundances, the body's temperature, and much other information.

When the radiation from a hot, opaque body (that is, radiation that is fairly intense and has a continuum spectrum) passes through a thin layer of gas, it

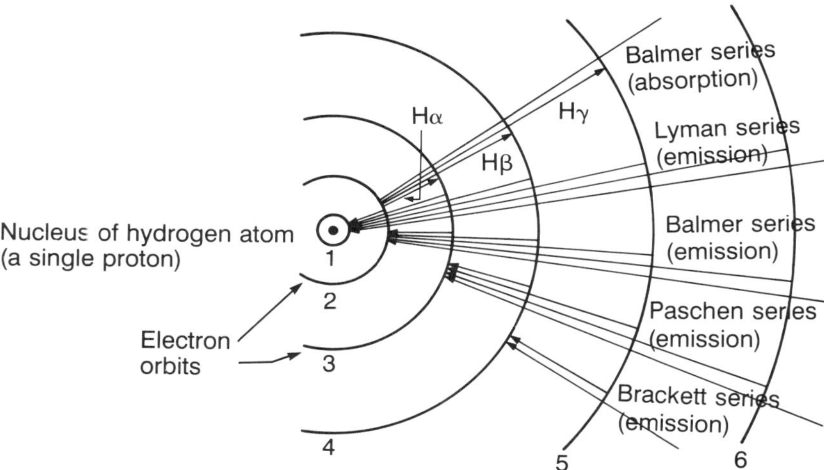

Figure I.19. Emission and absorption of electromagnetic radiation by the hydrogen atom. Six electron orbits are shown. Orbit 1 represents the ground state; the others are excited states. The inward-pointing arrows indicate possible jumps by the electron from higher to lower orbits, during which photons of distinct wavelengths are emitted. The outward-pointing arrows indicate possible jumps by the electron from the first excited state to higher orbits, during which photons of distinct wavelengths are absorbed. The names *Lyman, Balmer, Paschen*, and *Bracket* are labels for the series of emission and absorption lines (that is, emission and absorption spectra) resulting from the transitions indicated in the diagram. The arrows labelled Hα, Hβ, and Hγ refer to the Balmer series of hydrogen and correspond to prominent absorption lines seen in the solar spectrum, as shown in figure I.20. (*Note*: In reality, electron orbits are not quite as simple as indicated in this diagram.)

tends to kick electrons of the gas's constituent atoms into excited states. This process requires energy of specific, quantized values, corresponding to the energy differences between the electron's lower and higher orbits. Thus, only photons having wavelengths that correspond to these energy differences can kick the electrons into excited states. Consequently, these photons are absorbed (that is, they disappear due to the kicking), which creates a distinct pattern of valleys in the original continuum spectrum. This pattern is called an *absorption spectrum*. Figure I.20 shows that absorption spectra, too, are unique for each element. They are commonly observed in the electromagnetic radiation from the Sun. They are also observed in the radiations we receive from distant stars, from galaxies, and from gas clouds lying between these distant objects and us.

If you are somewhat uneasy about this rather brief discussion of continuum, emission, and absorption spectra, do not worry. This is a difficult subject and would require several chapters and advanced physical theory to be dealt with thoroughly. The main point is that by analyzing the spectra of the electromagnetic radiation coming from stars, interstellar gas clouds, galaxies, and other distant objects, astronomers are able to deduce a great deal of information. In the next five chapters I shall discuss some of this information.

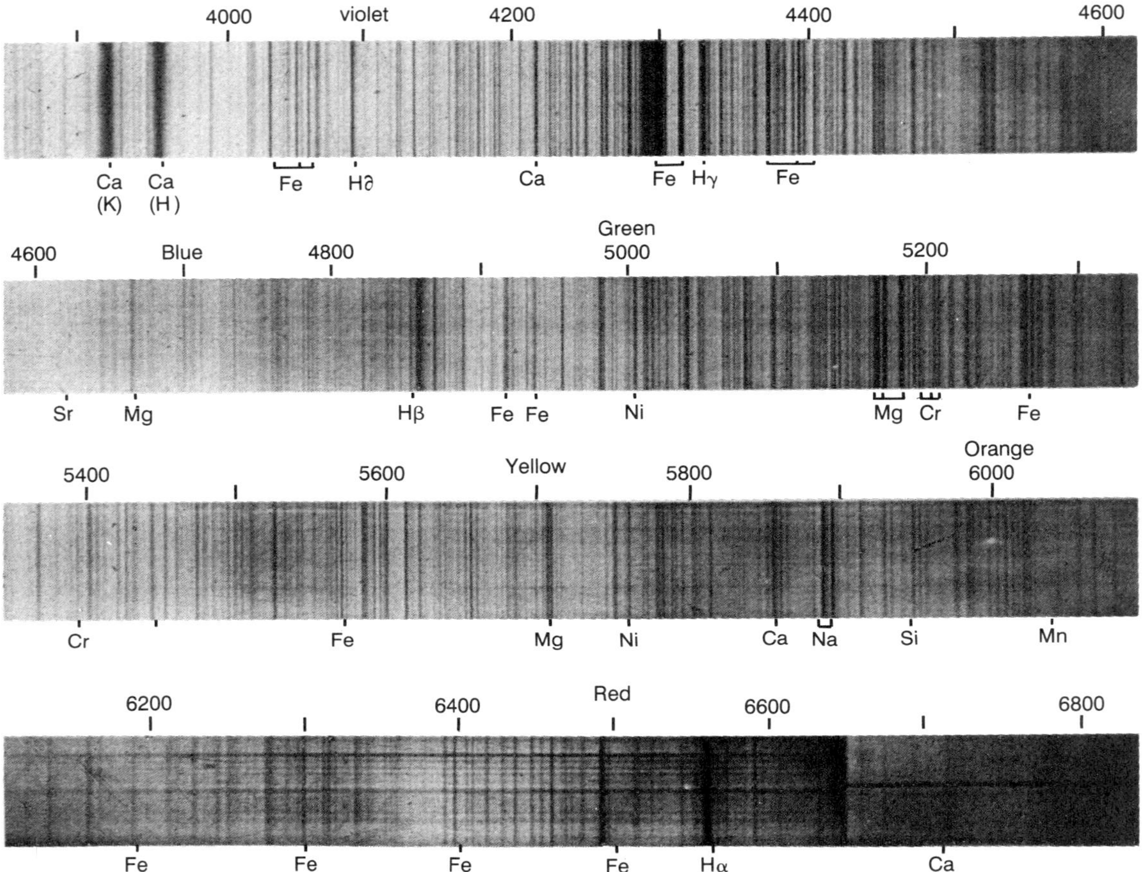

Figure I.20. The solar absorption spectrum. This spectrum is produced by continuous radiation from the Sun's interior as it passes through the solar atmosphere. Elements in the solar atmosphere selectively absorb radiation of specific wavelengths and create the pattern illustrated here. The numbers shown above the spectrum refer to wavelength in Ångstroms (Å; one Å equals 10^{-8} cm or 10^{-4} μm). The symbols shown below the spectrum refer to the elements responsible for the absorption lines. (Note the absorption lines of hydrogen [labeled Hα, Hβ, Hγ, and Hδ] and the H and K lines of calcium. They will be referred to in later chapters.)

SCIENTIFIC THEORY

In this Introduction to Part One I presented the five kinds of stable elementary particles that make up the matter and energy content of the Universe—electrons, protons, neutrons, neutrinos, and photons. I then introduced the four forces by which these particles interact with each other—the gravitational, electromagnetic, nuclear, and weak forces—and then described how protons and neutrons combine to form atomic nuclei and how these nuclei surround themselves with electrons to form atoms. Finally, I described electromagnetic radiation, which is really just another way (the wave model) of representing photons (the particle model).

Knowledge of the stable elementary particles and their interactions provides sufficient background to begin the study of the origin and evolution of the Universe. However, before we do so, let us concern ourselves with the question, "How closely do the theories discussed in this book correspond to the evolution that really happened?"

To answer this question, we need to recognize the basis of the theories being discussed. In brief, their basis is scientific knowledge that has been acquired over several centuries by many people. This knowledge is rooted in careful observations of natural phenomena as well as in countless experiments designed to help us understand specific aspects of nature. Scientists use the results of observations and experiments to construct theories that they hope will explain natural phenomena, relate different phenomena to each other, and in general contribute to an understanding of nature.

When scientific theories are first put forth, they are always tentative. They are then tested by additional observations and experiments, often carried out by many different scientists or teams of scientists and, whenever possible, under a great range of conditions. Depending on the results of the new observations and experiments, the original theories are refined, altered, or discarded.

This process of observation, experimentation, development of theories, communication of the theories to other scientists, testing, and critiquing lies at the heart of doing science. It is a very social process, one that involves many people with differing degrees of talent, expertise, background, temperament, style, and motivation. And, like all human endeavors, it progresses in very nonlinear ways. Quite often a given path turns out to be wrong, even though at first it appeared to be promising, and it needs to be either redirected or abandoned. Observational and experimental data are sometimes misinterpreted. Some individuals acquire high status and their opinions are accepted as established facts. However, misinterpretations and mere opinions usually do not survive very long in science. They do not last because scientific theories are continually reexamined, questioned in the light of new discoveries in other areas of science, compared with alternative theories, and tested through the use of new and often more sensitive observational and experimental tools. Ultimately, neither dogma nor opinion makes scientific theories acceptable. It is how well these theories correspond to what is observed in nature that determines their acceptance by the scientific community. That is why Martin Schwarzschild (1958) referred to observations as the " . . . pillars . . . on which we base our theories. . . . "

The correspondence of a given theory to nature is never perfect. At best the correspondence is limited by the unavoidable uncertainties present in all observational and experimental results. Additional uncertainties usually arise from incomplete observational and experimental data. They may also arise from our intellectual limitations in interpreting correctly the data we do have and from the particular models or biases we start with when carrying out our observations or setting up our experiments. Werner K. Heisenberg (1901–1976), one of the great physicists of our century, referred to these last sources of uncertainties when he said, "What we observe is not nature itself, but nature exposed to our method of questioning" (1958). Thus, in general, scientific theories span the whole spectrum from being very good, when they correspond extremely closely to all relevant observational and experimental data, to being rather tentative and speculative.

Many of the theories presented in this book are good ones and are generally accepted by the scientific community. But because many of the topics deal with events that took place long ago, at great distances, or under uncommon circumstances, a good part of the book's discussion is of a tentative nature. Thus, the answer to the question "How closely do this book's theories correspond to the evolution that really happened?" is that some (maybe many) of the discussions of evolution probably correspond quite closely to what really happened. In the case of others it is impossible to be so certain. No doubt, some of the theories will have to be significantly revised or even discarded as continued scientific inquiry provides new information.

In most instances in which a theory is clearly speculative, I point this out in the text. Even in cases that are not so qualified, I would like to caution the reader not to think that the theories represent the final word on the subject. Science is a multifaceted and rapidly developing human enterprise and no one can predict in which directions our understanding of nature will progress.

> If a man will begin with certainties, he shall end in doubts; but if he will be content to begin with doubts he shall end in certainties.
> **Francis Bacon** (1561–1626)
> *The Advancement of Learning, Book I*

Exercises

1 Describe how the macroscopic structure of matter, such as that of ice crystals, snowflakes, and diamonds, is determined by the microscopic properties of matter.

2 (*a*) Confirm through your own measurements that the density of liquid water is 1 g/cm^3.
(*b*) Why is this value so much less than the density of a proton or a neutron, which is roughly 10^{14} g/cm^3?
(*c*) Determine, also through your own measurements, the densities of a piece of wood, a rock, and a steel bolt.

3 Make a short list of applications of the gravitational and electromagnetic forces, in addition to the applications found in table I.2.

4 (*a*) Explain the difference between chemical and nuclear reactions.
(*b*) Why are nuclear reactions far more energetic, in general, than chemical reactions?

5 Explain the meanings of CO, CO_2, N_2, O_2, CH_4.

6 (*a*) Explain the meanings of the symbols $^{3}_{2}He$, $^{4}_{2}He$, $^{12}_{6}C$, $^{13}_{6}C$, $^{14}_{6}C$, $^{14}_{7}N$, $^{16}_{8}O$, $^{24}_{12}Mg$, $^{28}_{14}Si$, $^{31}_{15}P$, $^{32}_{16}S$, $^{56}_{26}Fe$, $^{197}_{79}Au$, $^{235}_{92}U$, $^{238}_{92}U$.
(*b*) Which of these symbols stand for isotopes of the same element, and which stand for different elements?

7 (*a*) Induce a wave in a taut rope. Measure the wavelength, amplitude, period (in seconds), frequency (in cycles per second), and speed of the wave.
(*b*) Repeat for a wave induced on a lake.

8 (*a*) What is meant by *electromagnetic spectrum*?
(*b*) Distinguish between X rays, UV radiation, IR radiation, microwaves, and radar waves, and give examples of sources of these kinds of radiation.

9 Look up the Greek alphabet in a dictionary and familiarize yourself with both its lower and upper case letters.

10 Look up the meanings of *micro-*, *milli-*, *kilo-*, and the other prefixes below in a dictionary and explain the following terms: micrometer, millimeter, centimeter, decimeter, decameter, hectometer, kilometer, megalight-year, milliliter, milligram, kilogram, nanosecond, microsecond, millisecond.

11 Use the metric units indicated in parentheses to measure: the thicknesses and diameters of United States coins (mm, cm); the width and length of a

dollar bill (cm); your height (cm, m); the heights of a few familiar mountains (m); the length of the Olympic marathon (km); the distances between several important United States cities (km); and the circumference of the Earth (km).

12 (*a*) What makes a scientific theory a "good" theory?
(*b*) How is such a theory developed?

Suggestions for Further Reading

Abell, G. O. 1982. *Exploration of the Universe*. 4th ed. New York: CBS College Publishing.* College-level astronomy text

Berman, L., and **J. C. Evans.** 1983. *Exploring the Cosmos*. 4th ed. Boston: Little, Brown and Co.* College-level astronomy text.

Boslough, J. 1985. "Worlds Within the Atom." *National Geographic*, May, 634–63. Current research into the ultimate structure of matter with giant machines and elegant mathematics.

Chaisson, E. 1981. *Cosmic Dawn: The Origins of Matter and Life*. Boston: Little, Brown and Co.* An enjoyable book for the general reader, covering much of the same material as this one.

Cloud, P. 1978. *Cosmos, Earth, and Man: A Short History of the Universe*. New Haven and London: Yale University Press.* An excitingly written book by an eminent geologist and paleontologist about what he has "learned from a lifetime of research, reading, and reflection about our planet, its history and cosmic connections, the development of life on it, and its capability to sustain our descendants."

Feinberg, G. 1967. "Ordinary Matter." *Scientific American*, May, 126–34. Ordinary matter consists of only three particles: electrons, protons, and neutrons. In addition to explaining the meaning of this statement, the article follows the history of human thought that has led to this contemporary model of matter.

Georgi, H. 1981. "A Unified Theory of Elementary Particles and Forces." *Scientific American*, April, 48–63. The author explains that at a range of 10^{-29} cm the world may be much simpler than the one we experience, with just one kind of elementary particle and one important force.

Goldman, T., R. J. Hughes, and **M. M. Nieto.** 1988. "Gravity and Antimatter" *Scientific American,* March, 48–56. "Would an antiapple fall to the ground at the same rate as an apple?" The traditional answer is "yes," but recent attempts to reconcile general relativity and quantum mechanics challenge this view.

Gribbin, J. 1981. *Genesis: The Origins of Man and the Universe*. New York: Delacorte Press. A book covering roughly the same topics as this one, in a style suitable for casual reading.

Hallett, J. 1984. "How Snow Crystals Grow." *American Scientist*, November–December, 582–89.

Huens, J.-L., and **T. Y. Canby.** 1974. "Pioneers in Man's Search for the Universe." *National Geographic*, May, 626–33.* Illustrated biographies of Copernicus, Galileo, Kepler, Newton, Herschel, Einstein, and Hubble.

Jeffries, A. D., P. R. Saulson, R. E. Spero, and **M. E. Zucker.** 1987. "Gravitational Wave Observatories." *Scientific American,* June, 50–58B. Observatories planned for the 1990s may be able to detect gravitational waves from extragalactic sources.

Learned, J. G., and **D. Eichler.** 1981. "A Deep-Sea Neutrino Telescope." *Scientific American*, February, 138–52.

McCusker, B. 1983. *The Quest for Quarks*. Cambridge University Press. A short, popularly written account of elementary particle structure, in particular about the question, "Are quarks the fundamental building blocks of protons and neutrons?"

Murdin, P., and **D. Allen.** 1979. *Catalogue of the Universe*. New York: Crown Publishers.* An atlas of exquisite photographs of the Solar System, the Milky Way, and galaxies, including up-to-date and concise descriptions.

Pauling, L., and **R. Hayward.** 1964. *The Architecture of Molecules*. San Francisco and London: W. H. Freeman and Co. A beautifully illustrated book about atoms, the periodic table, and many of the molecules and larger structures into which atoms can be arranged. The book is easy to read, topically arranged, and particularly suitable for the chemistry novice.

*The asterisk denotes books and articles that are general references.

Quigg, C. 1985. "Elementary Particles and Forces." *Scientific American*, April, 84–95.

Shapiro, S. L., R. F. Stark, and **S. A. Teukolsky.** 1985. "The Search for Gravitational Waves." *American Scientist*, May–June, 248–57.

Shimony, A. 1988. "The Reality of the Quantum World." *Scientific American,* January, 46–53. "Is a photon a particle or a wave." According to quantum mechanics, it depends on how and when you look at it.

Struve, O., and **V. Zebergs.** 1962. *Astronomy of the 20th Century*. New York: The MacMillan Co.* A comprehensive account of astronomical discoveries between 1900 and 1960. The book is written in a lively style and reflects the touch of Otto Struve (1897–1963), one of the foremost astronomers of the twentieth century. A must for anyone interested in the history of astronomy.

Zeilik, M. 1985. *Astronomy: The Evolving Universe*. 4th ed. New York: Harper & Row, Publishers.* College-level astronomy text.

Additional References

Heisenberg, K. 1958. *Physics and Philosophy*. New York: Harper & Row.

Schwarzschild, M. 1958. *Structure and Evolution of the Stars*. Princeton: Princeton University Press.

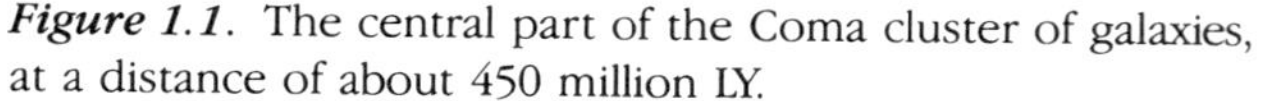

Figure 1.1. The central part of the Coma cluster of galaxies, at a distance of about 450 million LY.

1 Cosmology

HOW DID THE UNIVERSE begin? How will it end? Did it, in fact, have a beginning and will it have an end? What does it contain? What is our place—both in space and in time—in the Universe? These are questions humans have been asking in one way or another for millenia. They have asked them because the answers offer a perspective on how we—humans and the rest of Earth-bound life—fit into the larger scheme of things. They give meaning and content to our existence.

The answers found over the ages are expressed in art, myths, and religions. They have become part of our cultural heritage. In this and the chapters to come, we, too, shall seek answers to the above questions. We shall do this by turning to modern science, which is the product of our own culture. In the present chapter, we shall turn to *cosmology*, the *study of the large-scale structure and evolution of the Universe. Large-scale structure* means that this chapter is concerned with the cosmic distribution of matter, in which clusters of galaxies measuring many millions of light-years across are the basic units. Details on a finer scale, such as individual galaxies, stars, planets, and life are left for other branches of science and will be discussed in later chapters. *Evolution of the Universe* means that this chapter will try to explain how the distribution of the matter changes with time. In particular, we shall seek answers in this chapter to questions about the origin of the Universe and about its eventual fate.

In our study we shall find that our galaxy does not occupy any special place in the Universe—it is just one of billions of galaxies. The natural laws that operate in our part of the Universe and that have made life on Earth possible are, to the best of our knowledge, the same everywhere, and they probably have been so almost since the beginning. Our bodies are made of the same kind of matter and are subject to the same forces as the stars, the galaxies, and the Universe at large. In short, we believe that the laws of nature are universal and that they apply equally to the very near and very far, the small and large, animate and inanimate.

THE STRUCTURE OF THE UNIVERSE

To begin the study of cosmology, this section describes the large-scale distribution of matter in the Universe as it is seen today. How the matter originated and evolved will be discussed in later sections. Let us start by taking an imaginary journey from Earth to the distant galaxies and see what we find along the way. We will pretend that we travel in a spaceship that moves at the speed of light. Traveling at that speed allows us to tell how far we have come just by looking at a clock. Every second of time we will cover a distance of one light-second, which equals 300,000 km or seven and a half trips around the Earth. Every year we will cover a distance of one light-year (abbreviated LY), which equals 9.5×10^{12} km or 32,000 round trips from the Earth to the Sun. This imaginary journey will last billions of years of time and it will take us across billions of light-years of space.

The Milky Way Galaxy

Let us begin the journey. Within four minutes of lift-off from Earth we are already crossing the orbit of Mars and ten minutes later we are passing through the

Figure 1.2. The open double-star cluster h and Chi Persei, at a distance of about 7000 LY from Earth. Each of the two clusters contains close to 300 stars.

Asteroid Belt, between Mars and Jupiter. Soon the giant outer planets—Jupiter, Saturn, Uranus, and Neptune—briefly fill our viewing window. And in just a little over five hours we have left all of the Sun's nine planets behind. During the next few weeks and months we sweep through the outer fringes of the Solar System, coming upon an occasional comet and other debris left over from the Sun's birth. All along, the Sun has been growing fainter and is now just one of many bright stars in the sky.

It will be four years before we encounter the first stars beyond the Sun. They are a triple star system—Proxima Centauri and Alpha Centauri A and B. Alpha Centauri A resembles the Sun in size and temperature, while the other two are somewhat smaller and cooler. Every few years we pass another star. They brighten up as we approach them and then fade in the distance. After just 50 years we lose sight of our Sun. It has become too faint to be seen with the unaided eye.

The centuries and millenia of our journey tick away. We are passing thousands upon thousands of stars. Many are alone, others are double and triple star systems, and still others are grouped in clusters (figures 1.2 and 1.3). Some of the stars are very young and are still surrounded by the clouds of gas and dust (figure 1.4) from which they were born. Others are old and nearing the end of their active lives. Occasionally we pass a star that pulsates or one that explodes, spewing gas in all directions and momentarily blinding us with its brilliance.

Figure 1.3. The globular star cluster Messier 13, in the constellation Hercules. This beautiful cluster lies at a distance of about 25,000 LY and is the largest globular cluster in the northern sky. It contains perhaps half a million stars.

Figure 1.4. The Horsehead nebula. This dark and cold interstellar cloud of dust, in the shape of a horse's head and neck, protrudes into the bright emission nebula known as IC 434. It is about 1000 LY distant and lies in the constellation Orion.

Eventually the concentration of stars lessens and we break out of the obscuring layer of gas and dust associated with them. Looking back we see an enormous number of stars, star clusters, and clouds, arranged in a huge but very flat disk-like distribution. At its center the disk thickens into a very prominent and bright spherical bulge, like the hub of a giant wheel. From the bulge a distinct spiral pattern of luminous, dark matter winds outward to the edge of the disk. The luminous matter consists of clouds of gas brightly lit by massive stars, while the interspersed dark patches are cold, light-absorbing interstellar dust.

What we are seeing is the Milky Way galaxy—our home in the vastness of intergalactic space. It is a *spiral galaxy* like many others that we shall encounter on our journey. It measures about 100,000 light-years across and contains close to 150 billion stars. Apart from its central bulge or nucleus, its most distinctive feature is the pattern of spiral arms. In fact, our journey began from an inconspicuous spot at the inner edge of one of them, roughly 30,000 light-years from the center of the Galaxy.

The Local Group

As we leave our galaxy, we still pass an occasional star or star cluster, whose motions carry them far above the Galaxy's disk. But most of the Milky Way lies behind us. We are entering intergalactic space.

Ahead of us lies the Universe filled with many billions of galaxies. Each is like an island universe, separated from the others by many light-years of emptiness. Many of the galaxies are spirals. Others have spherical or elliptical structures and are called *elliptical galaxies*. Still others look highly asymmetric and distorted; they are known as *irregular galaxies*. The galaxies are not distributed randomly through space, but are grouped together in small and large clusters. The small clusters contain up to a few dozen galaxies, while the large clusters have thousands of them.

The Milky Way belongs to a small cluster of galaxies, known as the Local Group, that contains some thirty members. The two nearest members are the Large and Small Magellanic Clouds (figure 1.5), and it is in their direction that we are now steering our ship. Both clouds are composed of young blue stars, bright gaseous nebulae, and dark dust lanes, the same matter that dominates the spiral arms in our galaxy. However, that matter is not distributed in well-defined spiral arms. Powerful forces seem to distort it into rather disorganized shapes. In fact, the Small Magellanic Cloud appears to have been torn into two misshapen components, perhaps by a collision with the Large Magellanic Cloud. Furthermore, an extended stream of hydrogen gas trails behind the clouds, as if it had been pulled out by the forces of such a collision and by tidal interactions with the Milky Way. Because of the clouds' disorganized shapes, astronomers have classified the Magellanic Clouds as irregular galaxies. From Earth they are visible only from the Southern Hemisphere and Europeans did not become aware of them until they began their worldwide explorations in the sixteenth century. The two clouds were named in honor of the Portuguese sea captain Ferdinand Magellan (1480–1521), who led the first circumnavigation of the Earth.

Beyond the Magellanic Clouds and after roughly 750 thousand years we pass two dwarf elliptical galaxies – Leo I and Leo II (figure 1.6). They impress us with their near perfect symmetry and uniformity. Gas and dust seem to be absent in them, and their light comes entirely from faint, old stars. As our journey continues, we will find that by far the largest number of galaxies in the Universe are dwarf ellipticals.

We are now leaving the Milky Way's immediate neighborhood (see figure 1.7) and are changing course toward the most spectacular galaxy in the Local Group, the great spiral galaxy in Andromeda (figure 1.8). We pass this galaxy in about 2 million years. From Earth it appears as a small and faint haze of light when viewed with the unaided eye. However, from close up we are able to see many details that remind us of the Milky Way: an almost spherical and very bright nucleus, a great disk of stars, gas, and dust, and an impressive pattern of spiral arms. Like the Milky Way, it also has two nearby neighbors, the small elliptical galaxies M 32 and NGC 205.*

Figure 1.5. The Large (left) and Small Magellanic Clouds. These two irregular galaxies, at distances of roughly 160,000 LY and 200,000 LY, respectively, are our closest galactic neighbors. The bright object in the upper right is the star Achernar.

*M 32 and NGC 205 are labels astronomers use to refer to galaxies and other celestial objects. M 32 is the 32nd object listed in the *Messier Catalogue*, compiled by the eighteenth-century French astronomer Charles Messier (1730–1817). NGC 205 is the 205th object listed in the *New General Catalogue of Nebulae and Clusters of Stars*, compiled at the end of the nineteenth century by the Danish astronomer Johan L. E. Dreyer (1852–1926).

Figure 1.6. Galaxy Leo II. This dwarf elliptical galaxy (of type E1; see also figure 2.10) at a distance of about 750,000 LY is highly resolved into stars and except for its greater size resembles a globular star cluster.

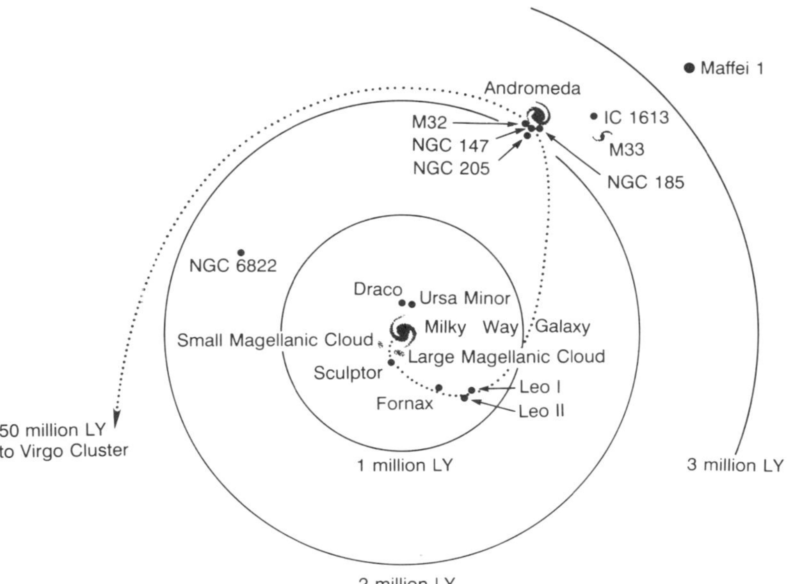

Figure 1.7. Flight path of the cosmic journey (dotted line; see text for details). The diagram illustrates the layout of the Local Group of galaxies in two dimensions. The third dimension—into and out of the page—is not shown.

Figure 1.8. The Andromeda galaxy. This galaxy (of type Sb) is tilted so that its outlines appear elliptical. Near its upper edge lies M 32, and slightly below it lies NGC 205. M 32 and NGC 205 are two dwarf ellipticals that are close companions of the Andromeda galaxy.

The Virgo Cluster

After we have inspected at close range the Andromeda galaxy, M 32, NGC 205, and several other galaxies in the same vicinity, we change course once again and leave the Local Group. Our next destination is the nearest large cluster of galaxies, the Virgo cluster. This cluster is still some 50 million light-years distant, and along the way we come upon several other smaller clusters similar to the Local Group. Each of them contains one or two large galaxies about which other smaller ones are gathered.

One of the first galaxies we pass is NGC 3115 (figure 1.9), a highly flattened elliptical. It closely resembles a spiral galaxy, except that gas and dust and, therefore, arms are absent. After roughly 16 million years we fly past a very unusual galaxy, Centaurus A. It has the spherical shape of an elliptical galaxy, but it also contains a disk of gas and dust. From Earth we see the disk edge-on, making the dark, light-absorbing dust contrast sharply against the bright stellar component (figure 1.10). A number of violent explosions have occurred in the recent past in the nucleus of Centaurus A, ejecting jets of gas at up to 5% of the speed of light. A pair of these jets, tossed out in opposite directions roughly 30 million years ago, are powerful sources of radio emission. Centaurus A's nucleus is also a strong and variable emitter of X rays, indicating that violent events are still taking place today. Astronomers suspect that the source of the radiation and

Figure 1.9. Galaxy NGC 3115. This highly flattened galaxy (of type E7 or type S0), at a distance of about 13 million LY, has a mass distribution intermediate between that of ellipticals and spirals.

Figure 1.10. Centaurus A (NGC 5128). This unusual galaxy lies at a distance of about 16 million LY, and is the nearest of the giant radio galaxies.

of the explosions is a gigantic *black hole** at the very center of the galaxy, which periodically tears up passing stars and draws their matter into its bottomless gravitational well.

After about 45 million years the frequency of galaxies increases as we approach the Virgo cluster (figure 1.11). This cluster contains thousands of members. They are of all sizes and types—spirals, ellipticals, irregulars—and they are rather randomly scattered about. Near the cluster's center lie several giant elliptical galaxies. They form the gravitational hub of the Virgo cluster, about which the other galaxies trace their orbits.

The largest of the elliptical galaxies near the center of the Virgo cluster is M 87 (figure 1.12). It is an impressive and remarkably symmetric galaxy, containing close to one hundred times as many stars as the Milky Way. This makes it the most massive galaxy known. The radiation emitted from the main body of M 87 is so intense that few details can be discerned. However, toward the outer fringes of the galaxy, the light gradually fades revealing a great number of star clusters. Each contains hundreds of thousands to millions of stars and resembles the globular star clusters in our galaxy (more about them in the following chapter). Short exposure photographs of the nucleus of M 87 show a jet-like structure, with a faint indication of a second one protruding in the opposite direction (see figure 2.19). The jets must have resulted from an explosion of enormous magnitude. Is the source of this violence a black hole, like the one

*A black hole is a star that collapsed under its own gravity. Its surface gravity is so strong that nothing, including light, can escape. Matter can, however, fall into it, releasing enormous amounts of energy. Black holes will be discussed in more detail in the next two chapters.

Figure 1.11. A part of the Virgo cluster of galaxies. With thousands of members, this cluster is the nearest rich cluster of galaxies and is believed to be the gravitational hub of the local supercluster of galaxies.

Figure 1.12. Galaxy M 87. This giant elliptical galaxy (of type E0), with a nearly perfectly spherical structure, is the brightest of the galaxies in the Virgo cluster and, like Centaurus A, a powerful emitter of radio waves.

believed to form the center of Centaurus A? Many astronomers are leaning toward this interpretation. They base their opinion on evidence that within less than 300 LY of M 87's very center exists a mass equal to at least 5 billion suns. Only a black hole can contain so much mass in such a small volume and also provide the energy to power the observed events.

It takes us well over 10 million years to thread our way through the heart of the Virgo cluster. Besides the kinds of galaxies already described, we pass some that are colliding or have recently (that means during the past billion or so years) undergone collisions. Many of these galaxies are grotesquely distorted by the gravitational forces they exert on each other. Often they are connected by long, thin bridges of gas and dust that they have torn off each other. Others are surrounded by gaseous and dusty rings. Some spiral galaxies are stripped entirely of their gas and dust by these interactions and have become what are known as *SO galaxies* (see figure 2.10). The Virgo cluster is a world of great fascination and violence.

As we travel on, the concentration of galaxies eventually begins to thin again and, as earlier, we encounter only now and then a small cluster of galaxies. About 100 million years after leaving the Milky Way their frequency diminishes still

Figure 1.13. Negative print of the cluster of galaxies Cl 0024 + 1654. This cluster is one of the most remote ones known, at a distance of about 7 billion LY. The light received on Earth today from this cluster was emitted roughly 2.5 billion years before the birth of the Solar System. Due to their great distance from Earth, the individual galaxies are barely distinguishable from the images of stars within our galaxy.

further and we find ourselves in mostly empty space. We have left our local cosmic neighborhood.

Superclusters of Galaxies

Looking back we see that we have come from a huge *supercluster of galaxies*, measuring more that 100 million light-years across. Its center is dominated by the Virgo cluster with its rich concentration of galaxies. Surrounding it like a halo is a sparse distribution of smaller clusters, one of which is the Local Group. These small clusters are weakly tied to the Virgo cluster by gravity and circle in giant orbits around it. The whole system contains millions of galaxies of all sizes and types.

The Virgo supercluster is not alone in the Universe. As our journey continues, we encounter such superclusters of galaxies again and again. The Coma cluster (figure 1.1) and Cluster Cl 0024 + 1654 (figure 1.13) are just two examples. No matter how far we travel or in which direction, the pattern remains similar. On the average, we cross another supercluster every 100 to 200 million years, although occasionally we see stretched-out aggregates of clusters of galaxies that are two or three times that size. The superclusters of galaxies form the basic structural units of the Universe, and between them exist great voids nearly as vast as the superclusters themselves.

Let us end our journey at this point and summarize what we have discovered: The matter of the Universe is concentrated in large superclusters of galaxies, each measuring, on the average, 100 to 200 million light-years across and containing millions of galaxies. There are differences in the exact number of galaxies, in the individual appearance of these galaxies, and in their spatial arrangement. However, if we disregard such fine-scale details, we find that the superclusters are very much alike. Furthermore, they are rather evenly distributed through space, as the map of nearly a million galaxies seen in the northern sky (figure 1.14) illustrates. We therefore conclude that on the largest scale the Universe is homogeneous and isotropic (*homogeneous* means that the Universe has the same structure and composition everywhere; *isotropic* means that it looks the same in all directions).

The Fourth Dimension

Our imaginary journey to the distant galaxies has left unanswered some compelling questions, such as "Where is the center of the Universe?", "How far is its edge?" and "Where exactly are we located?" We did not find answers to such questions because there aren't any. Such questions are perfectly valid when asked in the context of the three-dimensional Earth environment. However, our local experiences do not necessarily apply to the large-scale structure of the Universe, and frequently they lead us completely astray. In the study of the Universe as a whole, a fourth dimension—time—must be included along with the three spatial dimensions. The above questions then lose their meaning, for it turns out that in four-dimensional space–time the Universe has neither a center nor an edge. It only has a beginning in time and, possibly, an end. Some of these bewildering features of the Universe are discussed later in this chapter.

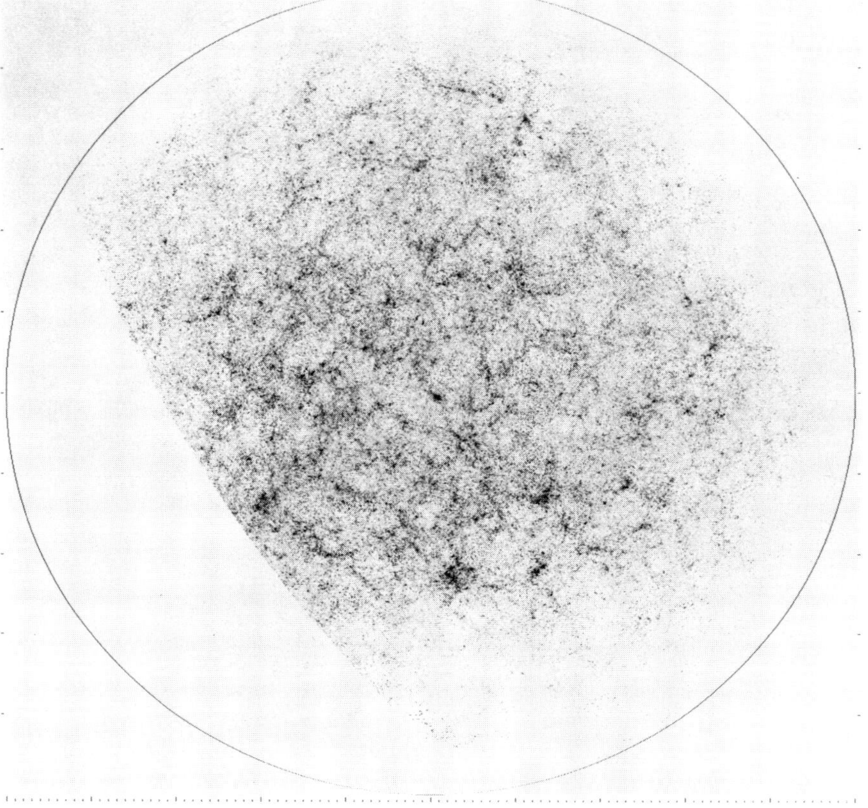

Figure 1.14. Map of the distribution of galaxies in the northern sky. Nearly a million galaxies are included. The map illustrates that on the large scale the distribution is uniform, though it shows clumpiness where galaxies are concentrated in clusters.

THE EXPANSION OF THE UNIVERSE

Up to 500 years ago, most humans believed in a geocentric Universe, with Earth at the center and the Sun, the planets, and the stars circling around it. The Polish astronomer Nicolaus Copernicus (1473–1543) revolutionized that point of view by placing the Sun at the center of the Universe and making Earth one of the planets. In the following centuries, astronomers gradually realized that the Sun is just one star among many millions in the Milky Way and that it is far from the center of the Galaxy. The prevailing view was that the Milky Way galaxy comprised the entire Universe. However, by the eighteenth century a few scientists and philosophers, such as William Herschel (1738–1822) in England and Immanuel Kant (1724–1804) in Germany, dared to consider that the Universe may extend beyond these limits. They suggested that the many faint nebulae visible through telescopes might be separate galactic systems.

Only in the early part of the twentieth century, with the development of modern reflecting telescopes with large light-gathering mirrors, were astronomers able to estimate accurately the distances to those nebulae. The new telescopes allowed detailed study of individual stars in many nebulae. From the faintness of their light, astronomers calculated that they were at distances of hundreds of thousands to many millions of light-years, which places them far beyond the Milky Way. This proved beyond the shadow of a doubt that these nebulae were separate galaxies* and that the Universe was much bigger than anyone had imagined. Despite some initial dispute among astronomers concerning this conclusion, by the mid–1920s it was generally accepted and modern cosmology was born.

The Hubble Constant

Among the pioneers who participated in these discoveries was the great American astronomer Edwin P. Hubble (1889–1953). After the true nature of the galaxies had been established, he turned his attention to their motions. He was motivated to do this by a puzzling report by Vesto M. Slipher (1875–1969) of the Lowell Observatory in Arizona that many of the faint nebulae are moving away from us with velocities of hundreds and thousands of kilometers per second (a velocity of 1000 km/sec equals 3.6 million km/hr or 2.2 million mi/hr). That seemed peculiar, for the stars in the Milky Way have velocities much smaller than that and some move away while others move towards us. In contrast, the velocities of nearly all galaxies, with the exception of a few nearby ones, are always directed away from us. Slipher had made his observations during the 1910s, before anyone was certain that these nebulae were really galaxies far beyond the Milky Way. Hence, he did not know what to make of their unusual velocities. However, by the mid–1920s Hubble did know that the galaxies were separate galactic systems and he began a systematic study of the relation between their velocities and their distances.

Working with the new 2.5-meter (100-inch) telescope at the Mount Wilson Observatory in southern California (figure 1.15), Hubble and his colleague

Figure 1.15. The 2.5-meter (100-inch) Hooker telescope at the Mount Wilson Observatory, California. This telescope, completed in 1917, was the first of the large reflectors that contributed to the modern perspective of the structure and evolution of the Universe.

*Upon the discovery that many of the faint nebulae were separate stellar systems beyond the Milky Way, astronomers began calling them *galaxies*. The word *nebula* is now used solely to refer to clouds of gas and dust that, along with stars and star clusters, are the major components of the spiral and irregular galaxies.

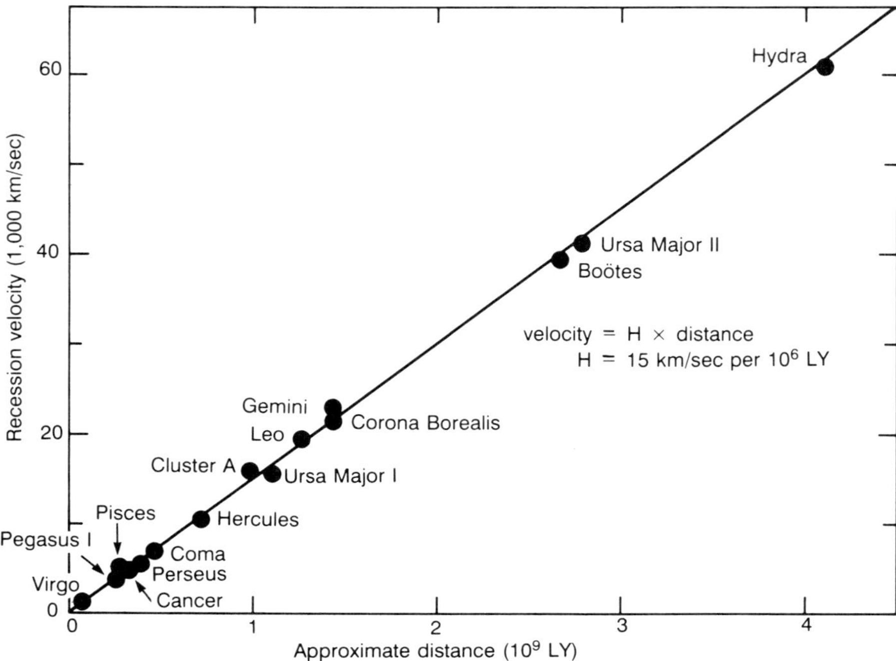

Figure 1.16. Hubble velocity–distance relation for 15 clusters of galaxies. The alignment of the data along a nearly straight line implies that recession velocity varies regularly with distance. For instance, galaxies at 1 billion LY have a recession velocity of 15,000 km/sec, those at 2 billion LY have a recession velocity of 30,000 km/sec, and so forth.

Milton L. Humason (1891–1972) let the faint light of distant galaxies pass through a spectrograph, an instrument that splits electromagnetic radiation into its spectral components (some spectrographs use prisms to accomplish this, see figure I.16). In almost every case they found that the light was shifted toward the color red. This meant that the galaxies are receding from us and confirmed Slipher's earlier result. (How recession velocities follow from redshifts is explained in box 1.1.) Furthermore, Hubble and Humason noticed that these recession velocities vary regularly with distances: Galaxies that have two, three, or ten times greater recession velocities than nearby ones are approximately two, three, or ten times farther away. A plot of recession velocity versus distance (see figure 1.17) clearly shows that, on the average, for every increase in distance by 1 million light-years there exists a corresponding increase in velocity by 15 km/sec. Today astronomers refer to the rate of increase of recession velocity with distance as the *Hubble constant* or, more briefly, as *H*.*

*Hubble and Humason's estimate for *H* was 170 km/sec per million LY. They obtained such a large value because in the 1920s and 1930s the distances to the galaxies were underestimated by roughly a factor of ten. The distances to the galaxies are still uncertain today and the modern estimate of *H* reflects this. It lies in the range from 12 to 30 km/sec per million LY. In this book, I have adopted the value of 15 km/sec per million LY.

BOX 1.1.

The Doppler Effect: An Explanation of the Cosmological Redshift

THE cosmological redshift refers to the displacement of light from distant galaxies toward longer wavelengths, as discovered by Vesto M. Slipher in the 1910s. Edwin P. Hubble interpreted this displacement as being due to recession velocities of the galaxies, and virtually all astronomers today agree. The larger the redshift, the faster the galaxies are receding.

The same phenomenon is observed in the case of stars in our Galaxy, except with them the shift in color is smaller because their motions relative to us are smaller. Furthermore, in the case of stars, both redshifts and blueshifts (that is, displacements toward shorter wavelengths) are observed. Stars whose light is redshifted move away from us and stars whose light is blueshifted move toward us. A striking example of this can be observed in the case of binary star systems, in which two stars orbit around a common center of mass and alternately approach and recede along our line of sight. As a result, the light from these stars is alternately blue- and redshifted, as was first discovered by the Austrian physicist Christian Johann Doppler (1803–1853). In his honor, red- and blueshifts of light, due to the motion of the source, are now known as the *Doppler effect*.

The Doppler effect is not limited to light, but occurs with all wave phenomena in which the source and observer are in relative motion along the line joining them. One example of the Doppler effect that we have all experienced is the shift in pitch of the whistle of a train passing at high speed. During the approach, the whistle has a higher pitch than normal. After the train has passed and is moving away, the pitch is lower than normal.

What is happening is this: Sound is a perturbation of the air and consists of pressure waves that spread out in all directions from a source. The spreading of a sound wave may be compared with the spreading of the circular rings of water waves from some source of perturbation on a lake, such as a rock thrown into the lake. The major difference between water and sound waves is the following: Water waves consist of up and down motions of the water surface. These motions are *transverse* to the direction in which the waves propagate and, hence, the water waves are called *transverse waves*. Electromagnetic waves are another example of transverse waves. Sound waves consist of back and forth motions of air molecules *along* the direction of wave propagation. Hence, sound waves are called *longitudinal waves* (see figure A).

Figure A. Comparison of transverse and longitudinal waves.

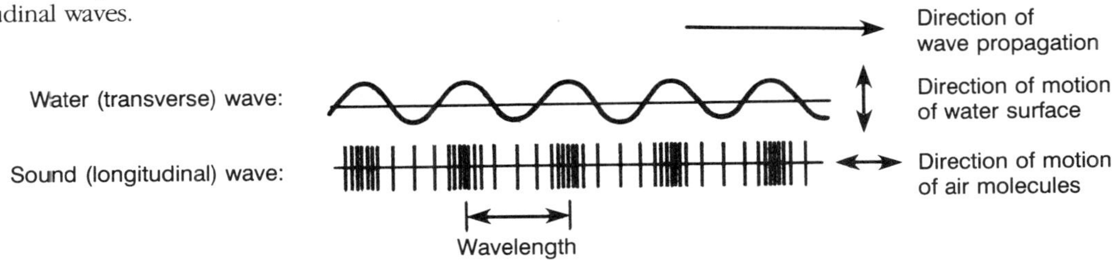

The pitch of sound is related to the rate at which the high pressure crests of the waves pass us. If they pass at a high rate, we hear a high pitch; if they pass at a low rate, we hear a low pitch. If the source of sound is moving toward us, successive pressure crests have less far to travel to reach us. Hence, the wavelengths are shortened and successive pressure crests pass us at a higher rate than those from a source at rest relative to us. That is why we hear the whistle of an approaching train at a higher pitch than normal. If the source of sound is moving away from us, successive pressure crests have further to travel to reach us. Hence, the wavelengths are stretched and successive pressure crests pass us at a lower rate than those from a source at rest relative to us. That is why the whistle of a receding train has a lower pitch than normal (see figure B).

The red- and blueshifts of light are subject to the same physical explanation as the one just given for sound. If a light source approaches, the wavelengths

BOX 1.1 continued

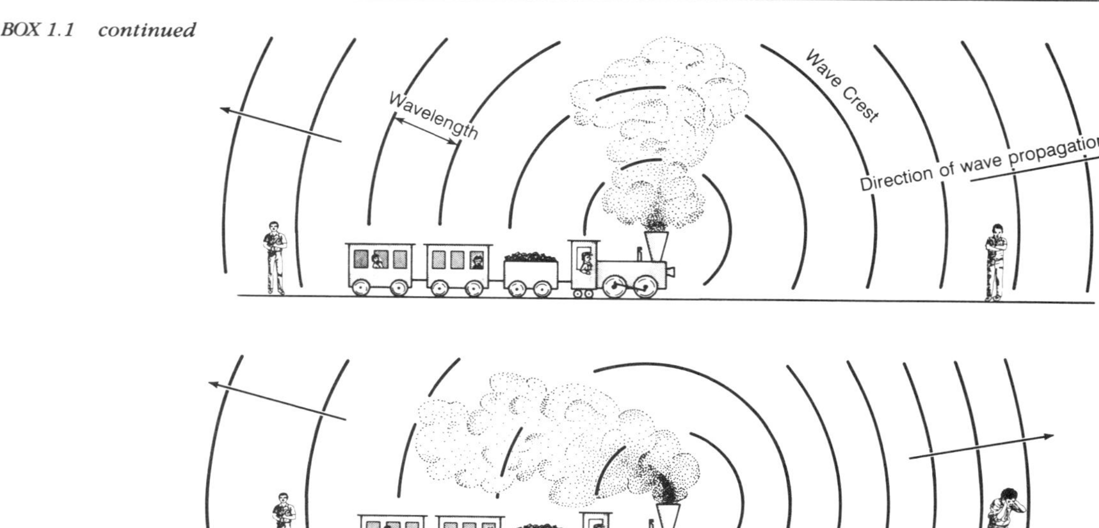

Figure B. The Doppler effect in the case of sound. At top, the train and the observers are standing still; both observers hear a normal pitch. At bottom, the observers stand still but the train moves. The observer whom the train is leaving hears a lower pitch; the observer whom the train approaches hears a higher pitch.

are shortened and we perceive a blueshift. If a light source recedes, the wavelengths are stretched and we perceive a redshift. This explanation of red- and blueshifts of light was first given by the French physicist Armand H. L. Fizeau (1819–1896).

The cosmological redshift is illustrated in figure C, a composite photograph of the spectra of light from five galaxy clusters. For instance the shift in wavelength of light from the Hydra cluster corresponds to a recession velocity of 61,000 km/sec and, assuming $H = 15$ km/sec per million LY, to a distance of 4 billion light-years. Light from the most distant quasar discovered to date has a wavelength shift corresponding to a recession velocity of 93% the speed of light and to a distance of approximately 12 billion light-years (see chapter 2). The most extreme example of the cosmological redshift is that of the cosmic background radiation. Today we observe this radiation at wavelengths of roughly 0.1 cm. However, when that radiation decoupled from matter (see text), its wavelengths were approximately 1000 times shorter. They were Doppler-shifted by the cosmic expansion of the Universe.

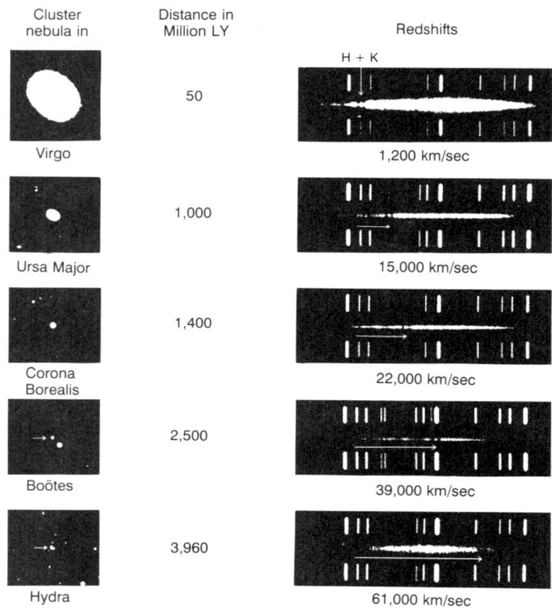

Figure C. The relationship of recession velocities and distances for selected galaxies in clusters. The recession velocities are deduced from the redshifts of the H and K lines of calcium in the spectra of the galaxies. Distances are deduced from the faintness of the galaxies in each cluster.

The Big Bang

The relation between recession velocities of galaxies and their distances has powerful implications for the origin and evolution of the Universe. It suggests that the entire Universe is in a state of expansion. This in turn implies that in the past the Universe was much more compact than it is today and that, possibly, at some long-ago time it came into existence in one gigantic explosion. The explosion sent matter flying in all directions. Matter that initially acquired large velocities relative to us is consequently very distant from us today, and matter that acquired a small velocity is comparatively still quite close. The situation may be compared with the explosion of a grenade. The energy of the explosion breaks the grenade into many fragments, some of which acquire high velocities and others low velocities. As the fragments fly apart, the ones with high velocities fly farther than those with smaller velocities. This analogy should, however, not be taken too literally, for there exists a definite center in the case of the explosion of the grenade. There is no such center in the case of the Universe (more about this apparent paradox later).

Observers in other galaxies (if, indeed, there are any) see the expansion of the Universe as we do, except from their vantage point it is our galaxy that is moving away from them. And the more distant we are from them, the faster they see the Milky Way receding. Hence, they share our impression that the Universe was born in a gigantic explosion—the *Big Bang*.

Age of the Universe: First Estimate

The relation between the velocities and distances of galaxies allows us to estimate the age of the Universe by the following reasoning. From the Hubble constant and also from the graph of figure 1.16 we know that, on the average, galaxies at a distance of 100 million light-years move away from us with a velocity of 1500 km/sec, those at 1 billion light-years with 15,000 km/sec, and so forth. If, for example, we assume that galaxies at 100 million light-years have been receding from us with a velocity of 1500 km/sec since the Big Bang, we can compute how long it took them to reach that distance:

$$\text{time} = \frac{\text{distance}}{\text{velocity}} = \frac{\text{100 million LY}}{\text{1500 km/sec}} = \text{20 billion years.}^*$$

Thus, we may conclude that the Universe was born approximately 20 billion years ago.

A later section will show that this simple arithmetic leads to overestimating the age of the Universe. Gravity is continuously slowing down the expansion of the Universe; hence, the value of the Hubble constant—that is, the rate of the Universe's expansion—must have been greater in the past than it is today. Consequently, the Universe probably reached its present size in less than 20 billion years.)

*This calculation is the same as that used to find how long it takes to drive 180 miles at 60 mi/hr:

$$\text{time} = \frac{\text{distance}}{\text{velocity}} = \frac{\text{180 mi}}{\text{60 mi/hr}} = \text{3 hr.}$$

The only added complication in the astronomical calculation is that light-years need to be converted to kilometers and seconds to years (see box I.2 for a list of conversion factors).

BOX 1.2.
The Meaning of Temperature

TEMPERATURE is a physical parameter that expresses how hot or cold a substance is. It also gives an indication of how much heat energy a substance possesses. The higher the temperature, the greater is the heat energy.

What is meant by heat energy? If the substance is a solid, the atoms and molecules are tightly bonded to each other. However, the atoms and molecules are not at rest, but oscillate back and forth relative to each other. The heat energy is the sum of the oscillation energies of all of the atoms and molecules that make up the solid. If the substance is a liquid or a gas, in which the atoms and molecules are free to move about, the heat energy is the sum of the energies that the atoms and molecules possess due to their various kinds of motions—translational motions (that is, motions from one place to another) and, in the case of molecules, rotations and internal oscillations. (Atoms are too small to have measurable motions of rotation and internal oscillation; but molecules may rotate and oscillate like weights connected by springs.)

To illustrate the correspondence between heat energy, temperature, and the structure of matter, consider what happens when we heat or cool a familiar object, namely an ice cube. At temperatures below the freezing point of water, the water molecules of the ice cube, which are tightly bonded together into a solid configuration, oscillate back and forth. The closer the temperature is to the melting point, the faster are the oscillations and the greater is the heat content of the ice cube. At the melting point, the oscillations are so strong that the bonds between neighboring molecules break. If we continue to add heat, molecule after molecule breaks free and the water changes from the solid to the liquid state. We say "The ice melts."

The water molecules of liquid water are not bonded to each other as they are in ice. They are free to move about, collide with each other, bounce off each other, speed up and slow down, rotate, and oscillate. They do this in a random, helter-skelter manner, which gives liquid water its fluid properties. If we heat the liquid water and raise its temperature, the various kinds of motions of the water molecules become faster. Eventually they become so fast that the water molecules form gaseous bubbles that rise to the liquid's surface (because of the bubbles' lower densities), and from there they fly off into the surrounding atmosphere. The water changes from the liquid to the gaseous state (that is, into steam), a process we call boiling. In the gaseous state, the water molecules are separated much farther from each other and move about even more freely than in the liquid state.

If we heat the steam, the molecular motions become still faster and the temperature climbs. Eventually, the collisions between the water molecules become so violent that they break up into their constituent hydrogen and oxygen atoms—the gaseous water changes into hydrogen gas and oxygen gas. If we continued to add heat, we would eventually reach the point beyond which even the atoms cannot survive. The collisions would knock the electrons off the atomic nuclei and produce a gas of free electrons and ions. This kind of ionization, due to thermal motions, occurs at temperatures corresponding to stellar atmospheres and interiors.

The process just described—from the melting of the ice cube to the boiling of water, the breakup of the water molecules into atoms, and the ionization of the atoms—requires the input of heat energy and occurs at progressively higher temperatures. If we go in the opposite direction by cooling the ice cube (that is, by taking away heat energy), the oscillations of the water molecules become slower and the temperature drops. Eventually, the molecules reach the state at which they oscillate at their lowest possible frequency (*note*: the oscillations can never come entirely to rest). According to well-tested laws of quantum mechanics, the oscillations cannot be slowed down further and no more energy can be taken away from the ice. The temperature has now reached its lowest possible value, called the ***absolute zero point***.

Temperature is usually measured in degrees according to scales whose zero points and gradations are defined arbitrarily. The most common temperature scales are the Celsius, Kelvin, and Fahrenheit scales.* The Celsius scale (formerly called the centigrade scale) is used in most parts of the world and in

*The three temperature scales are named after Anders Celsius (1701–1744, a Swedish astronomer), Lord Kelvin (1824–1907, a British physicist), and Gabriel D. Fahrenheit (1686–1736, a German physicist).

BOX 1.2 continued

many scientific applications; the Kelvin scale is preferred by physical scientists; and the Fahrenheit scale is used mainly in the United States (see figure A).

On the Celsius scale the zero point (written as 0° C) corresponds to the melting point of ice at sea level and 100° C corresponds to the boiling point of water at sea level. Absolute zero occurs at −273° C.

The Kelvin scale has the same size gradations as the Celsius scale, but its zero point is different. It coincides with absolute zero (0 K), and that is why this scale is preferred by physical scientists. It not only simplifies calculations involving temperature, but in this scale zero degrees refers to the lowest possible temperature to which material objects can be cooled. Degrees Kelvin (T_K) and degrees Celsius (T_C) are related according to the formula

$$T_K = T_C + 273.$$

At sea level the melting point of ice and boiling point of water occur at 273 K and 373 K, respectively.

On the Fahrenheit scale, the melting and boiling points of water fall at 32° F and 212° F, respectively. The difference between these two points is 180 degrees, which means that one degree Fahrenheit equals 5/9 degree Celsius or Kelvin (5/9 = 100/180). As a consequence, degrees Fahrenheit (T_F) and degrees Celsius (T_C) are related by the formula

$$T_C = \tfrac{5}{9}(T_F - 32).$$

Using this formula, we find that the zero point of the Fahrenheit scale (the lowest temperature Gabriel Fahrenheit could reach with a mixture of ice and common salt) equals −17.8° C, absolute zero occurs at −459° F, the normal human body temperature of 98.6° F equals 37° C, and −40° F is the same as −40° C.

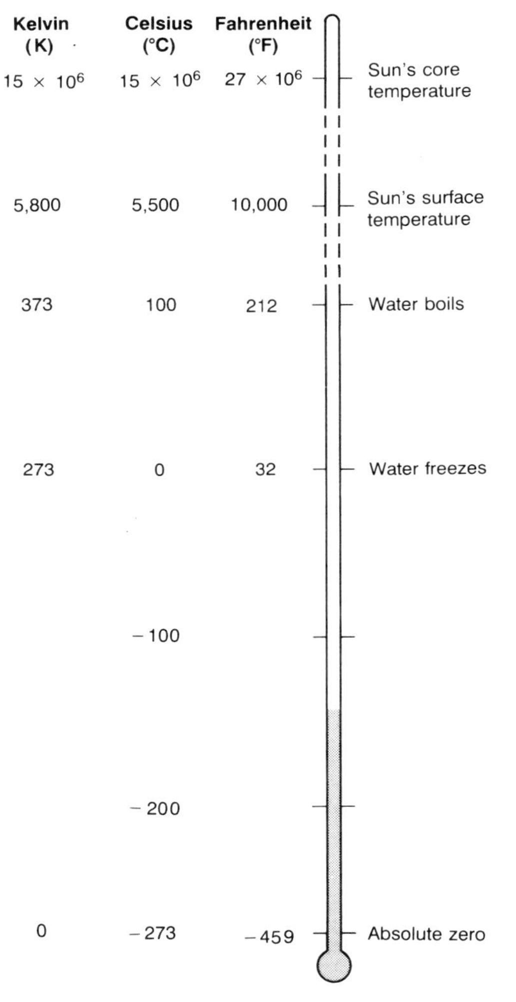

Figure A. Comparison of the three temperature scales in common use.

COSMOLOGICAL PRINCIPLE

Remember that during our imaginary journey through space we discovered that on the average *the Universe is homogeneous and isotropic*. This observation (which in reality, of course, was discovered by observations through Earth-based telescopes) plays such a fundamental role in the construction of theories about the origin and evolution of the Universe that it is referred to as the *Cosmological Principle*.

Does the expansion of the Universe not invalidate the Cosmological Principle? Will there not be changes with time that will destroy the homogeneity and isotropy? There will be changes with time, but because the Universe expands uniformly in all directions, those changes will be the same everywhere. A galaxy that today is twice as far away as another will still be that much farther away in a million or billion years because its recession velocity is also twice as great. Therefore, even though the Universe expands and ages, it remains homogeneous and isotropic.

To observe the Universe's homogeneity and isotropy we must compare different parts of it at the same time. When we look at galaxies through telescopes we don't do that because we see them not as they are today but as they were when the light that we pick up with our telescopes left them. The most distant galaxies visible through present-day telescopes are approximately 12 billion light-years away. This means that the light we receive from them was emitted 12 billion years ago and has been in transit every since. It gives us information about those galaxies at that distant time. According to the Cosmological Principle, however, the conditions in our local cosmic neighborhood 12 billion years ago were, on the average, the same as those that we observe by looking at galaxies 12 billion light-years away.

The delay in transmitting information across intergalactic space is due to the finite speed of light. A similar delay in the transmission of electromagnetic radiation occurs here on Earth, except that it is negligible because the distances are so small. Light or telephone signals take less than 2/100 of a second to travel from the Atlantic to the Pacific coast of the United States. This interval is too short to affect communication. However, when the astronauts landed on the Moon, noticeable pauses existed between the transmission of radio messages from Earth and the reception of answers. They were due to the 2.5-second travel time of radio waves from Earth to the Moon and back.

A message from the nearest star would take approximately four years to reach Earth and from the Andromeda galaxy it would take 2 million years. We do not expect that things would greatly change on Alpha or Proxima Centauri in just four years or in the Andromeda galaxy in 2 million years. However, with the distant galaxies this is not the case. In the course of several billion years and longer, significant evolutionary changes can take place in the Universe. Thus, observing distant objects gives astronomers an important tool for probing the Universe's past.

COSMIC BACKGROUND RADIATION

If we extrapolate the present structure and expansion of the Universe back in time, we find that the galaxies were closer together. Eight billion years ago they were approximately twice as close as they are today and 10 billion years ago they were three times as close. At the time of the Big Bang—well before the galaxies were formed—the matter must have been compressed to exceedingly high densities and heated to enormous temperatures. If the Universe started from such extreme conditions, some remnant of the early, intense heat might still be around today in the form of electromagnetic radiation and perhaps it can be detected.

This was the kind of reasoning that in the 1940s lead George Gamow (1904–1968), a Russian physicist who had emigrated to the United States, to calculate what the cosmic radiation left over from the Big Bang should be like today. He assumed that the radiation, if it existed at all, should be as much part of the Universe as is matter. Hence, it should participate in the expansion of the Universe and all its wavelengths should have become greatly stretched, just as the distances between galaxies became stretched. From the expansion rate of the Universe and its present density of matter, Gamow estimated that the present average wavelength of the cosmic radiation should be about 0.06 centimeters. This places it into the microwave range and gives it a temperature of about 5 degrees Kelvin (for a discussion of temperature scales see box 1.2). The radiation should be very faint, fill the entire Universe, and come to us equally from all directions (very much like white light comes equally from all directions during a whiteout in a snowstorm or heavy fog). Unfortunately, at the time no one took Gamow's bold predictions seriously, and no efforts were made to search for cosmic radiation.

The story continues in 1964 with two scientists of the Bell Telephone Laboratories, Arno Penzias and Robert Wilson. They were observing radio emission from our galaxy using a highly sensitive 20-foot radio antenna built for communication with the *Echo* and *Telstar* satellites. To their dismay, their instrument picked up low intensity microwave noise that resembled the static heard on a radio with poor reception. The noise was very weak and isotropic. At first they suspected that it was produced by faulty equipment, but careful checking showed that this was not the case.

When Penzias and Wilson discussed their problem with other scientists, they learned that Robert Dicke and James Peebles, two theoreticians at Princeton University, might have an explanation for their radio noise. Dicke and Peebles had taken up Gamow's old idea of cosmic background radiation and repeated his calculations. The two teams got together and upon comparing the observations with the theoretical predictions they realized that the microwave noise could only be the relic radiation from the Big Bang. Since then the *cosmic background radiation*, as it is now known, has been observed by radio astronomers around the world at wavelengths from fractions of a centimeter to 50 centimeters. It peaks near one-tenth of a centimeter, which is in the microwave range and corresponds to a temperature of about 3 degrees Kelvin, just a bit lower than Gamow's ingenious prediction 20 years earlier.

In measuring the cosmic background radiation, astronomers have paid much attention to its isotropy. Does it come equally from all directions as the Cosmological Principle predicts? The answer is no. In 1977 several researchers reported that radiation coming approximately from the direction of the Virgo Cluster is slightly blueshifted and radiation coming from the opposite direction is slightly redshifted. Because the two directions are exactly opposite each other and the blue- and redshifts are nearly equal in magnitude, only one conclusion can be drawn: The cosmic background radiation is isotropic, but Earth is traveling relative to the radiation. The speed of travel is about 390 km/sec, which is the combined result of the motion of the Sun within the Milky Way and the motion of the Milky Way through the Universe. If we subtract from this value the Sun's orbital velocity around the center of the Milky Way, we find that our galaxy is falling at approximately 600 km/sec towards the Virgo cluster, the gravitational center of our local supercluster of galaxies.

This result, and the otherwise near-isotropic nature of the cosmic background radiation, suggests that this radiation defines an absolute reference frame for the Universe, relative to which the motions of the galaxies and all other material objects can be measured. The frame is an expanding one, because the radiation is expanding right along with the expansion of the Universe. Any motion relative to the frame is a motion relative to the expansion of the Universe.

It is commonly believed that the superclusters of galaxies have no appreciable velocity relative to the frame defined by the cosmic background radiation.* However, individual galaxies move relative to the frame, because each of them has an appreciable velocity within the supercluster to which it belongs.

The discovery of the cosmic background radiation must be ranked among the great achievements of twentieth century science. In 1978 the Swedish Academy recognized this and awarded the Nobel Prize to Penzias and Wilson, commenting that "their discovery made it possible to obtain information about cosmic processes that took place a very long time ago, at the time of the creation of the universe."

PRIMORDIAL ABUNDANCE OF THE ELEMENTS

Since the discovery that the Universe is expanding, the Big Bang theory has had a number of rivals, such as the steady-state theory.† However, none of the other theories is able to explain both the expansion of the Universe and the existence of the cosmic background radiation. Nor are they able to explain an additional discovery astronomers made during the past two decades: the discovery of the primordial abundance of the elements.

The primordial abundance of the elements refers to the composition matter acquired when it was created by the Big Bang. Astronomers have estimated this abundance by examining the spectrum of light emitted by the oldest stars in our and other galaxies. They found that out of every 100 atoms, approximately 93 are ^{1}H and seven are ^{4}He. By mass, this amounts to approximately 76% ^{1}H and 24% ^{4}He. (^{4}He has four times the mass of ^{1}H, hence the difference in the percentage values by number and by mass.) Elements heavier than helium are present in only trace amounts. As explained in the final section of this chapter, this ratio of helium to hydrogen in the oldest stars is consistent with the Big

*The traditional view that superclusters of galaxies are distributed homogeneously and that they have no appreciable velocities relative to the frame defined by the cosmic background radiation has recently been questioned by astronomers in the United States and England. Their measurements indicate far greater clumping and larger drift velocities than previously suspected (*Scientific American*, June 1986, pp. 66, 67; *Science News*, July 5, 1986, p. 7). No convincing theoretical explanations of these results have been offered so far.

†The steady-state theory of the Universe was first proposed in the late 1940s by Hermann Bondi, Thomas Gold, and Fred Hoyle. According to it, the Universe is expanding forever, yet it never changes. Matter is created continuously and new galaxies are formed at just the rate at which old galaxies recede due to the expansion of the Universe. A steady-state Universe has, therefore, neither a beginning nor an end.

Bang model of creation. The Universe started out explosively from a very hot and dense state and quickly cooled as it expanded. The hot and dense conditions lasted long enough for some hydrogen to fuse into helium, but not so long as to allow the production of significant amounts of the heavier elements. They were made much later in the interiors of massive stars.

COSMOLOGICAL THEORY

So far we have been largely concerned with observational information about the Universe: the distribution of galaxies, the expansion of the Universe, the cosmic background radiation, and the primordial abundance of the elements. These observations led to the conclusion that the Universe was created some 20 billion years ago by a gigantic explosion—the Big Bang—and has been expanding ever since. We now face the obvious question: Will the Universe go on expanding forever or will it eventually fall back onto itself? To answer this question, observational information must be merged with theoretical predictions.

The expansion of the Universe may be compared to the flight path of a rocket launched from the surface of a planet. Whether or not the rocket is able to escape the gravitational pull of the planet depends on the planet's mass and radius and on the rocket's velocity at take-off. In the case of the Earth, the rocket's velocity must exceed 11 km/sec (about 40,000 km/hr or 25,000 mi/hr) to escape. If it is less than that, the Earth's gravity is sufficient to reverse the rocket's motion. The rocket will rise to a maximum height and then fall back to Earth and crash.

Likewise, the eventual fate of the Universe depends on how much matter it contains, on its present size, and on how fast the matter is expanding. The more matter there is, the greater are the gravitational forces holding the Universe together, and the more likely it is that the Universe is finite and will eventually collapse back on itself. The greater the present size of the Universe, the further are the galaxies and the rest of the matter separated from each other and the weaker is the gravitational attraction. This increases the likelihood that the Universe is infinite and will expand forever. Finally, the greater the velocity of expansion, the harder it is for gravity to slow the expansion to zero and to reverse it. Again, this contributes toward making the Universe infinite and ever-expanding.

The task of this section is to determine the relative contribution of each of these three factors—cosmic mass, size, and expansion velocity—to the evolution of the Universe and then to use them to decide on theoretical grounds what the destiny of the Universe will be. Before going on, it is important to understand two differences that exist between the launching of a rocket and the expansion of the Universe.

First, in the case of the rocket, there are two separate entities—rocket and Earth. The rocket is flying away from Earth. In the case of the Universe, there is just one entity—the Universe. It is not individual galaxies that fly away from the Universe. It is the entire Universe and all it contains that fly apart.

Second, when calculating the flight path of a rocket it is safe to assume that space is described by the three-dimensional geometry of Euclid (third century B.C.), with time as a separate parameter. Furthermore, the physics of the flight path may be assumed to be the classical one developed by Isaac Newton in the seventeenth century.* Neither of these assumptions is appropriate when dealing with the Universe at large. Here space and time must be linked into the four-dimensional geometry of space–time and the physics must be based on the General Theory of Relativity. This theory, developed early in the present century by the German-American physicist Albert Einstein (1879–1955), is considered one of the great intellectual accomplishments of our time.

Einstein's General Theory of Relativity

In Einstein's theory, massive objects warp the geometry of space–time and this warping gives rise to gravitational forces. The greater the warping, the greater are the gravitational forces. The Earth and Sun warp space–time only slightly because they are not very massive and they are rather extended. In their vicinity classical physics is a good approximation. However, in the case of the large-scale structure of the Universe or near highly compact objects like neutron stars and black holes, the curvature of space–time is appreciable. Classical physics breaks down and general relativity applies.

General relativity predicts a number of effects that classical physics cannot explain (see figure 1.17). It predicts that light travels through the Universe and past stars not in straight lines, but along curved paths; that time does not tick away at equal rates everywhere, but ticks more slowly where gravity is strong; and that light emitted from a star or galaxy loses energy and its wavelengths are stretched to longer values (which means that the wavelengths are shifted toward the red part of the electromagnetic spectrum, an effect that is called the *gravitational redshift*). The greater the warping of space–time, the greater are these effects. Since Einstein first published this theory in 1915, each of these predictions has been observed and verified as occurring in nature.

Perhaps the most dramatic confirmation of the predictions of general relativity has been the discovery in 1979 of the gravitational lens effect (figure 1.18). The light from two distant quasars (objects at great distances, characterized by luminous, point-like sources of light—see chapter 2), QSO 0957 + 561 A and B, was found to originate from a single quasar and to be bent by the mass of an intervening galaxy so as to produce two separate images 6.15 seconds of arc apart (figure 1.19). That the two images (A and B) are, in fact, of the same source is demonstrated by their identical redshifts (velocity of recession = 200,000 km/sec, corresponding to a distance of about 9 billion LY) and identical spectral features. Five other examples of gravitational lensing of distant quasars have been discovered since 1979, and many more are suspected to exist.

*Sir Isaac Newton (1642–1727) is considered one of the fathers of modern science. Among many other contributions, he invented the reflecting telescope, discovered the splitting of white light into its color components by prisms, suggested the corpuscle (photon) theory of light, was one of the co-developers of calculus, derived the classical theory of gravitation, and laid the foundation of mechanics with his three celebrated laws of motion (see box 4.2). In 1687 Newton published *Philosophiae naturalis principia mathematica* (Mathematical Principles of Natural Philosophy), a book considered by many to have had the single most important influence in the development of modern science.

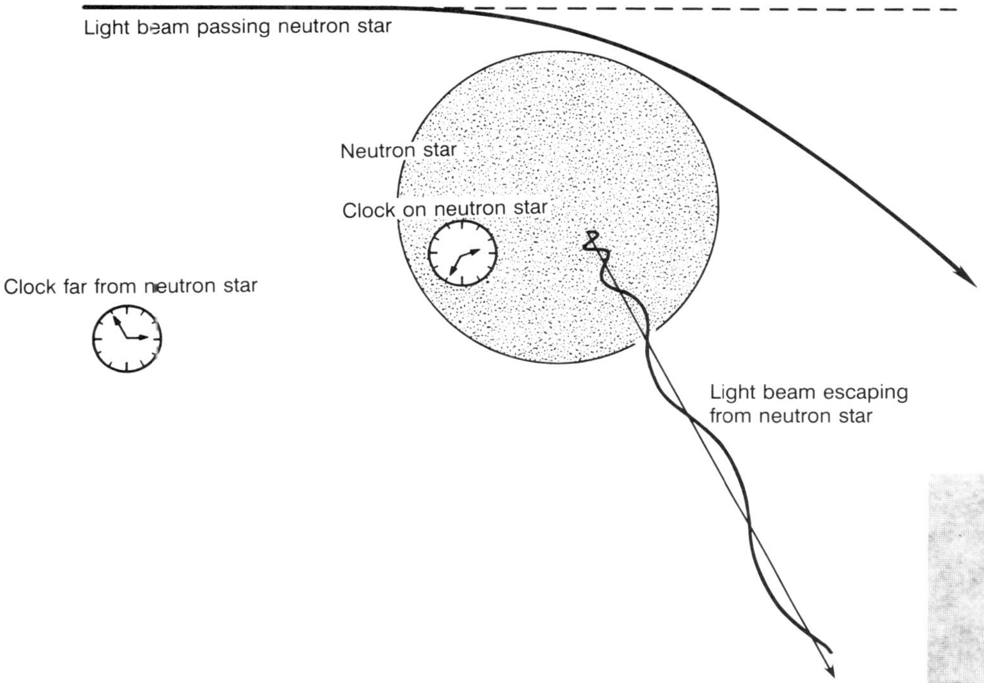

Figure 1.17. Illustration of three of the predictions of the General Theory of Relativity: 1. Light travels past a star not in a straight line, but along a curved path. 2. Clocks tick more slowly on a star than in free space. 3. The wavelengths of light emitted from a star become stretched (called the *gravitational redshift*). According to the General Theory of Relativity, all three effects are caused by the warping of space-time in the vicinity of the star.

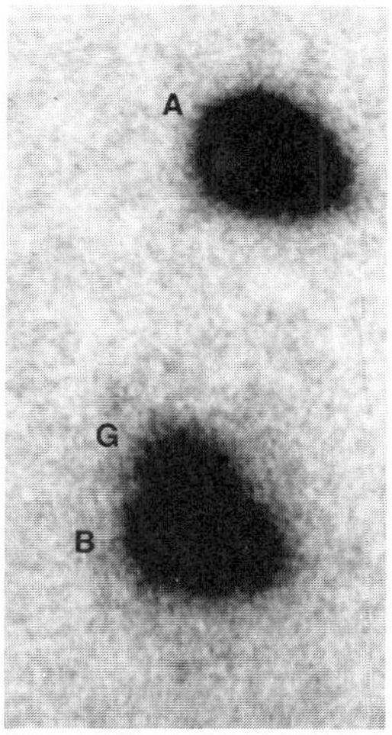

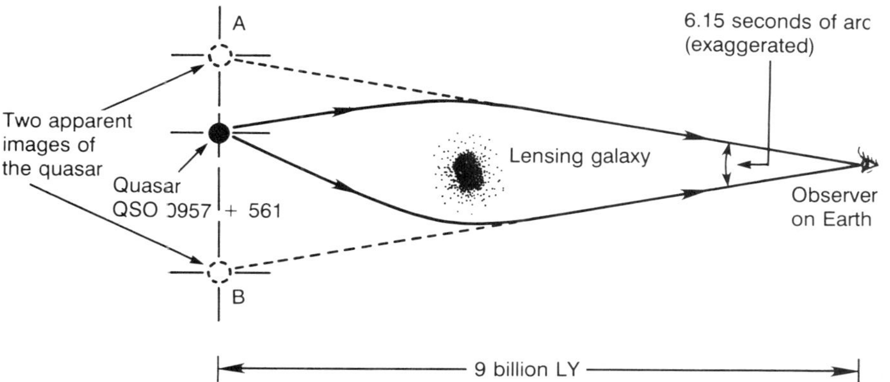

Figure 1.18. One of the most dramatic confirmations of Einstein's General Theory of Relativity, the gravitational lens effect. The light from a quasar (QSO 0957 + 561) was found to be bent by the gravitational field of a massive intervening galaxy so as to produce two images.

Figure 1.19. Quasar QSO 0957 + 561 A and B, an example of the gravitational lens effect. Three images, labeled A, B, and G, are shown. Images A and B are of the same object, namely the quasar. Image G is that of an intervening galaxy whose gravitational field produces the splitting of the quasar image.

Friedmann's Three Models of the Universe

After developing his General Theory of Relativity, Einstein and others applied it to the study of the large-scale structure and evolution of the Universe. The first major success in these efforts came in the early 1920s when the Russian mathematician Alexander A. Friedmann (1888–1925) obtained three different solutions to Einstein's equations. Each solution predicts that the Universe began with an explosion, but they differ on the subsequent evolution. In the first solution, the Universe expands for some time, reaches a maximum size, and then contracts. This model predicts a Universe that is *finite and closed*. In the other two solutions, the Universe expands forever, making it infinite. In one case, the Universe is still expanding even after all the galaxies and the rest of its content have become infinitely dispersed. This model universe is said to be *open*. In the other case, the Universe is just barely able to disperse to infinity, its expansion velocity slowing to zero in the process. This model universe lies at the border between the other two and is called the *flat universe*.

Einstein developed his General Theory of Relativity in order to describe in a consistent manner physical phenomena occurring near stars and other gravitating bodies. It is remarkable that this theory, derived from local phenomena, allowed Friedmann to make predictions about the evolution of the entire Universe, including the prediction of the cosmic expansion several years before Hubble confirmed it by observation. This result attests to the broad applicability of the theory, as well as to the genius of its inventor.

The theoretical predictions of Friedmann are most clearly illustrated by a graph of the scale* of the Universe versus time (see figure 1.20). All three models start from a very compact state—the Big Bang origin of the Universe. For some time thereafter they evolve nearly identically, and their graphs show this. However, as the rate of expansion is slowed by the decelerating effect of gravity, differences appear. The graph of the closed universe reaches a maximum height and then descends. Eventually, the graph dips all the way down as this universe collapses into the compact state of its birth. There may occur a bounce from which the closed universe re-emerges for yet another cycle in an ongoing evolution, though this is not a prediction that follows from general relativity. The other two models of the Universe expand to infinity, as shown by their open-ended graphs. The graph of the open universe remains steeper and lies above those of the other two. This indicates that its expansion velocity is the largest of the three models and that it is still expanding even after it has become infinitely dispersed. The graph of the flat universe becomes less and less steep until, after an infinite time, it becomes horizontal while simultaneously reaching infinite height. This means that the flat universe slows more and more until the expansion rate approaches zero while simultaneously dispersing to infinity.

Now we must decide which of Friedmann's three models corresponds to the actual Universe. As stated earlier, this decision needs to be based on observa-

*In this context it might at first appear that "size of the Universe" should be plotted versus time. However, "size" loses its meaning if the Universe is infinite. Therefore, the new concept "scale of the Universe" is introduced. It refers to the distance between *any* two superclusters of galaxies, as, for instance, the distance between the Virgo and the Coma clusters. As the Universe expands and the distance between the two superclusters increases, the scale increases proportionately. It does not matter whether the Universe is finite or infinite.

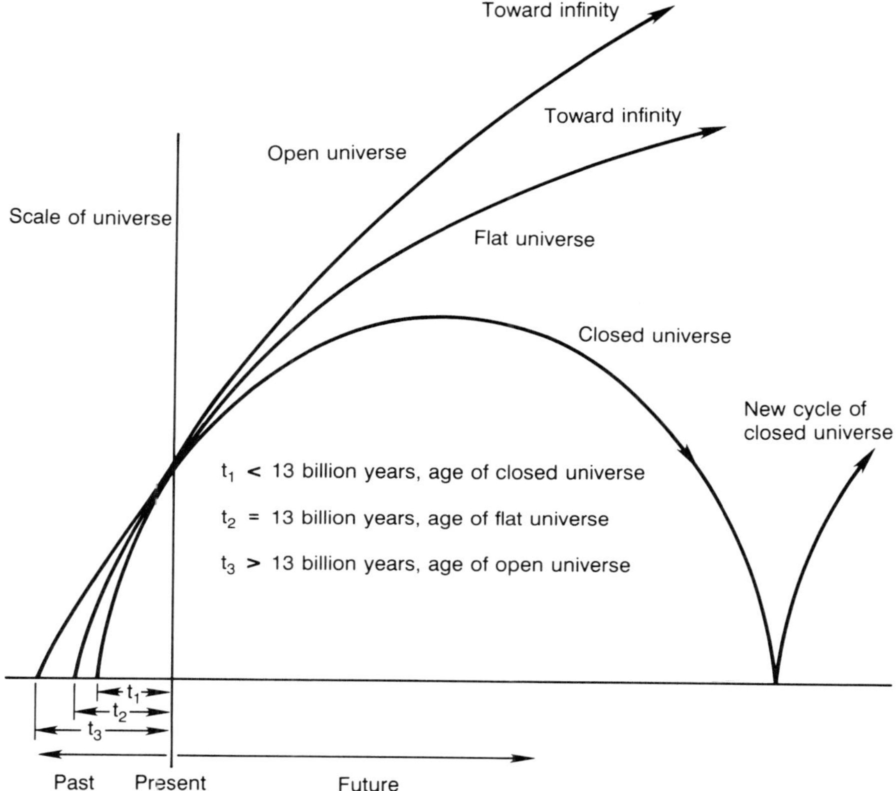

Figure 1.20. Evolution of Friedmann's three models of the Universe. The distances between clusters of galaxies (labeled *Scale of Universe*) remain finite in the closed universe, but they become infinite in the flat and open universes. The ages indicated are based on H = 15 km/sec per million light-years.

tional data and on theory. The observational data pertinent to the problem are the present average density of matter in the Universe and the present rate of cosmic expansion. The theory is Einstein's General Theory of Relativity. It will tell us whether or not the cosmic expansion rate is sufficient to overcome the gravitational pull of the matter.

The present rate of expansion of the Universe is given by the Hubble constant — 15 km/sec per million light-years. For this rate of expansion Einstein's theory predicts that the Universe is finite and closed if the average density of matter is greater than 5×10^{-30} g/cm^3; it is infinite and open if the average density is less than that value; and it is infinite and flat if the average density is equal to it.

The value of 5×10^{-30} g/cm^3, which corresponds to about three hydrogen atoms per cubic meter, is called the *critical density* of the Universe.* To decide

*The critical density is derived from theory and based on the value adopted for the Hubble constant. For a value of H twice as great as the one used above, namely H = 30 km/sec per million light-years, the critical density is 2×10^{-29} g/cm^3. In other words, if the Universe expands twice as fast as we have assumed, the density required to close it is four times greater.

in what kind of universe we live—open, flat, or closed—we need to compare the average density of the actual Universe with the critical density. The average density of matter is the amount of mass contained in a volume large enough to embrace a representative sample of the Universe. Such a volume must extend over several superclusters of galaxies. By adding up the masses of the galaxies of the superclusters in our vicinity of the Universe and dividing by the corresponding volume, it turns out that the average density of matter is approximately 7.5×10^{-32} g/cm^3, or four to five hydrogen atoms per 100 cubic meters of space. This is roughly 70 times less than the critical density and suggests that the matter in the Universe is insufficient for reversing the cosmic expansion. The Universe is infinite and open.

Could it be that our estimate of the average density of matter in the Universe is too low? Might there not exist somewhere in the vastness of space matter not yet discovered that is sufficient to close the Universe? For example, there are many more ways for matter to be invisible, or at least hard to see, than there are for it to be visible. Visible matter is either in the form of luminous stars or gas clouds illuminated by stars, but dark or hard-to-see matter may be in such different forms as rocks, planet-like objects, faint burnt-out stars, black holes, dark gaseous halos surrounding galaxies, intergalactic clouds of neutral hydrogen that never condensed into stars or galaxies, or even neutrinos (if they possess mass).

Although dark matter cannot be seen directly, its presence can be deduced from its gravitational effect on visible matter. This is done by measuring the velocities of stars and gas clouds within galaxies and of galaxies within galaxy clusters. The velocities tell how much mass (luminous and dark) is present to keep the fast moving objects bound by gravity and prevent them from flying off.

Measurements of stellar motions within galaxies indicate that, depending on the galaxy type, their true masses are anywhere from a factor of 3 to 80 times greater than the masses of the visible matter, with a factor of 6 to 7 being most common. If we include this invisible matter of the galaxies in our estimate, we obtain very approximately a value of 5×10^{-31} g/cm^3 for the average density of the Universe. This value falls still short by a factor of 10 of the critical density and again suggests that the Universe is infinite and open.

Measurements of the velocities of double galaxies around each other and of galaxies within clusters of galaxies indicate that additional matter exists in the space between the galaxies, not accounted for by the galaxies themselves. Without that matter, the clusters would have become dispersed long ago. This is the *missing mass*, whose existence was first suspected nearly 50 years ago by Fritz Zwicky (1898–1974) of the California Institute of Technology. Of course, the mass is not really missing, it is just not visible. As suggested above, this dark mass might be in the form of black holes or intergalactic gas clouds. Some of it could be in the form of small faint galaxies or globular star clusters that have escaped detection over intergalactic distances. Another contribution to the missing mass could arise from the hypercharge force that has recently been suggested by Ephraim Fischbach and his colleagues as a fifth fundamental force of nature (see the Introduction to Part One). The missing mass might also exist in the form of neutrinos, those elusive particles that, according to current theories, have a mass of about 1/100,000 the mass of an electron. There may, in fact, be so many neutrinos that, despite their tiny mass, they add up to ten times the ordinary mass found in galaxies. If this is correct, the average density of matter in the Universe would be very close to the critical density. That is a very interesting result. It means that the Universe is very nearly flat. However, at present the estimate of the Universe's average density is too uncertain to allow us to be more specific. We do not know whether the Universe is exactly flat, slightly open, or slightly closed.

Age of the Universe: Revised Estimate

The models derived by Friedmann allow us to estimate the age of the Universe more accurately than we did earlier. The estimate of 20 billion years was based on the assumption that the Universe has always been expanding at its current rate of 15 km/sec per million light-years. The steepness of the graphs in figure 1.20 shows, however, that this rate was much greater in the past. Hence, it took the Universe less time to reach its present scale than we estimated. If the Universe is flat, its age turns out (assuming $H = 15$ km/sec per million light-years) to be two-thirds of our earlier result—13 billion years. If it is open, its age is somewhat greater. And if it is closed, its age is a little less. Because it seems that the Universe is very nearly flat, 13 billion years is the estimate for the age of the Universe assumed in this text.

GEOMETRICAL INTERPRETATION OF THE MODEL UNIVERSES

Much can be learned by considering the geometrical characteristics of Friedmann's three models. To do this we must draw diagrams of their space–time structures. However, it is not possible to draw diagrams of four-dimensional structures. What we can do is "cut through"* one of the space dimensions and then draw the remaining three—one time and two spatial dimensions. Presumably, physical events happening along the space dimension deleted by cutting are, on the average, like the events happening along the two space dimensions that remain. Hence, what we learn from three-dimensional space–time diagrams should be correct in principle, if not in detail.

Flat Universe

The simplest geometry is that of the flat universe: it is the Euclidean geometry with which we are all familiar from daily life. Cutting through one space dimension results in three-dimensional space–time. The two remaining spatial dimensions are a flat surface that extends to infinity in all directions, and time, which runs perpendicularly to that surface.

*By "cutting" we reduce the number of dimensions of a real object or a geometrical figure. For instance, by cutting through a three-dimensional object, we obtain a two-dimensional surface; by cutting a two-dimensional surface, we obtain a one-dimensional line; and by cutting a line, we obtain a zero-dimensional point.

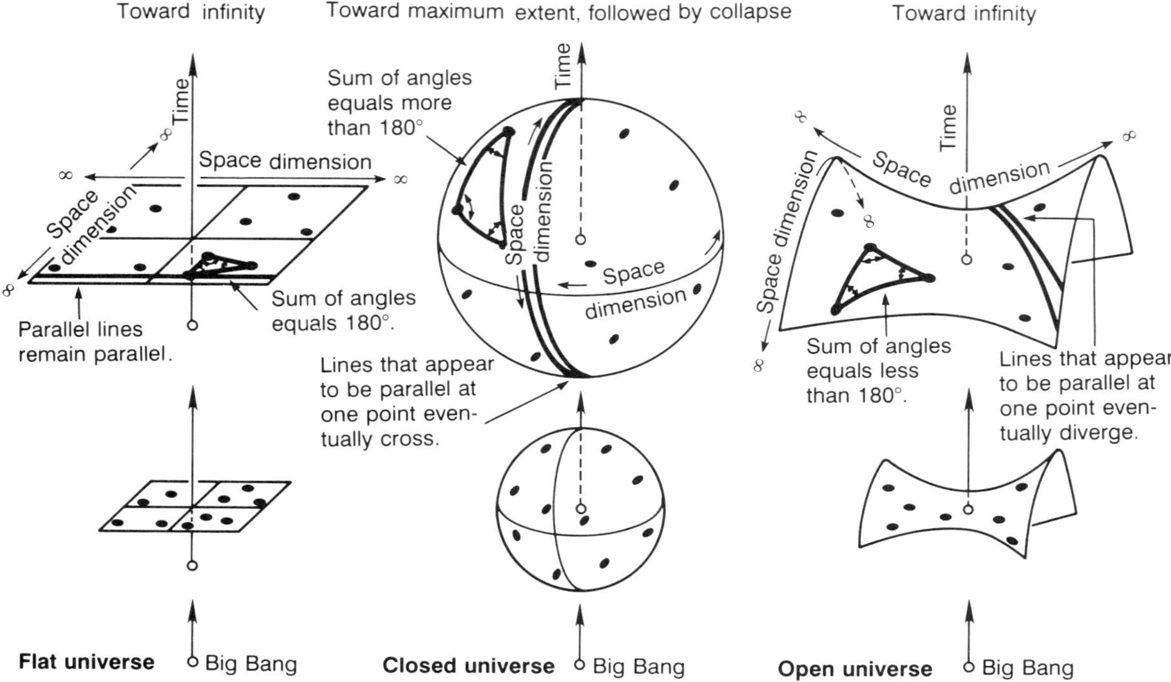

Figure 1.21. Geometrical interpretation of the flat, closed, and open models of the Universe. Two space and the time dimensions are drawn. The third space dimension has been deleted. All three models begin from a highly compact state—the Big Bang (shown by open circles). As time progresses, all three models expand and clusters of galaxies (shown by dots) recede from each other.

Let us focus on a finite square that lies on that flat surface obtained by cutting and see how it and its contents evolve (see figure 1.21, flat universe). Dots on the square represent clusters of galaxies. As time goes on, the square moves along the time axis and expands; the dots move farther and farther apart; and after an infinite length of time, all the dots become dispersed to infinity. These changes in the size and internal structure of the square simulate the aging and expansion of the flat universe. If we go back in time, we find that the square started out from a highly compact state, in which all the dots—the galaxy clusters—were crunched together and the Universe was born.

We selected the square quite arbitrarily. There exist infinitely many other squares exactly like it, one next to the other and filling the space of the flat universe. No matter how far we go in any direction, we never come to an end or to an edge. We will also never find a center, for that would mean that one square distinguishes itself from all the others. The only edge that we ever find in the flat universe is an edge in time—namely the beginning.

In Euclidean geometry the shortest distance between two points is a straight line, parallel lines always remain parallel, and the interior angles of triangles add up to 180°. Furthermore, light travels along straight lines. However, these features are unique to the flat universe and are not shared by the other two models.

Closed Universe

The geometries of the other two model universes are non-Euclidean. Their space–time is not flat but curved. Let us consider the closed universe (figure 1.21). By cutting through one of its spatial dimensions, we obtain a finite sphere (think of a balloon). The radius of the sphere corresponds to time and its surface to curved two-dimensional space. Again, we let dots on the surface represent clusters of galaxies. As the closed universe evolves, the sphere expands until it reaches a maximum size and then it contracts again (like blowing air into the balloon and then letting it out). During the expansion phase all the dots move away from each other, and during the contraction phase they approach each other. Eventually, the sphere collapses into a very compact state and all the dots are crushed together. This represents the unavoidable, catastrophic end of the closed universe, the so-called big crunch. Going back in time, we find that the closed universe was born from a similar, equally compact state in space–time.

This universe, too, has neither a spatial edge nor a spatial center, for they do not exist on the surface of a sphere. The point at the center of the sphere refers not to space, but to space–time. It corresponds to the two unique events in the evolution of the closed universe: its creation and its end.

Our geometrical intuition does not apply to the global structure of the closed universe. As the diagram illustrates, in this universe the shortest distance between two points is a great circle,* there are no parallel lines (that is, all lines cross), and the interior angles of triangles add up to more than 180°. Furthermore, light rays will not travel along straight lines, but follow the curvature of the universe, which is represented by great circles.

Open Universe

Finally, consider the geometry of the open universe (figure 1.21). Cutting through one of the space dimensions yields a saddle-shaped surface that extends to infinity in all directions. As time marches on, the saddle expands, its curvature lessens, and the dots—representing galaxy clusters—move apart. After an infinite length of time, all the dots will be separated from each other by infinite distances. Going back in time, the section of the saddle shown in the diagram becomes smaller and the dots move closer together. During the Big Bang, they were compressed together into the same highly compact state that we encountered at the beginning of the other model universes.

As in the other universes, neither an edge nor a center exists in the space dimensions of the open universe. No matter how far we go in any direction, we will find galaxy clusters. An edge exists only in space–time, namely at the beginning.

Again, our geometrical intuition fails in the open universe. The shortest distance between two points is a curve (although it is not a great circle), there are no parallel lines (that is, all lines cross or diverge), the interior angles of triangles add up to less than 180°, and beams of light follow curved paths.

*A great circle is a circle formed on the surface of a sphere by the intersection of a plane that passes through the center of the sphere. For example, the equator of the Earth is a great circle. All longitudinal circles are great circles, but latitudinal circles (except the equator) are not.

In principle, we should be able to tell in which kind of universe we live by observing the geometry of our cosmic environment. Do the interior angles of triangles add up to 180°, or are they greater or smaller? Are there parallel lines, or do all lines cross or diverge? What is the shortest distance between two points? Here on Earth—by drawing lines on paper or surveying the countryside—Euclidean geometry holds as far as we can tell. However, that does not help us decide in which kind of universe we live. Earth extends over such a small portion of the Universe that even if space were curved it could not be distinguished locally from flat space. This is analogous to looking at a microscopic section of the surface of a large balloon or saddle. It, too, cannot be distinguished from a section of a flat surface. Likewise, the section of the Universe that we can observe with precision with present-day telescopes is too small to permit us to discover the nature of the cosmic geometry. It could be flat, spherical, or saddle-shaped. All we can say is that if space is curved, it is not curved very much. Again, as when we compared the average density of the Universe with the critical density, we have failed to discover the Universe's structure and its ultimate fate.

THE BIRTH OF THE UNIVERSE

The currently available evidence suggests that the Universe came into existence in one gigantic explosion some 13 billion years ago, as discussed in the preceding sections. During the earliest phase of this explosion, the temperatures and densities were so extreme that neither space and time, nor matter and energy, nor the laws of physics were as we know them today. In order to get some idea of what the conditions may have been like then, we need to start from the known conditions of today and extrapolate backward in time. The best presently available theories concerning the behavior of matter at high energies along with the general theory of relativity shall be our guides.

Such an extrapolation is daring, to say the least. It may even be grossly misleading. Physicists have learned from hard experience that extrapolating beyond solid observational and experimental evidence has generally led them astray. However, at present there are no observational or experimental data of the first instant of creation, and possibly, there never will be. Extrapolation is the only means of such an investigation. Besides, to do so is an exciting intellectual adventure.

Our extrapolation will tell us that the initial extreme conditions passed in rapid succession through a number of transitions, during which the Universe acquired the characteristics that still distinguish it today. A single *superforce*, which ruled in the beginning, split into the four forces we are accustomed to. Matter was spontaneously created and acquired the enormous expansion velocities that we still observe. There occurred a brief burst of primordial nucleosynthesis, during which hydrogen (H) was converted into helium four (^{4}He). One by one, the lightest of the elementary particles—gravitons, neutrinos, and photons—decoupled from the rest of the particles and began to evolve independently. This entire sequence of events, with the exception of photon decoupling, was over in about 5 minutes. The photons decoupled approximately at age 500,000 years and subsequently evolved into today's 3 K cosmic background radiation. From then on, matter (that is, hydrogen and helium) evolved on an independent course as well and began to condense into the lumpy distributions — clusters of galaxies, galaxies, stars, and planets (around at least one star) — that we find in the Universe today.

A more detailed discussion of these events is best divided into two parts: the first 10^{-6} second and the evolution from age 10^{-6} second forward. Only during the first microsecond were conditions so severe that any treatment of them is unqualifiedly speculative and, no doubt, will be subject to major revisions in the future. In contrast, by age 10^{-6} second the content of the Universe and the physical laws governing it are believed to have been the same as they are today. Consequently, discussion of that period is much less in question, though uncertainties do remain.

Theoretical Background

Before discussing the Universe's birth, a few remarks are necessary concerning the theories upon which this description is based. One theory has already been introduced — the General Theory of Relativity — though to be fully applicable to the Universe's first instant of existence Einstein's formulation of relativity needs to be modified. In particular, Einstein's theory needs to be replaced by a quantum theory of gravitation. Unfortunately, no one has yet succeeded in working out such a theory in detail, and this constitutes the greatest source of uncertainty in the description below.

Another theory that we need goes by the poetic name of *quantum chromodynamics* or QCD. According to this theory, protons and neutrons are not really elementary, but are composed of more fundamental particles called *quarks* (see the beginning of the Introduction to Part One).

A third theoretical component of the description is the *grand unified theories* or GUTs (plural, because there are many grand unified theories, none of which is at present universally accepted). With GUTs, physicists seek to find relationships between the forces of nature and unify them into one. For example, the theories predict that at temperatures above approximately 10^{15} K two of the standard forces — the weak and the electromagnetic forces — become unified into the *electro-weak force*. At the same time, electrons and neutrinos lose their identity and behave in identical ways. They are then referred to collectively as *leptons* and they interact by the electro-weak force. Above approximately 10^{27} K, the electro-weak and nuclear forces become unified into the *grand unified force*. Finally, above approximately 10^{32} K, the grand unified force and gravitation are expected to become unified into a single *superforce*.

The concept of a theory that unifies the forces and, simultaneously, describes the fundamental nature of the elementary particles is not new. For example, after completing his General Theory of Relativity, Einstein attempted to unify gravity with the other forces. However, it was not until the late 1960s that real progress in this regard was made. The first to succeed in unifying the electromagnetic and weak forces were Steven Weinberg, Sheldon Lee Glashow (both of the United States) and Abdus Salam (of Pakistan), and in 1979 they were awarded the Nobel prize in physics for this work. Then in the 1970s, Howard Georgi (of the United States), Glashow, and others began working out theories (namely GUTs) unifying the electro-weak force with the nuclear force, an effort that continues today. No one has yet succeeded in unifying the grand unified force with gravity, in part because there is no quantum theory of gravitation. Therefore, the earlier reference to the existence of a single superforce above approximately 10^{32} K needs to be accepted with particular caution.

The application of GUTs to the early Universe described below was first made in 1980 by Alan H. Guth of the Massachusetts Institute of Technology and has been developed further by him and others since then. Guth called his theory the ***inflationary model*** of the Big Bang because it predicts an enormously fast initial expansion of the Universe, during which most of the matter that is still with us today came into existence.

First Microsecond

In the discussion of the birth of the Universe that begins here, the focus is on the sequence of transitions mentioned earlier. These transitions, in a step-by-step fashion, are believed to have marked the evolution from the initial extreme conditions to the more familiar ones of today (see figure 1.22).

Age 10^{-43} second. Age 10^{-43} second is the earliest point in time about which physics has anything to say. Neither matter and energy nor the laws of physics as we know them today existed then. Space–time was so severely curved that black holes were continually forming and bursting. They were microscopic in size, with diameters 10^{20} times smaller than a proton, and they lasted on the average about 10^{-43} second. Many black holes may have come into existence and then burst; we don't know. Along with the creation and destruction of black holes, space and time were also continually being broken and reconnected. Thus, space and time were not continuous. They were "quantized" analogous to the way today's photons and other elementary particles are quantized. Under these conditions even the concept of causality, namely such words as *earlier* and *later*, had little meaning.

At that early instant of time, the temperature was approximately 10^{32} K and the force that governed the black holes was the superforce introduced earlier. This force constituted a unification of our present-day gravitational, weak, electromagnetic, and nuclear forces. Sometime between ages 10^{-43} and 10^{-35} second, this era of black hole creation and destruction ended. The black holes that existed then were expanding, the temperature was dropping, and the superforce split into gravity and the grand unified force.

Stop and consider. Space–time was continually being broken and reconnected between ages 10^{-43} and 10^{-35} second. Does reference to these times make much sense when time itself was discontinuous and causality had little meaning? There is at present no clear answer. There may well have existed an infinite stretch of time before 10^{-43} second. Or that point in time may have been preceded by earlier cycles of oscillating universes. Or any number of conditions that, with our limited understanding, we cannot even conceive of may have existed then. We just don't know.

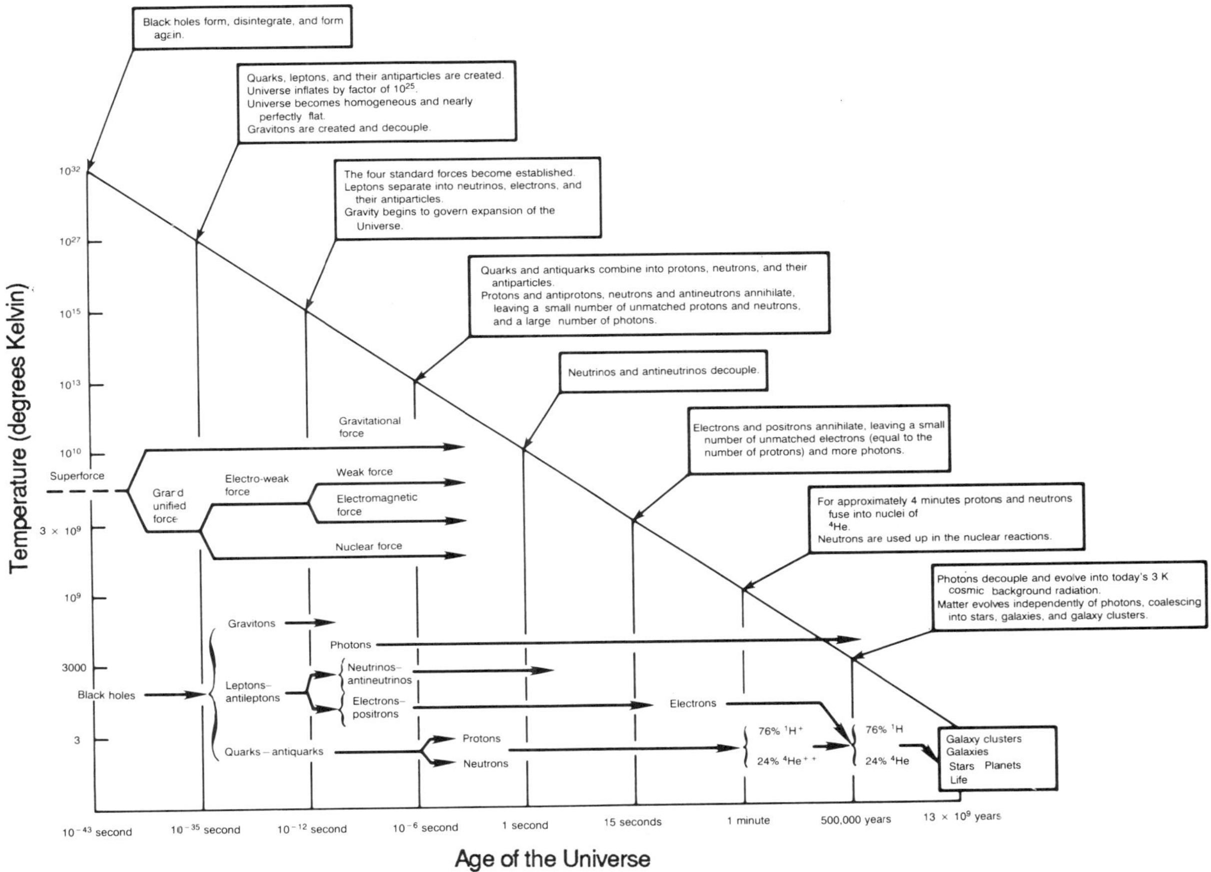

Figure 1.22. Key events during the birth of the Universe according to the inflationary model of the Big Bang.

Age 10^{-35} second. By age 10^{-35} second, the temperature had fallen to about 10^{27} K and the grand unified force separated into the nuclear and electro-weak forces. Within each of the black holes, this transition was accompanied by the spontaneous and rapid creation of quarks and leptons, as well as antiquarks and antileptons. We may compare this creation to the spontaneous formation of raindrops in the present Earth atmosphere, when water vapor cools below the condensation point, though the energies involved were far more extreme. According to GUTs, the quarks, leptons, and their antiparticles came into existence because the vacuum of that time contained energy (unlike today, when a vacuum contains no energy and, hence, cannot produce matter). In the language of physics, the creation of the particles constituted a phase transition from pure energy into matter.

The rapid creation of quarks, leptons, and their antiparticles led to enormous pressures, which forced the black holes to expand at superfast rates. This was the expansion that Alan Guth called *inflation*. Even though each of the black holes, along with the disconnected regions of space-time associated with them—called *domains*—expanded extremely quickly, they moved away from each other still more rapidly. Consequently, they never became connected and, in the course of time, got more and more out of touch. We do not know what became of any of these domains except one. It continued to expand and evolved into the Universe in which we live, as described in this book. The other domains may have become annihilated. Or they may still exist as separate universes, forever inaccessible to us. We shall never know.

Age 10^{-12} second. By about age 10^{-12} second, the temperature had dropped to 10^{15} K and the last of the unified forces divided. The electro-weak force separated into the electromagnetic and weak forces. Thus, all four of the forces that we experience today had become established.

Quarks, leptons, and their antiparticles were now no longer created spontaneously, and the epoch of inflationary expansion ended. Inflation had increased the size of the Universe by roughly a factor of 10^{25} from its embryonic state at age 10^{-35} second. Furthermore, the distribution of the particles throughout the Universe was nearly perfectly homogeneous, a condition still observed today on the largest scale.

Even though inflation had come to an end, the Universe continued to expand simply because of the momentum it had acquired during inflation. However, the expansion was not unopposed. The global attractive force of gravity—that is, the force of gravity exerted by *all* matter and energy in the Universe—began to slow the expansion. As a result the large-scale evolution of the Universe followed—and still follows today—one of the three curves derived by Alexander Friedmann. As already noted, we do not know which of the three curves it was, and GUTs give us no answer either. We only know that the cosmic expansion rate at that early time must have been extremely delicately balanced against the global gravitational attraction to ensure that the Universe would coast to its present size. Any deviation from this finely tuned balance would have become greatly magnified with time. If the expansion rate had been only a fraction greater, the Universe would be totally dispersed today. Had it been slightly smaller, the Universe would long ago have recollapsed into its initial, highly condensed state. Put differently, our Universe is nearly perfectly flat, as recognized in the previous section.

With the separation of the electro-weak force into the weak and electromagnetic forces, the leptons and antileptons no longer remained single species of particles. They separated into electrons and positrons (electron antiparticles), and neutrinos and antineutrinos. The electrons and positrons interacted by the electromagnetic force, while the neutrinos and antineutrinos interacted by the weak force. No corresponding change occurred in the case of quarks and antiquarks.

In addition to these particles and antiparticles, which are characterized by having mass, there also existed massless particles, namely gravitons and photons. The gravitons had been created during the Universe's rapid inflationary acceleration, and they continued to be created now, though at an ever diminishing rate (see the beginning of the Introduction to Part One and box I.1 for a brief discussion of gravitons).

The particles that filled the Universe at age 10^{-12} second were in a chaotic state. There was no structure or order. The particles were flying around at high speed in all directions, continually colliding and interacting violently with each other. Upon colliding, the particles and their antiparticles annihilated each other, creating high-energy photons, which, in turn, quickly decayed and changed back into matter and antimatter particles.

The only particles that did not participate in these collisions were the gravitons. They passed right through the other particles without interacting directly. They were decoupled from the rest of the matter and the photons. This happened because they interacted only by the gravitational force, which is the weakest of the four forces. However, despite being decoupled, the gravitons still felt the global gravitational attraction of the rest of the matter and energy in the Universe. Consequently they traveled along curved paths as dictated by the geometry of space–time and, presumably, they still do so today. Should we ever succeed in detecting them, they would constitute the most ancient direct evidence of the birth of the Universe.

Age 10^{-6} second. At age 10^{-6} second the temperature dropped below 10^{13} K and the quarks and antiquarks combined into protons and antiprotons as well as neutrons and antineutrons. Simultaneously, the protons and neutrons collided with the antiprotons and antineutrons and became annihilated, producing a huge number of photons. At the existing temperature, the photons were no longer energetic enough to create either protons and antiprotons, neutrons and antineutrons, or quarks and antiquarks. Thus, the back-and-forth reactions between heavy particle–antiparticle pairs and photons came to an end. The heavy particles and their antiparticles annihilated each other without being recreated by the photons. (This was not true of the electron–positron pairs. They continued their back-and-forth reactions with the photons for about another 15 seconds; see below.)

Not all of the protons and neutrons were annihilated, however. At age 10^{-35} second, slightly more quarks had been created than antiquarks—approximately one unmatched quark for every billion quark–antiquark pairs. These unmatched quarks combined into protons and neutrons that did not find antiprotons and antineutrons with which to annihilate. Hence, they survived. These excess protons and neutrons (together with excess electrons; see below) constituted the matter that eventually evolved into the stars, galaxies, and clusters of galaxies that mark the Universe today. Had matter and antimatter particles been created in exactly equal numbers, they would have completely destroyed each other and left the Universe in a rather impoverished state, with photons, neutrinos, and gravitons as its only content.*

*We happen to live in a Universe made of matter consisting of protons, neutrons, and electrons. In addition there are neutrinos and antineutrinos (which may have a tiny mass), photons, and gravitons (the last two are massless). Generally, there are no antiprotons, antineutrons, and positrons in our present-day Universe. However, during such violent events as supernova explosions, galaxy explosions, and Earth-based experiments with high-energy particle accelerators, particle–antiparticle

To summarize, the Universe at age 10^{-6} second was in a state of rapid expansion. It contained protons, neutrons, electrons and positrons (both in nearly equal numbers), neutrinos and antineutrinos, photons, and gravitons. The forces governing the interactions among the particles were the nuclear, electromagnetic, weak, and gravitational forces.

All the particles were in continuous chaotic motion, incessantly colliding and interacting with each other. The only exception was the gravitons, which were decoupled from the rest. The electrons and positrons were still continually annihilating each other and creating photons. The photons, in turn, quickly decayed and changed back into electrons and positrons.

None of the collisions produced nuclei of helium or heavier elements, nor did protons and electrons combine to form atoms. At the high prevailing temperatures, the collisions were still too violent, by many orders of magnitude, for that. Had nuclei of heavy elements or atoms formed, they would have been instantaneously broken up into their component particles by further collisions.

After the First Microsecond

Space and time, matter and energy, and the four forces all came into existence during our Universe's first microsecond of existence. Let us now examine what happened next.

Age 1 second. By age 1 second, the temperature had dropped to about 10^{10} K. As well as causing this drop in temperature, the cosmic expansion had reduced the neutrinos' energies and changed their interaction properties such that they ceased to feel the presence of the other particles. Thus, neutrinos and antineutrinos became the second species of particles to decouple from the rest. Like the gravitons, they still were part of the Universe and subject to its global gravitational attraction. Presumably, these primordial neutrinos are still racing through the Universe today, though they have not yet been detected. Should they be detected in the future, they will be the second oldest evidence from the Big Bang. In fact, there may be so many of them that their combined mass closes the Universe, as noted earlier.

Age 15 seconds. As the Universe aged and expanded further, its radiation and matter continued to cool. By age 15 seconds, the temperature had fallen below 3 billion degrees Kelvin and the photons no longer had the energy to produce electron–positron pairs. The electrons and positrons experienced the same fate that had befallen the protons and neutrons earlier. They annihilated each other without being recreated by the photons. The only exception was the small excess number of electrons with which the Universe was born. Thus, the

pairs are created—electron–positron pairs, proton–antiproton pairs, neutron–antineutron pairs, and still other, more exotic pairs. When these particles collide with their respective antiparticles (electrons with positrons, protons with antiprotons, and so forth) they annihilate each other and produce high-energy photons, just as during the first microsecond of the Big Bang. Furthermore, these high-energy photons are observed to decay into particle–antiparticle pairs. The antiparticles have the same masses as their respective particles. However, the positron is positively, and the antiproton is negatively, charged. The antineutron has zero charge, just like the neutron.

last vestige of primordial antimatter (with the exception of antineutrinos) disappeared. The Universe now consisted of equal numbers of electrons and protons, of neutrons, neutrinos, antineutrinos, photons, and gravitons. The neutrinos, antineutrinos, and gravitons were decoupled from the rest.

Age 1 minute. The time between ages 1 and roughly 5 minutes was the epoch of primordial nucleosynthesis.* During this interval, the temperature dropped from about 1.3 billion to 600 million degrees Kelvin (which brought it into the range of 40 to 80 times the central temperature of the Sun). Collisions between the particles became much less violent than they had been earlier. Consequently, collisions between protons and neutrons produced stable nuclei that were not torn apart again by further collisions. In particular, protons and neutrons fused to form nuclei of hydrogen-2 ($^2H^+$, also called deuteron), helium-3 ($^3He^{++}$) and helium-4 ($^4He^{++}$):†

$$\begin{array}{llll} p^+ & + n & \rightarrow {}^2H^+ & + \gamma \\ p^+ & + p^+ & \rightarrow {}^2H^+ & + e^+ + \nu \\ p^+ & + {}^2H^+ & \rightarrow {}^3He^{++} & + \gamma \\ {}^3He^{++} & + {}^3He^{++} & \rightarrow {}^4He^{++} & + 2p^+ + \gamma \end{array}$$

(The same nuclear reactions, with the exception of the first one, occur today in the interiors of the Sun and of many other stars. In fact, they are the major source of the stars' radiant energy [see chapter 3].)

By the time the Universe was five minutes old, the temperature had dropped below 600 million degrees Kelvin and nucleosynthesis came to an end. All the neutrons had become used up in the nuclear reactions. The matter now consisted of roughly 76% hydrogen (^1H), 24% helium-4 (^4He), and traces of deuterium (^2H) and helium-3 (^3He), all in fully ionized form. (Percentages refer to mass.) Virtually no heavier elements had been manufactured. The primordial abundance of the elements remained unchanged until the birth of the first generations of stars hundreds of millions of years later. Only then did the buildup towards the heavier elements—carbon, nitrogen, oxygen, the metals, and the rest—continue.

With continued expansion the temperature kept falling further—to 10 million degrees, 1 million degrees, and lower. Although these temperatures were below the threshold for nuclear reactions, they were easily high enough to keep the matter fully ionized. Hydrogen still existed as free protons and free electrons. As soon as a proton captured an electron to form a neutral hydrogen atom, it was torn apart again into a proton and an electron either by a photon or by collision with some other particle. The same was true of helium. It existed in the form of free helium nuclei and free electrons.

**Nucleosynthesis* refers to the production of elements heavier than ^1H by nuclear reactions.

†A deuteron consists of a proton and neutron tightly held together by the nuclear force. It is the nucleus of deuterium, the isotope hydrogen-2 (^2H). The symbols p^+, $^2H^+$, $^3He^{++}$, and $^4He^{++}$ refer to the nuclei of hydrogen-1, hydrogen-2, helium-3, and helium-4. The right-hand superscripts indicate that these *nuclei* are singly or doubly charged. The symbols ^1H, ^2H, ^3He, ^4He (without the right-hand superscripts) refer to the *neutral atoms* hydrogen-1 and hydrogen-2 (deuterium) and helium-3 and -4. (For further details, see the Introduction to Part One.)

All the particles—except the neutrinos and gravitons—were still colliding with each other incessantly. In particular, the photons were unable to travel very far before striking an electron, a proton, or a helium nucleus. This made the Universe opaque. It resembled thick fog, except that it was much hotter.* Because of the constant collisions, the photons and the material particles shared their energies of motion; and because temperature is a measure of this energy, photons and matter shared a common temperature.

Age 500,000 years. Up to roughly 500,000 years, the state of the Universe was as just described. By then the temperature had fallen to approximately 3000 K. As a consequence, the energy of the photons had become so greatly reduced that they were less and less able to keep the nuclei and electrons from combining. Soon they were totally unable to do so and matter changed into neutral atomic form. This also changed the appearance of matter from opaque to transparent, and the photons did what the gravitons and neutrinos had done earlier. They ceased to interact with the matter and the Universe became transparent, as it still is today.

The decoupling of the photons from the matter is the last transition included in this description of the birth of the Universe. From here on, photons and matter evolved independently, each according to its own physical laws. The matter evolved into the structures we see all around us today—galaxy clusters, galaxies, stars, planets, and life. The photons evolved very little. Like the gravitons and neutrinos, the photons remained part of the Universe and continued to feel the global gravitational attraction. However, they merely partook in the cosmic expansion. They did not acquire any structure. Their energies were reduced more and more, their wavelengths became longer and longer, and the temperature characterizing them kept falling. Today we observe them as the 3 K cosmic background radiation, a faded remnant from the birth of the Universe.

One final comment is in order. As noted earlier, the 3 K cosmic background radiation observed today is nearly perfectly isotropic. For example, if we measure the intensity of this radiation from one direction and then from the opposite direction, we find that the two results are identical to roughly one part in 10^4. This is rather surprising. The two opposite locations were approximately 10^7 LY apart when the radiation that we measure today was emitted, at age 500,000 years (the age of photon decoupling). In 500,000 years no signal—not even light, the fastest mode of communication—could have traveled 10^7 LY to "instruct" the two places to emit the same intensity background radiation. Hence, these two locations should not have been causally linked. Yet they emitted radiation of nearly identical intensity.

*Fog consists of microscopic water droplets. It is opaque because light rays passing through continually collide with the droplets and are scattered. They cannot follow a straight path. The same was true of the photons in the early Universe—they kept colliding with and were scattered by electrons, protons, deuterons, and helium nuclei. It was also true of the gravitons and neutrinos before they decoupled from the other particles.

Apparently, the isotropy had been established much earlier, at age 10^{-35} second, when the Universe was much more compact and, despite its young age, all of its parts were in communication and causally linked. During the subsequent inflationary expansion, when its size increased by 10^{25} orders of magnitude, the various parts of the Universe got causally out of touch with each other. However, they all inherited the same conditions. That is why at age 500,000 years the intensity of the cosmic background radiation was virtually the same everywhere and why it still is so today. It is also the reason why the distribution of matter at that time was nearly the same everywhere and why, on the largest scale, the distribution of clusters of galaxies is still homogeneous and isotropic today.

Exercises

1 What is meant by "large-scale structure of the Universe"?

2 Using the fact that light travels 300,000 km per second and the number of seconds per year, calculate how far light travels in one year. Compare your answer with the value of one light-year given in the text.

3 (*a*) Listen to a car or train passing you at high speed. You'll notice a distinct change in the sound from high to low pitch. Explain this change in terms of changes in the wavelength and frequency of the sound waves as they arrive at your ear.
(*b*) Consider the following experiment: A friend throws tennis balls toward you at a constant rate (say, one ball every second). Catch the balls while first running toward and then away from your friend. How do the rates at which you catch the balls differ from the rate at which your friend throws them?
(*c*) How does this experiment relate to your answer to (*a*) of this exercise?
(*d*) How does it relate to observations of stars and galaxies?

4 Compare the meanings of *homogeneous, heterogeneous, isotropic*, and *anisotropic*, and find examples from everyday life illustrating these terms.

5 Look up yesterday's maximum and minimum temperatures reported in your local newspaper. If you live in the United States, the temperatures are probably given in degrees Fahrenheit. Convert them to degrees Celsius and Kelvin.

6 Throw a ball up vertically. Observe how gravity decelerates the ball, momentarily stops it, and then accelerates it downward.
(*a*) How does this simple experiment correspond to the evolution of the closed universe? How is it an oversimplification of that evolution?
(*b*) How would you have to change the experiment to make it correspond to the evolutions of the flat and the open universes?

7 What is meant by the "missing mass of the Universe"? Is the mass really missing?

8 What would the age of today's Universe be if H = 30 km/(sec MLY) and if (*a*) the Universe's rate of expansion has been constant since the Big Bang and (*b*) the rate of expansion has slowed down due to the Universe's self-gravity. (Assume that the Universe can be represented by a Friedmann model.)

9 Inflate a spherical balloon. Cover its surface with dots to represent superclusters of galaxies. Continue to inflate the balloon and observe how the distances between the dots increase.
(*a*) Is there any central dot?
(*b*) Is there an edge?
(*c*) How can you use this simple experiment to model the evolution of the closed Universe? In this model, which characteristics of the balloon correspond to time; to space?

10 According to the Big Bang theory, why does the primordial abundance of the elements consist mainly of hydrogen and helium, and virtually of no elements heavier than helium?

11 You probably have experienced a whiteout, either in a snowstorm or heavy fog, and know that it is easy to lose your balance or get lost under that condition.
(*a*) Why is this so?
(*b*) How does the condition of a whiteout correspond to conditions in the early Universe, before the decoupling of radiation and matter?

Suggestions for Further Reading

Adair, R. K. 1988. "A Flaw in the Universal Mirror." *Scientific American*, February, 50–56. The Universe exists because of a slight asymmetry in a "mirror" called CP invariance, which gave matter the edge over antimatter during the Big Bang. (Also see the article by F. Wilczek referenced in this chapter.)

Davies, P. 1985. "Relics of Creation." *Sky and Telescope*, February, 112–15. About progress in understanding nature's fundamental forces at the subnuclear level and how this enables cosmologists to study the first few moments of the Universe's existence.

DeWitt, B. S. 1983. "Quantum Gravity." *Scientific American*, December, 112–29. The author explains that in a quantum-mechanical theory of gravitation, space–time would be subject to continual fluctuations, and even the distinction between past and future might become blurred.

Dicus, D. A., et al. 1983. "The Future of the Universe." *Scientific American*, March, 90–101. A forecast for the expanding Universe through the year 10^{100}.

Dyson, F. J. 1954. "What Is Heat?" *Scientific American*, September, 58–63. The author defines heat as "disordered energy" and explains what this statement means. He also describes the range of temperatures found in nature, entropy, and other concepts from the science of thermodynamics.

Einstein, A. 1961. *Relativity: The Special and the General Theory; a Popular Exposition*. New York: Crown Publishers. An easy-to-read little book about the theory of relativity, revealing the clarity of thought and deep physical intuition that distinguished Einstein.

Gale, G. 1981. "The Anthropic Principle." *Scientific American*, December, 154–71. Traditionally it has been argued that life evolved on Earth because conditions in the Universe were favorable. The anthropic principle argues the reverse: The presence of life may "explain" the conditions in the Universe.

Gamow, G. 1956. "The Evolutionary Universe." *Scientific American*, September, 136–54. A lucid account of the geometrical interpretation of the space–time structure and evolution of the Universe.

Gore, R. 1983. "The Once and Future Universe." *National Geographic*, June, 704–49. A well-illustrated summary of our current understanding of the structure and evolution of stars, nebulae, galaxies, quasars, and the Universe, including brief descriptions of some of the modern instruments used in the study of these objects.

Gott, J. R., III. 1983. "Gravitational Lenses." *American Scientist*, March–April, 150–57.

Gregory, S. A., and **L. A. Thompson**. 1982. "Superclusters and Voids in the Distribution of Galaxies." *Scientific American*, March, 106–14.

Gribbin, J. 1987. "The Structure of the Universe." *New Scientist*, 29 October, 58–62. The distribution and motions of matter in the Universe are not as homogeneous as once thought. According to newly proposed theories, primordial neutrinos, cold dark matter, cosmic strings, and "strange attractors" may be responsible for the Universe's inhomogeneities.

Guth, A. H., and **P. J. Steinhardt**. 1984. "The Inflationary Universe." *Scientific American*, May, 116–28.

Harrison, E. R. 1981. *Cosmology, the Science of the Universe.* Cambridge, U.K.: Cambridge University Press. An authoritative and remarkably clear exposition of a difficult subject, intended for the serious reader, but not necessarily one with a background in the natural sciences.

Hawking, S. W. 1984. "The Edge of Spacetime." *American Scientist*, July–August, 355–59. This article, written by one of the most imaginative cosmologists of our time, addresses the question, "What happened at the beginning of the expansion of the Universe?"

Hodge, P. 1984. "The Cosmic Distance Scale." *American Scientist*, September–October, 474–82. Two distinct approaches to the measurement of cosmic distances have produced two different values for the size of the Universe.

Hoskin, M. 1986. "William Herschel and the Making of Modern Astronomy." *Scientific American*, February, 106–12.

Hoyle, F. 1956. "The Steady-State Universe." *Scientific American*, September, 157–66.

Price, R. H., and **K. S. Thorne**. 1988. "The Membrane Paradigm for Black Holes." *Scientific American*, April, 69–77. The interaction of a black hole with its environment may be understood by thinking of the hole as a thin, electrically conducting spherical membrane.

Rowan-Robinson, M. 1985. *The Cosmological Distance Ladder: Distance and Time in the Universe*. San Francisco: W. H. Freeman and Co. In this important book, the author critiques the methods used to establish the cosmological distance scale. He disagrees with the derivations of the two most-often quoted values of the Hubble constant, 15 and 30 km/sec per million light years, and suggests instead the value of 20 km/sec per million light years with an uncertainty of about ±20%.

Schneider, S. E., and **Y. Terzian.** 1984. "Between Galaxies." *American Scientist*, November–December, 574–81. The authors discuss how the search for elusive "missing matter" is changing some of the most basic ideas about the nature of the Universe.

Schramm, D. N. 1983. "The Early Universe and High-Energy Physics." *Physics Today*, April, 27–33. The author discusses what the new particle field theories predict about the early history of the Universe, and he explains that comparisons of these predictions with astronomical data may offer the only possible tests of these theories.

Schramm, D. N., and **G. Steigman.** 1988. "Particle Accelerators Test Cosmological Theory." *Scientific American*, June, 66–72. A readable review of the currently proposed families of elementary particles, the creation of these particles during the Big Bang, the unity of nature's four fundamental forces, and the experimental testing of whether there is a limit to the number of families of elementary particles.

Shaver, P. 1987. 'Ever More Distant Quasars?" *Nature* 330: 426 (3 December). The discovery in 1987 of 6 quasars with redshifts exceeding 4.

Silk, J., A. S. Szalay, and **Y. B. Zel'dovich.** 1983. "The Large-Scale Structure of the Universe." *Scientific American*, October, 72–80.

Sulak, L. R. 1982. "Waiting for the Proton to Decay." *American Scientist*, November–December, 616–25. In the grand unified theories of elementary particle structure the proton is not stable, but is predicted to decay on a time scale of about 10^{30} years. This article describes experiments designed to detect this decay.

Trefil, J. S. 1983. "How the Universe Began." *Smithsonian*, May, 32–51. "How the Universe Will End." *Smithsonian*, June, 72–83. A readable, two-part account of the recently developed grand unified theories of particle interactions and the predictions that follow from them concerning the origin and end of the Universe. The second article suggests that the Universe's future will be determined by the "missing mass" that is thought to exist in galaxies and galaxy clusters.

——. 1980. "Einstein's Theory of General Relativity Is Put to the Test." *Smithsonian*, April, 74–83. A summary of recent and still-in-progress experiments designed to test various predictions of Einstein's General Theory of Relativity, including the possibility that gravity is weakening over time.

Turner, E. L. 1988. "Gravitational Lenses." *Scientific American*, July, 54–60.

Vilenkin, A. 1987. "Cosmic Strings." *Scientific American*, December, 94–102. Thin strings of energy, created during the first second of the Big Bang and resembling crinkles in the space-time structure of the Universe, may have caused the observed clustering of galaxies.

Weisskopf, V. F. 1983. "The Origin of the Universe." *American Scientist*, September–October, 473–80. A respected scientist's discussion of recent theoretical developments linking cosmology and particle physics.

Wilczek, F. 1980. "The Cosmic Asymmetry Between Matter and Antimatter." *Scientific American*, December, 82–90. Evidence from cosmology and particle physics suggests an explanation of why the Universe today consists almost entirely of matter.

Wilson, R. W. 1979. "The Cosmic Microwave Background Radiation." *Science* 205: 866–74. Wilson's Nobel prize lecture on the discovery and status of understanding of the cosmic background radiation, delivered on December 8, 1978, in Stockholm.

Figure 2.1. Galaxies NGC 5194 and 5195. The larger of these two galaxies, NGC 5194 (type Sc), also known as the Whirlpool Galaxy, is one of the most magnificent spirals in the sky. It is connected by an extension of one of its arms to its smaller companion, an irregular galaxy.

2 Galaxies

THE PREVIOUS CHAPTER ended with a description of the birth of the Universe and its evolution to age one-half million years. At that time the temperature had dropped to 3000 K and the primordial photons had stopped interacting with the matter. Both the photons and the matter were smoothly distributed throughout space, but only the photons have remained so. During the intervening 13 billion years the matter acquired the structures and forms that surround us near and far—from atoms, molecules, and dust grains to stars, galaxies, and galaxy clusters.

How did matter come to organize itself in this astounding way? How, in addition, did it evolve into life on a tiny planet called Earth? These are among the central questions challenging scientists today.

THE FIRST STRUCTURES IN THE UNIVERSE

When matter set out on its independent evolution it consisted of approximately 76% hydrogen and 24% helium by mass. It was in gaseous form and smoothly distributed, and its initial, nearly infinite density had been lowered by the cosmic expansion to about 10^{-21} g/cm^3. However, this smoothness was not quite perfect; it contained weak, random fluctuations in density.

Such fluctuations are a common phenomenon and occur today everywhere in our environment. For instance, from a distance the surface of a lake or the ocean may appear to be smooth, but there are almost always small ripples or waves on the surface. Even on a perfectly windless day, when there are no perceptible waves, fluctuations on the molecular level still exist, so that at any instant the water molecules are concentrated a bit more in some places than others. Within fractions of a second the concentrations shift to different locations, and they continue to do so in an ever-changing, random pattern. Likewise, when water freezes and the molecules become firmly bonded into a hexagonal pattern (see the Introduction to Part One), the resulting crystal structure is not perfect. Microscopic variations exist in the density and structure of the ice, and the molecules oscillate back and forth perpetually.

Random fluctuations in density cannot be avoided in nature. When the gaseous matter and photons decoupled in the early Universe, weak random fluctuations in the density of matter existed everywhere, extending over a wide range of scales. From the present-day clumping of matter into superclusters of galaxies, we may assume that the largest fluctuations stretched over regions of about 10^{15} to 10^{16} $M_\odot$.* Superimposed on them were fluctuations over smaller regions, and they contained still smaller fluctuations. This state of alternating high and low concentrations of matter in the early Universe is analogous to today's ocean waves. They, too, consist of a hierarchy of structures—very long waves that have smaller ones of varying lengths superimposed on them.

Initially the fluctuations were not very strong. Nowhere did the density vary from its average by more than a fraction of a percent. This follows from the near isotropy of today's cosmic background radiation. Careful measurements of the intensity of this radiation from different directions in the sky show variations of

*The symbol $M_\odot$ (read "solar mass") is a common one in astronomy and refers to a mass equal to that of the Sun.

only about one part in ten thousand. The initial fluctuations in the matter density are expected to have been comparably small, because until decoupling, the matter and photons were interacting strongly.

Formation of Primordial Clouds

Upon decoupling, the primordial photons retained their almost perfectly smooth distribution, but the matter did not. Wherever there existed small enhancements in density, the attractive force of gravity counteracted and tended to slow the expanding matter. At first it succeeded in this only over the largest scales. Over smaller scales, the pressure exerted by the matter's constituent particles was too strong to permit gravitational contraction. As a result, regions extending over approximately 10^{15} to 10^{16} $M_{\odot}$ of matter expanded more and more slowly and eventually began to contract. This led to the formation of huge primordial clouds of hydrogen and helium—the first large-scale structures in the evolving Universe (see figure 2.2). Between these clouds, voids developed that were as large or larger than the clouds themselves. (Note that

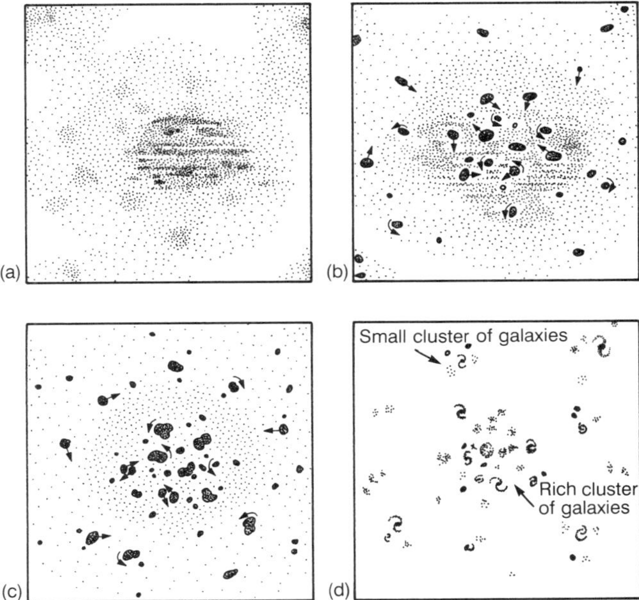

Figure 2.2. Model of the birth of galaxies and clusters of galaxies. (*a*) Huge primordial clouds of hydrogen and helium condense from density fluctuations present in the young Universe. (*b*) Local gravitational forces break up the primordial clouds into gaseous fragments. In the process, the fragments acquire large velocities relative to each other and many begin to rotate. (*c*) The fragments collide and merge into larger, more massive fragments. The first stars are forming and the fragments turn into galaxies. (*d*) Within roughly 2 to 3 billion years the birth of galaxies is largely completed. The original primordial clouds have evolved into superclusters of galaxies, each containing millions of member galaxies.

whenever matter gets compressed, the pressure rises and opposes the compressional forces. Under the conditions existing in the early Universe, the pressure was successful in opposing gravity on all but the largest scales. The concept of pressure is discussed in box 3.2).

Although there is no direct observational knowledge about the formation of the primordial clouds, it seems reasonable to assume that their contraction did not proceed smoothly and uniformly, and few if any of the primordial clouds contracted into spheres. As first suggested by Yokov B. Zel'dovich, professor of astrophysics at Moscow University, the compression of the matter generally tended to occur more rapidly along one of the three spatial axes than along the other two. As a result, the first large-scale primordial clouds are believed to have taken on the shapes of huge, flattened pancakes. Where these pancakes intersected, long, thin filaments formed. Thus, the structure and distribution of the primordial clouds was probably highly irregular and interconnected. Three-dimensional maps of the present-day distribution of superclusters of galaxies, which evolved from the primordial clouds, reflect this (see figure 2.3).

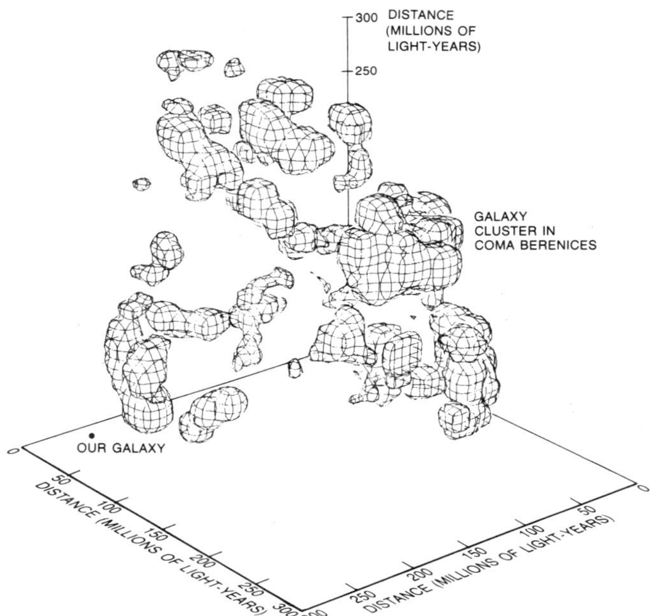

Figure 2.3. Three-dimensional map of the distribution of bright galaxies within 250 million LY of the Milky Way, as seen in the northern sky. The map outlines the central regions of the superclusters of galaxies, and it illustrates the interconnectedness of the distribution of matter on the largest scale. The map is based on a survey of 1801 galaxies. It was constructed by Carlos Frenk and Simon White of the University of California at Berkeley from a map prepared by Marc Davis and his colleagues at the Center for Astrophysics of the Harvard College Observatory and the Smithsonian Astrophysical Observatory. From J. Silk, A. S. Szalay, and Y. B. Zel'dovich, "The Large-Scale Structure of the Universe," *Scientific American*, October 1983, 78. Used by permission.

While the primordial clouds were forming, the Universe continued to expand. By contracting the gas into clouds, gravity succeeded in *locally* counteracting the cosmic expansion. However, *globally* the expansion of the Universe was not affected. The clouds participated in the cosmic expansion by moving farther and farther away from each other, a phenomenon still observed today among the superclusters of galaxies.

At present astronomers do not understand why the primordial clouds condensed over regions as large as 10^{15} to 10^{16} $M_\odot$ (values deduced from the masses contained in today's superclusters of galaxies). Simple theoretical estimates give masses that are one to two orders of magnitude smaller. For example, if the maximum initial density fluctuations of the matter, as allowed by the variations in the intensity of today's cosmic background radiation, are used in the mathematical modeling of the Universe's early evolution, structures of at most 10^{14} $M_\odot$ are obtained.

So how did the primordial clouds of hydrogen and helium acquire their large masses? At present, astronomers take very seriously the suggestion that the neutrinos created during the Big Bang were responsible, provided that they possess a tiny mass (see the Introduction to part 1). As noted in chapter 1, the primordial neutrinos decoupled from the photons and the rest of the matter when the Universe was only about one second old. From then on, they followed their own evolutionary course. Their initial density distribution, too, was marked by small fluctuations that, under the influence of gravity, began to grow with time. However, because this growth had already begun at age one second, when the Universe was still much more compact than at age 500,000 years (when the matter and photons decoupled), the force of gravity was stronger and more time was available to amplify the fluctuations. Therefore, the regions over which the neutrinos became concentrated were considerably larger than the structures of 10^{14} $M_\odot$ obtained for the evolution of the matter alone. These large regions constituted gravitational potential wells into which the matter was pulled, upon its decoupling from the photons. In this manner, the primordial clouds of hydrogen and helium, encompassing 10^{15} to 10^{16} $M_\odot$, are thought to have formed.

Though astronomers are still debating whether neutrinos indeed played a role in the formation of the first large-scale structures of the Universe, it is a plausible hypothesis and, in broad terms, the theoretical consequences are consistent with observations. Furthermore, neutrinos may constitute the dark matter that stellar and galactic motions indicate is present within galaxies and in intergalactic space, as noted in chapter 1. Another hypothesis is that this dark matter might also consist, at least in part, of black holes or some other, even more esoteric, form. We just don't know. Surely, a number of surprises still await us in our search to understand the formation of the Universe's first structures of large scale.

Breakup of Primordial Clouds

As the primordial clouds formed, they inherited all the weak, small-scale density fluctuations that had characterized the matter when it decoupled from the photons. Gradually, the local gravitational contractions that began on the largest scales affected the small-scale density fluctuations as well. In many regions where the density within the clouds was greater than the average, gravity managed to contract the gas further and to intensify the concentration of matter. This increased the local gravitational force still more and led to the breakup of the primordial clouds into large numbers of cloud fragments.

At present, we do not understand the details of this breakup. A good guess is that the fragments acquired masses in the range of 10^5 to 10^8 $M_\odot$ and, possibly, even more. The smaller ones were probably far more numerous than the larger ones. Furthermore, the fragments were strongly concentrated toward the centers of the primordial cloud concentrations where the density of matter was greatest and, hence, gravity acted most strongly.

The formation of the cloud fragments may be compared to the condensation of water vapor into rain droplets in the clouds of the Earth's atmosphere. The major differences between the two kinds of condensations are in scale, time, and the forces involved. The primordial cloud fragments extended over masses of millions of suns, their formation took hundreds of millions of years, and the participating force was gravity. In contrast, rain droplets weigh a fraction of a gram, they condense spontaneously, and the force responsible is the electric force, which pulls water molecules together.

The primordial clouds of hydrogen and helium were the first structures of large scale to form in the evolving Universe. Their breakup into fragments created second-order structures on the next smaller scale. Next, the fragments broke up further into stars and interstellar clouds, producing third-order structures on a still smaller scale.

BIRTH OF GALAXIES AND CLUSTERS OF GALAXIES

The breakup of the primordial clouds into fragments was chaotic and violent. At first the local gravitational forces, which produced the fragments, were rather weak and the contractions were slow. As the matter became more clumped, the gravitational forces increased in strength and the gaseous material fell together faster and faster. Eventually it reached supersonic speeds. Shock waves, like those from supersonic jets but much stronger, were created and thrust their way through the clouds. They acted like the blades of a giant mixer, agitating the matter and giving it rapid, turbulent motions. Whirlpools were stirred up, extending over thousands and millions of solar masses of gas, and windstorms began to blow with velocities of tens and hundreds of thousands of kilometers per hour. The fragments were accelerated to large velocities relative to each other and many acquired rapid rotational motions.

This violent activity occurred during the first few hundred million years, when the Universe was much more compact than it is today. Consequently the fragments had very little elbow room, particularly because many of them had large velocities. They started to collide, but not in a simple way like two tennis balls that bounce off each other. The collisions were inelastic and complicated. Fragments of gas containing masses equal to millions of suns were plowing into each other at thousands of kilometers per hour. On impact the gas of each fragment tried to maintain its own momentum and to penetrate into the oncoming fragment. The ensuing struggle compressed and heated the gas. New shock waves were created that spread through both fragments and caused further turbulence and compression. The friction between the oppositely directed streams of gas gradually slowed the motions and the fragments merged into one larger fragment.

Quite often this sequence of events was not permitted to run its full course. Before one merger was completed, collisions occurred with other fragments, creating further compression, heating, shock waves, turbulence, and merging. In the course of dozens to hundreds of collisions and mergers, the masses of the surviving fragments increased manyfold and their numbers decreased correspondingly.

Birth of the First Stars

The shock waves created by the collisions compressed the gas of the fragments. This happened to such a degree that huge regions, involving up to millions of solar masses and scattered throughout the fragments, became gravitationally unstable and collapsed into compact clouds within which the first generations of stars began to form. In some fragments star formation occurred in rapid bursts and most of the gas was quickly converted into stars. In others, which possessed a great deal of internal turbulence and motion, the collapse into stars was less efficient and much of the matter remained in gaseous form.

The birth of the first generations of stars had two far-reaching consequences. Up to now the fragments had been rather cold and dark objects,* except for regions where collisions and shock waves had temporarily heated the gas. This changed rather quickly when the newly formed stars began to pour forth their radiant energy and illuminate the gaseous matter surviving in their vicinity. In a semiregular pattern, along the paths where shock waves had compressed the gas and induced gravitational collapse, light after light turned on as stars and groups of stars were born. For the first time since decoupling from the primordial photons, the matter of the Universe lit up and became visible.

The second consequence of star formation resulted from the accompanying change in the density of matter. The gas that collapsed into stars was no longer spread out over vast regions of space as a diffuse and amorphous mass. It became concentrated into relatively small volumes. The stars' diameters were up to a million times smaller than the distances separating them. Consequently, when two fragments collided the stars did not generally hit each other; they were too compact for that.† They merely interacted by the long-range gravitational force and altered each other's motions until, after some time, their orbits became intermingled and the two groups of stars merged into one larger stellar system (see figure 2.4).

*Before the birth of stars, the fragments were cold and dark because they radiated away their heat energy very efficiently in the infrared and radio parts of the electromagnetic spectrum. This kind of heat loss is crucial for the formation of stars, as will be seen in chapter 3.

†A difference by a factor of 10^6 between the sizes and separations of the stars in galaxies is comparable to scattering a handful of sand so that the average distance between one sand grain and the next is one to two kilometers. If two batches of sand, dispersed to this extent, were thrown toward each other, the probability that two grains collide would be vanishingly small.

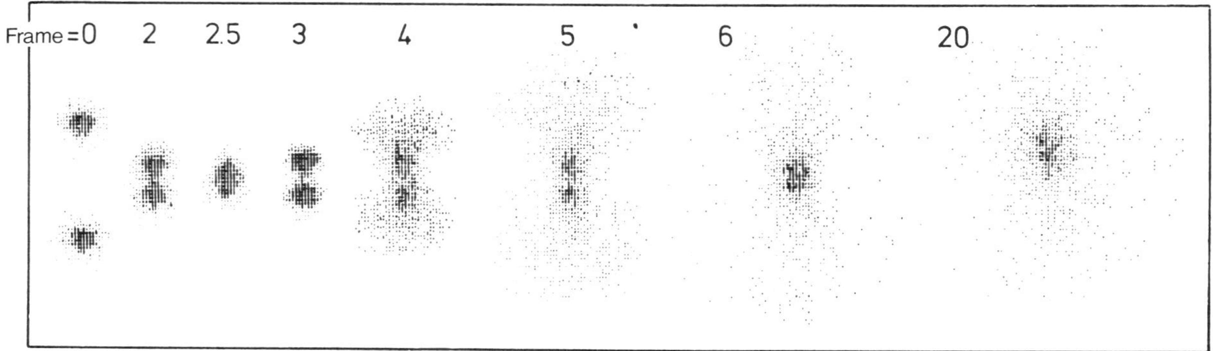

Figure 2.4. Computer simulation of the merger of two elliptical galaxies, leading to the formation of a daughter galaxy that is much more extended than either of the original ones. In nature, such a merger takes hundreds of millions of years. In this simulation, each galaxy consists of 1000 stars, though in reality elliptical galaxies contain millions to many billions of stars. From Joseph Silk, *The Big Bang* (San Francisco: W. H. Freeman & Co., 1980), 171. Used by permission.

Birth of Galaxies

In contrast to the stars, the uncondensed gaseous components of two colliding fragments interacted as they always had. They plowed into each other, became compressed and heated, shock waves and turbulence were created, and new

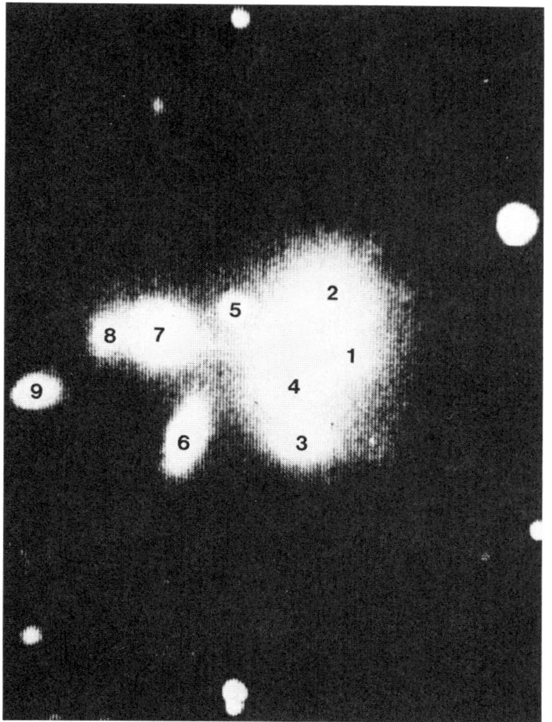

Figure 2.5. Nine small galaxies in the process of coalescing into a single galaxy. The motions of the nine galaxies suggest that they are the nuclei of once independent galaxies that collided and, due to their mutual gravity, are now merging into a single system. Within roughly a billion years, the nine galaxies are expected to be no longer identifiable as separate entities and the coalescence will have produced a new giant elliptical galaxy.

instabilities formed, leading to more star formation. In some collisions the gaseous components stayed with the stars. In other collisions the gas was cleanly swept away from the stars. Thus, a broad spectrum of end products resulted, from stellar systems consisting of only stars to clouds comprised of mostly gas.

As this sequence of collisions and mergers of fragments and of conversion of gas into stars progressed, the surviving systems looked more and more like star clusters and galaxies (figure 2.5). The fragments that were most successful in capturing and assimilating others grew into large galaxies with masses of 10^{11} $M_\odot$ and more. Those that managed to retain their gaseous material became spiral galaxies, like our own or the Andromeda galaxy (see figure 1.8). Fragments that lost their gas or converted most of it into stars, evolved into giant ellipticals, like galaxy M 87 in the Virgo cluster (see figure 1.12). Fragments that grew less successfully became dwarf ellipticals and globular star clusters. Finally, fragments that collided with or were captured into close orbit about another more massive galaxy experienced strong tidal perturbations and ended up with rather distorted shapes (figures 2.6 and 2.7). They turned into irregular galaxies, such as the Large and Small Magellanic Clouds (see figure 1.5).

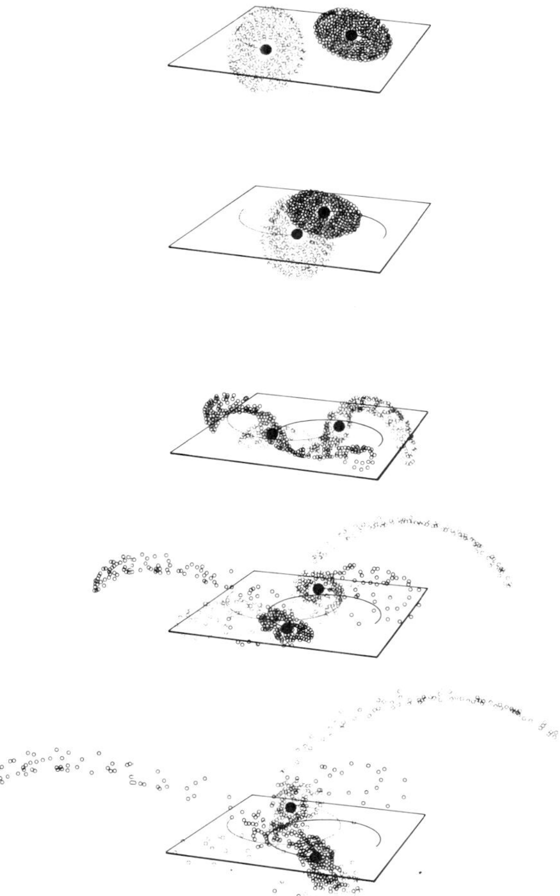

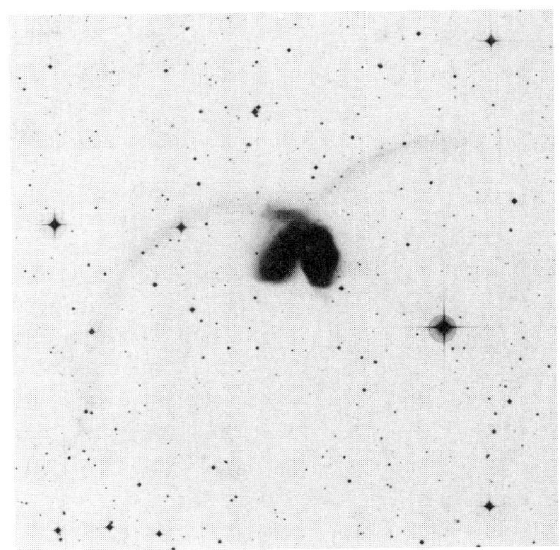

Figure 2.6. **(*Left*)** A sequence of five frames from a computer-generated motion picture, simulating the gravitational distortions resulting from the close encounter of two galaxies. The initial conditions were chosen so as to reproduce the tidal distortions observed in galaxies NGC 4038 and 4039 (also known as the Antennae, figure 2.7). From George O. Abell, *Exploration of the Universe*, 4th ed. (Philadelphia: Saunders College Pub., 1982), 618. Used by permission.

Figure 2.7. **(*Above*)** The Antennae, a pair of spiral galaxies, NGC 4038 and 4039, seen shortly after a very close encounter.

Figure 2.8. Stephan's quintet. This interesting group of five galaxies was named after their discoverer, a nineteenth-century astronomer observing from Marseilles, France. Four of the galaxies are in close proximity of each other, as indicated by the luminous bridges of gas connecting them and by their nearly identical redshifts. They are the compact center of a small cluster of galaxies at a distance of approximately 250 million LY. The fifth galaxy (an elliptical), in the lower right part of the photograph, is a foreground galaxy and not associated with the others.

Most of the newly formed galaxies found themselves in the vicinity of others and were bound to each other by the force of gravity. This clustering was particularly pronounced near the central regions of what used to be the primordial clouds (figure 2.8). Today that is where the rich clusters of galaxies with thousands of members are found, such as the Virgo and Coma clusters. Farther out the galaxies clustered into sparser groupings with only a handful to a few dozen members. Our Local Group is an example of such a small cluster. Most of these outlying galaxy clusters are weakly tied by gravity to a rich central cluster. Together they form the superclusters of galaxies, which we visited on our imaginary journey in chapter 1.

Note that this description of the origin of galaxies and galaxy clusters merely represents some of the current thinking of astronomers. It is far from the final word on the subject. The description is largely based on theoretical deductions from observations of cloud collisions that presently happen on much smaller scales in our galaxy, on observations of collisions and near-encounters among some nearby galaxies (see figures 2.5 and 2.7), and on computer simulations (see figures 2.4 and 2.6). It is not based on observations of the actual events that produced the galaxies and galaxy clusters.

In fact, Valérie de Lapparent, Margaret J. Geller, and John P. Huchra of the Harvard-Smithsonian Center for Astrophysics have suggested a scenario of galaxy cluster formation quite different from the one described here. From a study of some 1100 galaxies distributed in a three-dimensional wedge of sky 6° × 117° and extending to a distance of about 650 million LY (see figure 2.9), they deduce that galaxies are not grouped into distinct superclusters with voids separating them from each other. Instead, they find that galaxy clusters are much more interconnected and appear to have a shell-like or bubble-like distribution. The bubbles are roughly spherical or elliptical, with typical diameters of 150 million LY, though the largest bubble in the survey is twice that size. The galaxies are located predominantly on the surfaces of the bubbles, while the interiors of the bubbles are largely void of matter. Huchra suggests that "these bubbles fill the Universe just like suds filling the kitchen sink."*

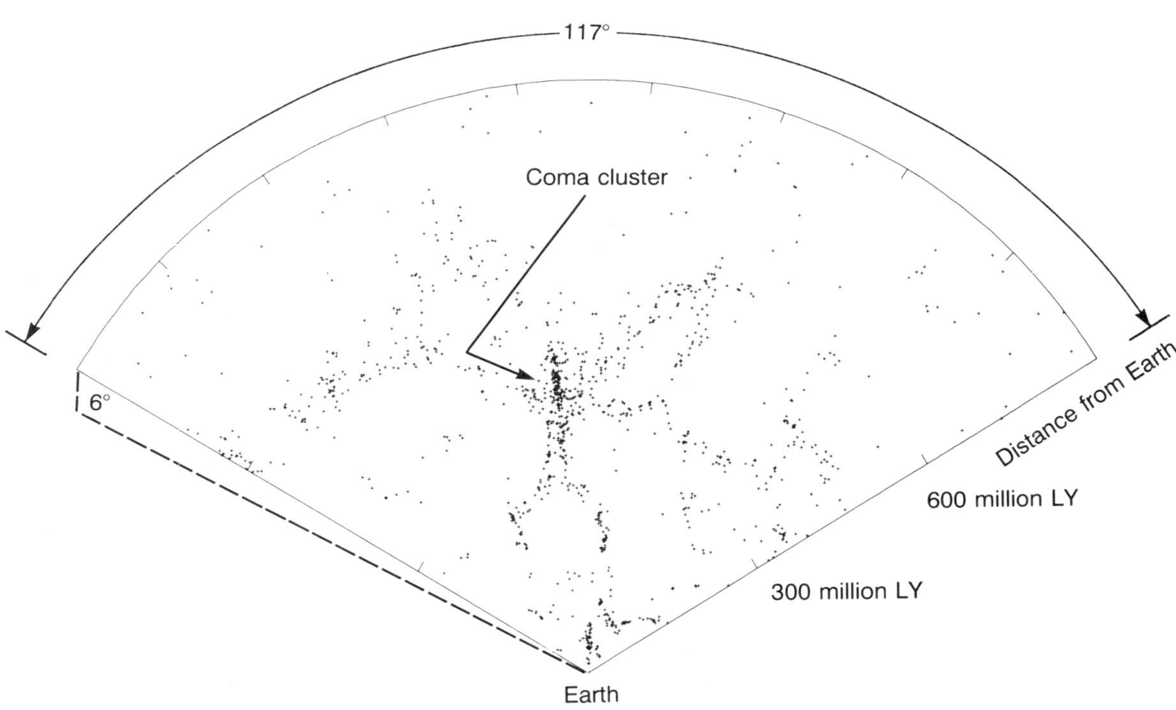

Figure 2.9. Distribution of nearly 1100 galaxies in a three-dimensional slice (6° by 117°) of the Universe in the direction of the Coma cluster, as compiled by Valérie de Lapparent, Margaret J. Geller, and John P. Huchra. The diagram suggests that the large-scale distribution of galaxies in the Universe is bubble-like. The Coma cluster of galaxies is the richest supercluster system in this survey and appears to lie at the intersection of several bubbles.

**Science and the Citizen* section, *Scientific American*, March 1986, pp. 59, 62.

If Huchra and his colleagues are right, then gravity may not have been the dominant force in initiating the formation of galaxies and galaxy clusters, as is traditionally assumed. The dominant force may have been shockwaves that resulted from explosions of enormous magnitudes within a few hundred million years after the Big Bang. What caused the explosions is not known, but according to one theory they were supernova explosions of early generations of massive stars. The shockwaves pushed the primordial matter apart locally and created the observed bubble-like voids. Where the shockwaves from neighboring bubbles collided, the matter became compressed (probably unevenly), cloud fragments resulted, and galaxies began to form, perhaps in the manner discussed in this section.

To decide how galaxies and galaxy clusters really formed—by one of the scenarios described in this chapter or, perhaps, by yet another one—observations like those by Huchra and his colleagues need to be extended over larger parts of the sky. Furthermore, galaxies 10 and more billion LY distant need to be observed in order to see them in their formative stages. However, light from such distant objects is, with few exceptions, too diffuse and faint to be detected with the present generation of telescopes. This should change when the Hubble Space Telescope, with a 2.4-meter mirror, is launched into Earth orbit aboard a space shuttle, probably some time in 1988 or 1989. (The launch was originally scheduled for 1985, rescheduled for 1986, and then delayed further by the explosion of the space shuttle *Challenger* on January 28, 1986.) Because this telescope will be located far above the atmosphere, objects 50 times fainter than are presently observable with ground-based optical telescopes should come within our view, allowing us to see and to study first hand some of those long-ago events that created the large-scale structures in the Universe.

THE STRUCTURE OF GALAXIES

Let us now turn from the topic of galaxy birth to that of galaxy structure. We begin with the normal galaxies, namely those that have been able to reach some semblance of equilibrium and do not experience major perturbations from either outside or within.

The most widely used scheme for classifying the normal galaxies is one introduced by Edwin Hubble, who lined them up along a tuning fork pattern according to their structural appearance (see figure 2.10). The handle of the fork is occupied by the *elliptical galaxies*, starting with the perfectly spherical ones and ending with the highly flattened ones. They are labeled E0 through E7, depending on their degree of flattening. The point where the prongs of the fork divide is marked by the *S0 galaxies*. They resemble the spirals except that they contain no gas or dust. Along the upper fork lie the *spiral galaxies* in order of tightness of the winding of their arms, from the most tightly wound (type Sa) to the most loosely wound (Sc). Along the lower fork are situated the *barred spirals*. They are like the ordinary spirals, except that they have a luminous bar running through their nucleus. This scheme of classifying galaxies, the *Hubble sequence*, is summarized in figure 2.10. It is further illustrated by figures 1.6, 1.8, 1.9, 1.12, 2.1, 2.11, and 2.13–2.15. Table 2.1 summarizes the range of physical characteristics of the galaxies, including the irregulars.

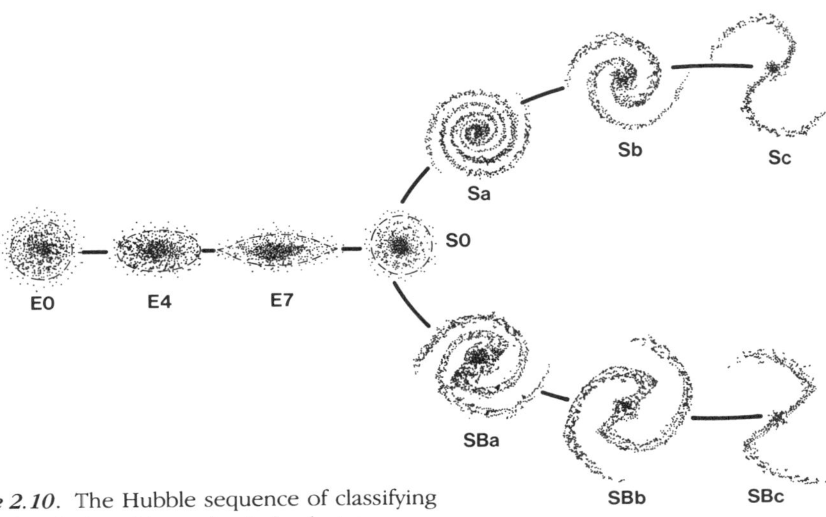

Figure 2.10. The Hubble sequence of classifying elliptical, spiral, and barred spiral galaxies.

Elliptical Galaxies

As table 2.1 indicates, the elliptical galaxies (figure 2.11) possess by far the greatest range in mass, size, and luminosity. The smallest of them resemble the globular star clusters, with approximately 1 million stars. The largest ones contain up to 10^{13} stars, nearly one hundred times more than there are in the Milky Way. Elliptical galaxies contain very little gas or dust. They either converted

Figure 2.11. Galaxy M 86, an elliptical of type E3. This galaxy does not have quite the perfectly spherical shape of M 87 (see figure 1.12); hence the designation E3.

Table 2.1. Range of Physical Characteristics of Elliptical, Spiral, and Irregular Galaxies

	Ellipticals	Spirals	Irregulars
Mass (solar masses)	10^6 to 10^{13}	10^9 to 10^{12}	10^8 to 10^{11}
Diameter (thousands of LY)	1 to 500	20 to 200	5 to 30
Radiant energy output (solar units)	10^6 to 10^{11}	10^8 to 2×10^{10}	10^7 to 2×10^9
Content	Old stars, little or no gas and dust	Old and young stars, much gas and dust	Old and young stars, much gas, dust may be present or absent

NOTE: Adapted from G. Abell, *Exploration of the Universe*, 4th ed., (Saunders College Publ., 1982), 615.

most of the gas and dust into stars or lost that material during encounters with other galaxies, probably during their formative years. As a result, star formation has ceased in them long ago, and the only stars present today are old ones.

Spiral Galaxies

Of all the galaxies, the spirals are of greatest interest to us. First, our galaxy—the Milky Way—is one of them. Second, they contain much gas and dust, from which stars are still forming today. By studying how this is happening we can learn about the origin of the Sun and the planets, including Earth.

Spiral galaxies span a considerably narrower range in mass, size, and luminosity than the ellipticals. The Milky Way is one of the largest with approximately 150 billion stars, a diameter of more than 100,000 light-years, and a luminosity roughly 10 billion times the Sun's. The structure of spirals is much more complicated than that of ellipticals (see figure 2.12). It consists of two main

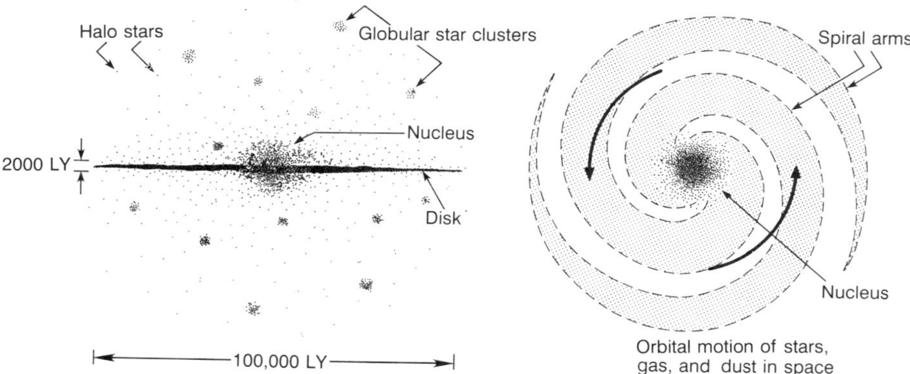

Figure 2.12. Schematic diagram of the main structural features of a spiral galaxy, showing the nucleus, halo, disk, and spiral arms (left, edge-on view; right, face-on view). The dimensions given are those of the visible part of the Milky Way. If the dark matter that is suspected to exist in the halo of spirals were included, the diameter would increase by roughly a factor of three.

Figure 2.13. Galaxy NGC 4565. This galaxy, seen edge-on, is of type Sb. Its distribution of gas, dust, and stars is similar to ours.

components, a spherical and a disk component. The *spherical component* possesses only old stars, many of which are grouped into globular star clusters (see figure 1.3). (The meanings of "old" and "new" stars are discussed in chapter 3.) Both the individual stars and globular star clusters show a strong concentration toward the galactic center, the *nucleus*; but they also extend in a sparser distribution outward to great distances, the *halo*. In addition, the halo of the spherical component contains a great amount of dark matter that extends far beyond the visible stars and star clusters. In fact, the motions of stars and interstellar clouds in the outlying regions of spiral galaxies (see below) indicate that this dark halo may constitute the major part of the masses of these galaxies. The *disk component* of most spirals is rather flat, with a ratio of thickness to diameter comparable to two or three 12-inch phonograph records stacked together (figure 2.13). The disk contains much gas and dust, clumped into large clouds, as well as many young stars. The brightest and youngest of the stars lie

Figure 2.14. Galaxy M 81. This beautiful galaxy, with a pronounced central bulge and graceful arms dotted with bright emission nebulae, is of type Sb. Dark, light-absorbing dust lanes are clearly distinguishable along the arms. The Milky Way, viewed from afar, is believed to resemble this galaxy.

mostly along the *spiral arms*, where they have recently been born and where they act as luminous markers defining the spiral pattern (figures 2.14 and 2.1).

The flatness of the disk component is due to gravity and the disk's rotation about the galactic center. Gravity concentrates the matter of the disk into the flat, relatively thin distribution, and rotation gives rise to centrifugal forces, which prevent the matter from falling toward the galactic center. The rotation however, is not uniform like that of a record. The galactic nucleus with its great concentration of mass acts as a gravitational hub, keeping the stars, gas, and dust of the disk in orbit. Consequently, the inner parts of the disk rotate more quickly than the outer parts, much as the Sun's inner planets revolve more quickly than the planets farther out and for the same physical reason. However, toward the outer regions of the disk, the rotational velocities do not drop off as they do in the case of the planets circling the Sun. They don't because of the presence of the halo, which, as noted above, contains a great deal of mass (in contrast to the Solar System, in which nearly all of the mass is concentrated in the Sun).

The arms of spiral galaxies do not rotate rigidly along with the disk. If they did, the differential rotation of the disk would long ago have wound them up into an indistinct blur. Instead, the arms—that is, the very bright, young stars that define the spiral pattern, along with clouds of gas and dust that give rise to them—are constantly being renewed.

This renewal is believed to take place as follows. When the interstellar gas and dust, in their orbits about the center of the galaxy, enter a spiral arm they collide with matter already in the arm. This compresses the gas and dust into huge clouds of many thousands of solar masses. Still other clouds form when shock waves from supernova explosions (see below and chapter 3) expand into interstellar gas and dust and compress them. In many of the clouds thus formed, the force of gravity becomes sufficiently strong to collapse some of the gas and dust into stars (for details on star formation see chapter 3). This happens in much the same way as it did billions of years earlier in the compact gas clouds of the protogalactic fragments, except on a much smaller scale. Usually hundreds of stars are born simultaneously and for some time they orbit around the galactic nucleus as loosely bound groups. They are called *open star clusters*, in distinction to the much older and richer globular star clusters. Figures 1.2, 3.3, 3.6, and 3.7 show open star clusters in various stages of their evolution.

Among the newly formed stars, a few are always much more massive than the Sun. Such stars use up their nuclear fuel at a furious rate. Within just a few million years, and before their motions have even carried them through the arm in which they were born, they literally burn themselves out. While they last, these massive stars shine up to a million times more brightly than the Sun and mark the location of the spiral arms. In the end, they explode as supernovae and throw much of their matter back into space. During such explosions, enormous amounts of energy are released and shock waves are created, which contribute to the formation of new generations of stars and the maintenance of the spiral arms.

Barred Spiral Galaxies

A third group of galaxies recognized in the Hubble sequence are the barred spiral galaxies (figure 2.15). They resemble normal spirals, except that their

Figure 2.15. The barred spiral galaxy NGC 1300, of type SBb.

arms start not from the nucleus but from the ends of a luminous bar of stars, gas, and dust, which passes right through the galactic center. Computer simulations show that a disk of orbiting stars, gas, and dust quite readily assumes a bar-like structure, suggesting that, possibly, all spiral galaxies periodically develop bars and then lose them again.

S0 Galaxies

The S0 galaxies, which lie at the origin of the fork in the Hubble sequence, are similar to the spirals as far as the spatial distribution of stars goes, but they contain little gas or dust. Hence, they have neither spiral arms nor clouds in which stars are forming. Very likely S0 galaxies were once normal spiral galaxies, but lost their gas during past collisions with other galaxies. This hypothesis is supported by the fact that they are most abundant in the central regions of rich clusters of galaxies, where collisions capable of sweeping away the gas and dust occur most often.

Irregular Galaxies

The irregular galaxies are galaxies that do not readily fit into the Hubble sequence. They come in the most unusual and contorted shapes (figures 2.16, 2.17). Many of them are close satellites of larger galaxies and their distorted

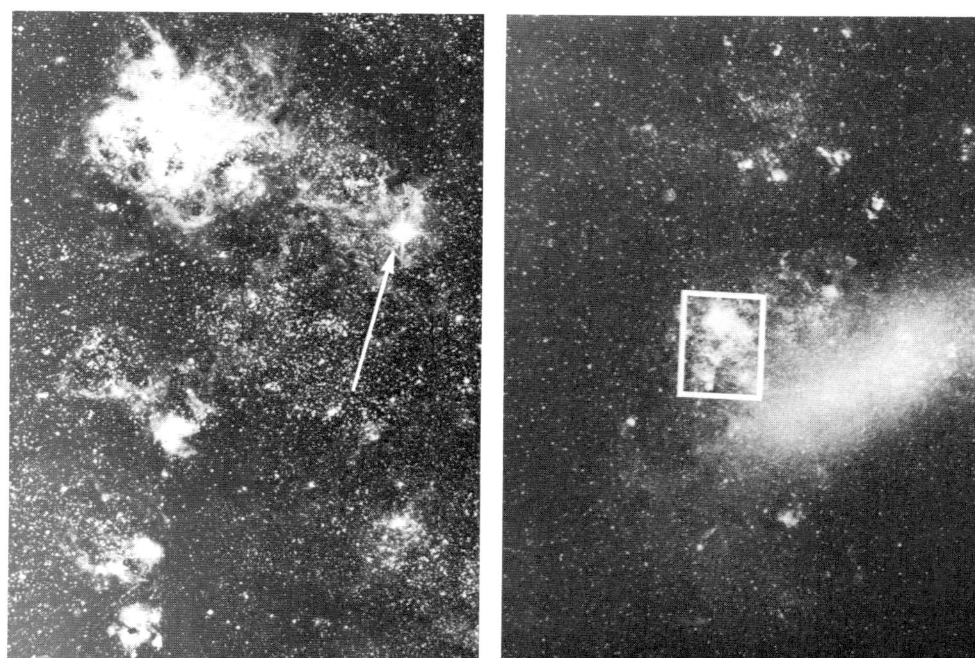

Figure 2.16. The Large Magellanic Cloud (on the right), one of the two companion galaxies of the Milky Way. (See also figure 1.5.) The region within the rectangle is enlarged on the left and shows the supernova (arrow) that was sighted on February 23, 1987.

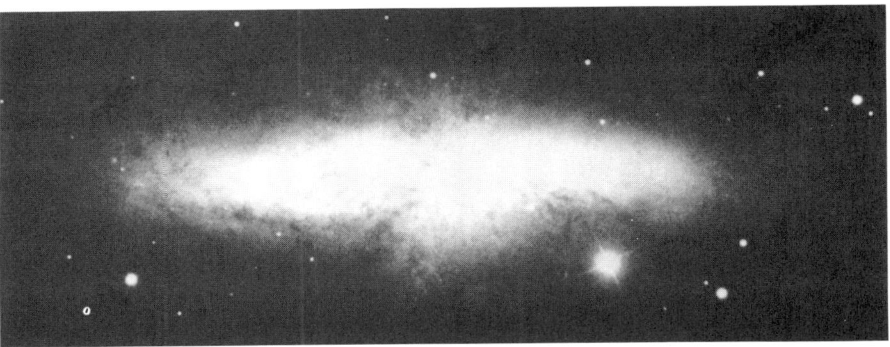

Figure 2.17. Galaxy M 82. Though this galaxy is classified as an irregular, it probably is a disk galaxy (possibly a spiral) seen edge-on that a few million years ago experienced an explosion of enormous magnitude at its center. The explosion accelerated millions of solar masses of gas to velocities of up to 500 km/sec and produced the chaotic structure seen in the galaxy's central regions. Incidentally, Galaxy M 82 lies in the vicinity of the spiral galaxy M 81 (see figure 2.14) and is connected to it by a faint bridge of gas, the result of a close encounter between the two galaxies about 100 million years ago.

shapes are caused by tidal forces, much like the Earth's oceans are distorted by the Moon's tidal force. Others undergo internal upheavals from explosions of gigantic magnitude. Still others either are colliding or have recently collided with another galaxy and are suffering from this experience.

Active Galaxies

A final group of galaxies, unknown in Hubble's time and hence not included in his classification, are the radio galaxies, Seyfert galaxies, and the quasars. Collectively they are known as *active galaxies* because all three are characterized by unusually energetic activities.

Figure 2.18. Radio telescope at Bad Münstereifel-Effelsberg, near Bonn, Federal Republic of Germany. The dish of this telescope is 100 meters across, making it the world's largest steerable radio antenna. Radio telescopes allow astronomers to extend their observations to the long-wavelength part of the electromagnetic spectrum (millimeters to many centimeters).

The *radio galaxies* emit much of their energy as radio waves (figure 2.18). In many cases the radio emission comes from pairs of sources placed hundreds of thousands to millions of light-years apart on opposite sides of the galactic nuclei. Exceptionally powerful explosions in the galactic nuclei must have ejected matter and thus created these double radio sources.

The previous chapter introduced two of the most interesting radio galaxies—M 87 and Centaurus A (figure 1.10). M 87, seen in figure 2.19, is the giant elliptical galaxy near the center of the Virgo cluster. The faint, lumpy jet shooting out from its nucleus is probably the result of a series of explosions. The radio emission of this galaxy is strongly concentrated toward the core of the galaxy and gradually tapers off farther out. Radio contour maps show that the tapering-off is elongated in opposite directions from the galactic nucleus and is aligned along the axis defined by the jet. This suggests that two jets are involved, which is only vaguely evident from the optical photographs. It is believed that most of the radio waves emitted by galaxies like M 87 and Centaurus A are *synchrotron radiation*, which originates from electrons accelerated to nearly the speed of light and forced by magnetic fields to travel in tight spiral paths.

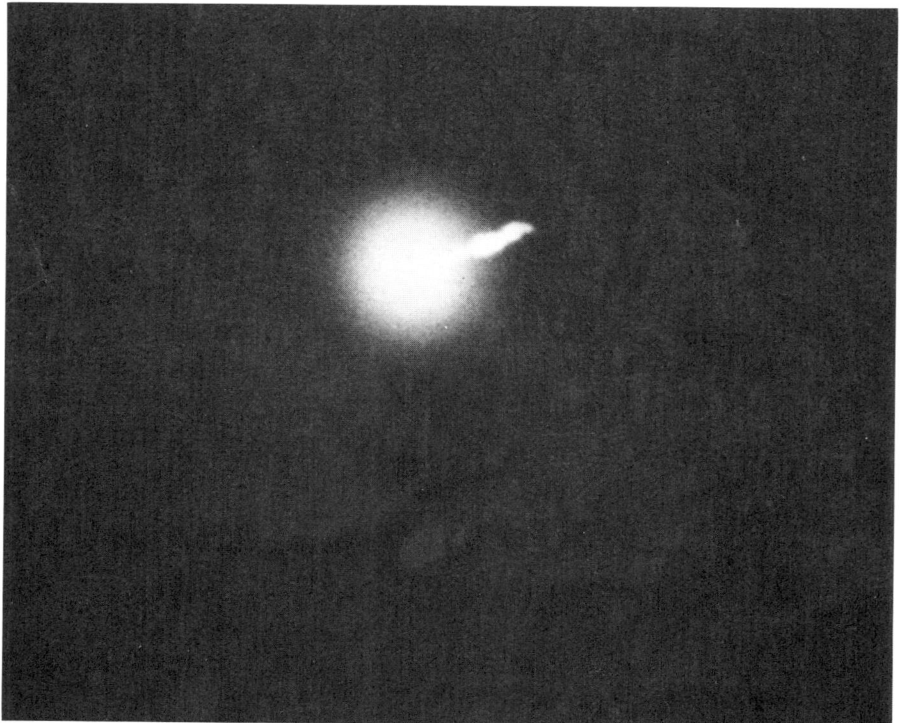

Figure 2.19. The giant elliptical Galaxy M 87 (of type E0, also known as Virgo A) in the Virgo cluster (see also figure 1.12).

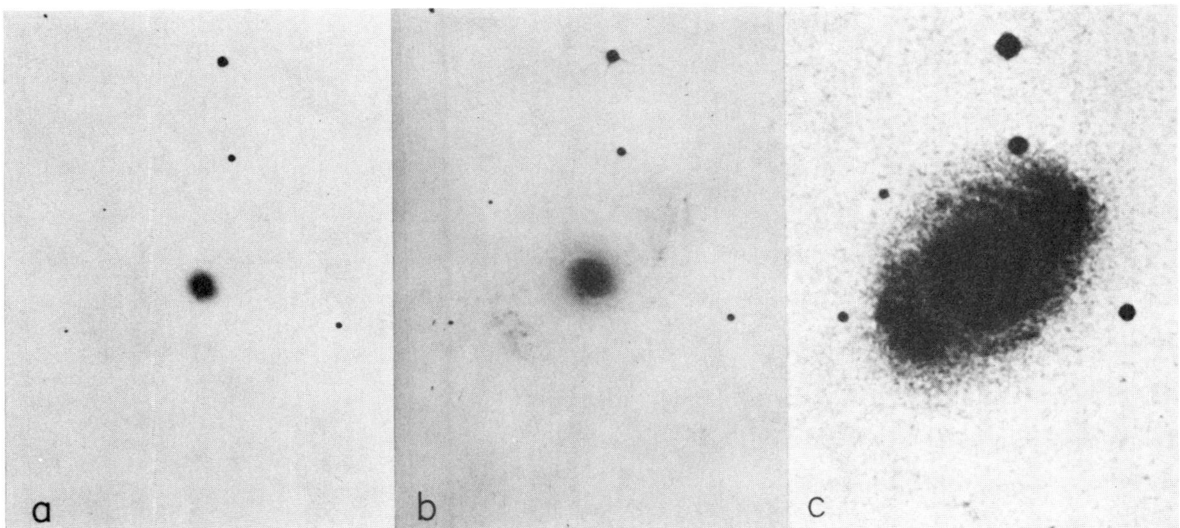

Figure 2.20. A sequence of three negative prints of progressively longer exposures of Seyfert galaxy NGC 4151. In the (a) photograph, the galaxy's nucleus is visible as a compact, star-like object. In the other two photographs, the galaxy's spiral structure can be seen.

The *Seyfert galaxies** are intrinsically much brighter than ordinary galaxies, with an energy output exceeding that of the giant elliptical galaxies by at least a factor of ten to a hundred (figure 2.20). The energy is emitted not only in the optical part of the spectrum, but also as infrared, ultraviolet, and X radiation. Often these galaxies are powerful sources of radio waves as well. Most of this energy originates from a very compact region, measuring only a few light-weeks or light-months across.

Quasars are another group of objects that many astronomers today suspect to be galaxies also—or, more accurately, the central nuclei of galaxies. They have the point-like appearance of stars, yet their spectra are clearly nonstellar. In the early 1960s, when they were first discovered, astronomers classified them as star-like or *quas*i-stell*ar* and contracted these two words into *quasar*. In many respects, their energy output resembles that of Seyfert galaxies, though most of them are brighter still. They lie at very great distances where, if they are indeed galaxies, only their compact central nuclei would tend to show up on photographs (figure 2.21).

Initially, astronomers were quite uncertain about the nature of quasars, not only because of their star-like appearance, but also because of the unusual features in their spectra. They contain emission and absorption lines never seen in astronomical objects before. In 1963, three years after the discovery of the first quasar, Maarten Schmidt of the Hale Observatory recognized that these

*The Seyfert galaxies are named after Carl K. Seyfert who, as a postdoctoral fellow at the Mount Wilson Observatory in California, described their properties in 1943.

Figure 2.21. Quasar 3C 273. This quasar is the brightest one visible from Earth, with a recession velocity of 15% of the speed of light and a corresponding distance of 3 billion LY. Radio observations indicate that it consists of two sources, the fainter of which shows up on photographs as a jet, reminiscent of that of M 87.

spectral features originate in the UV part of the electromagnetic spectrum and are redshifted into the optical range. They appear so unusual because UV radiation is normally absorbed by the Earth's atmosphere and, hence, in the early 1960s astronomers were not familiar with its spectral features. In the case of the quasars, the redshift moves the UV radiation into the optical range where, like light from the Sun, it readily penetrates to the ground. For example, for quasar 3C 273, whose light Schmidt analyzed, the redshift amounts to 0.158, and for the quasar with the largest redshift discovered to date, it is 4.43. In comparison, stars in our galaxy have redshifts of only about 10^{-4}, with some being positive and others negative, and clusters of galaxies typically have redshifts in the range of 0.01 and 0.2.*

For two decades astronomers have debated what the source of the quasars' unusually large redshifts could be. Are they caused by intense gravitational fields, by very high recession velocities, or by yet other processes that we presently do not understand? If they are due to high recession velocities, are those velocities related to the cosmic expansion as are the recession velocities

*Astronomers refer to the ***redshift*** of a star or a galaxy by the letter z and define it as the ratio $(\lambda - \lambda_0)/\lambda_0$, where λ_0 is the wavelength of the radiation as it is emitted by the source and λ is the redshifted wavelength as observed on Earth. Redshifts of 4.43, 0.2, 0.01, and 10^{-4} correspond to recession velocities of 93%, 18%, 1%, and 0.01% of the speed of light, respectively.

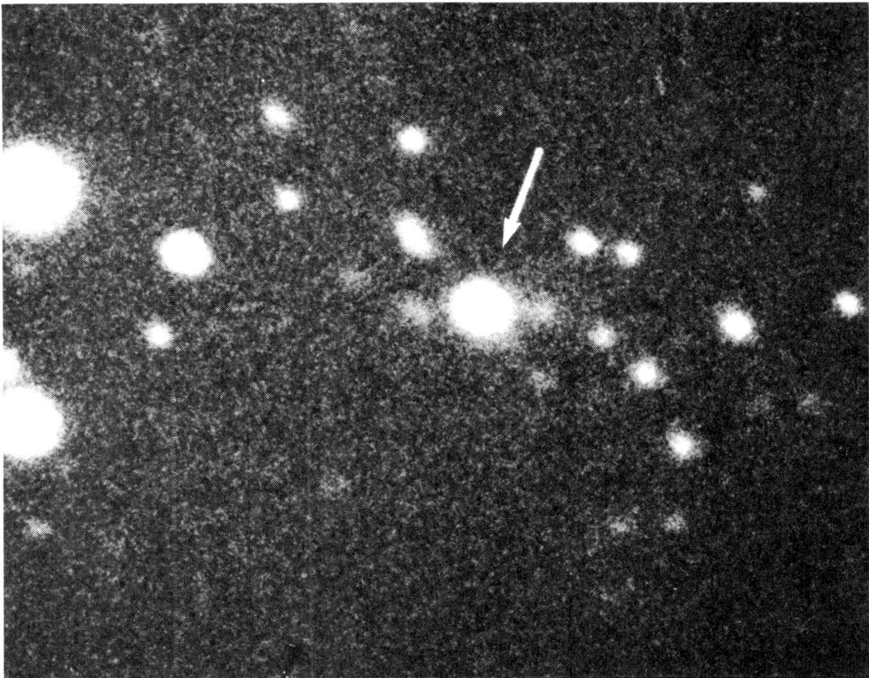

Figure 2.22. Quasar 3C 206. This quasar is surrounded by a cluster of galaxies with about 200 members. The quasar and galaxies have similar redshifts (z = 0.203 to 0.206, corresponding to a distance of about 3.7 billion LY), strongly suggesting that the quasar is a member of the cluster.

of distant clusters of galaxies? Or are they locally produced velocities? For most astronomers the controversy has now been resolved by the discovery that some of the quasars are members of clusters of normal galaxies (figure 2.22) and others show absorption lines originating from intervening intergalactic gas. These discoveries leave little doubt that the quasars' redshifts are due to large recession velocities and the expansion of the Universe. The recession velocities measured to date range from 5 to 93% of the speed of light, placing the quasars at distances from 1 to 12 billion light-years (providing that Hubble's velocity–distance relationship applies to the quasars and our notions of the large-scale structure of the Universe are correct). Thus, the most distant quasars emitted their radiation when the Universe was only about 1 billion years old.

Note, however, that a minority of astronomers still questions that the currently available data on quasars proves beyond the shadow of a doubt their cosmological nature. They suggest that quasars may be nearby objects, whose large redshifts have nothing to do with the cosmic expansion and whose origin and physical structure we do not presently understand. Therefore, they prefer not to think of quasars as the luminous nuclei of galaxies, but restrict their definition of quasars to *intense, point-like sources of light (and, in some cases, of radio waves, X rays, and gamma rays) that are characterized by large redshifts of the lines in their visible spectra.*

Black Holes: Engines of Active Galaxies

What could the energy source of quasars, Seyfert galaxies, and radio galaxies be that produces such powerful explosions and creates a brightness of up to a thousand times that of our galaxy, all from a volume not much bigger than the Solar System? One rather intriguing answer, which many astronomers are coming to accept, is that the sources are gigantic black holes, with masses of millions or even billions of suns, at the centers of the active galaxies. The energy is produced when gas clouds and stars, in the course of their orbital motions about a galaxy, pass too close to the black hole (see figure 2.23). Intense tidal forces tear the gas and stars apart. The gases, which may be at tens or hundreds of millions of degrees Kelvin, are then pulled inward to form a thin *accretion disk* around the black hole. Gradually they spiral closer and closer to the black hole and, eventually, are irretrievably drawn into it. During the final descent, the matter is accelerated to nearly the speed of light and enormous amounts of energy are released, far more than is available from nuclear reactions or any other brown source. The outbursts are repeated as new gas and stars approach the black hole and succumb to its inescapable gravitational pull.

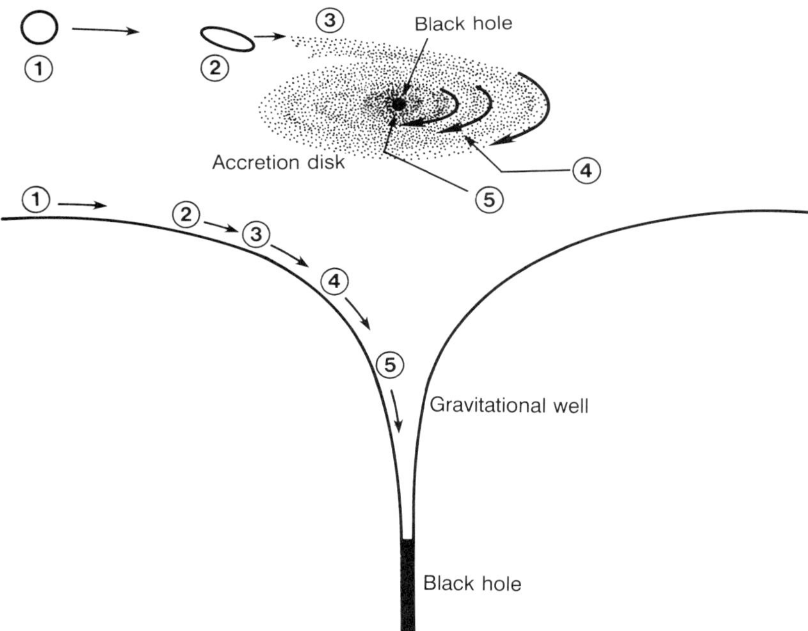

Figure 2.23. Proposed model of the engine powering the active galaxies. According to this model, the engine is a gigantic black hole lying at the centers of quasars, Seyfert galaxies, and radio galaxies. Stars (1) that pass too close to the black hole are tidally torn apart (2), spread out into an accretion disk (3), and finally are drawn into the bottomless gravitational well of the black hole (4, 5), releasing enormous amounts of energy. If the Sun, for example, were to meet this fate, the energy released would be 100 times more than it radiates in the course of its main-sequence lifetime of 10 billion years (see chapter 3).

The black hole model explains another feature observed in many active galaxies, namely the jets of gas and radio lobes. These features could result from the high pressures created by heating and compression when the gas spirals through the accretion disk toward the black hole. The pressures become so great that they squirt part of the gas outward at nearly the speed of light in opposite directions along the disk's spin axis (the directions of least resistance). Besides creating jets and radio lobes, this kind of ejection would also lead to the emission of gamma rays, X rays, and synchrotron radiation (due to relativistic electrons of the jets interacting with magnetic fields surrounding the black hole), which are observed from near the cores of many active galaxies.

It has been suggested that the sequence of events just described, which spells the demise of entire stars, is the engine powering more than just the active galaxies. Possibly, even ordinary galaxies, including our own, possess black holes at their centers and periodically flare up as matter falls into them.

BUILDUP OF HEAVY ELEMENTS

When the galaxies formed, matter consisted of approximately 76% hydrogen (by mass), 24% helium-4, traces of deuterium and helium-3, and virtually nothing else. By the time the Solar System was born, 4.5 billion years ago, the amount of hydrogen in our Galaxy had been reduced by about 2 to 3%, helium-4 had increased slightly, and roughly 1 to 2% of the matter had been converted into heavier elements—carbon, nitrogen, oxygen, neon, magnesium, silicon, iron, and the rest. Similar changes had occurred in all other spiral, barred spiral, S0, and most irregular galaxies. In elliptical galaxies and in the halo of the spiral galaxies, the changes were considerably less. Only one one-hundredth to one-tenth of 1% of their matter had been changed into heavy elements.

Significance of Heavy Element Buildup

Is such a small change as 1 to 2% buildup of heavy elements in the spiral galaxies significant? The answer is yes. First, it reveals a great deal about the formation and early evolution of the galaxies. Second, and more important for us, the heavy elements made a planet like Earth and life on it possible. The core of our planet consists largely of iron and nickel; its mantle and crust are made of silicon, oxygen, magnesium, aluminum, and many other heavy elements; our present atmosphere contains mostly nitrogen and oxygen; and life is based on the chemistry of carbon. Thus, the buildup of the heavy elements, even though it amounted to only 1 to 2%, was crucial for our existence.

The dependence of our planet's structure and of life's development on the heavy element buildup is discussed in later chapters. Here let us focus on what this buildup tells us about the making of the galaxies. Look at a few more facts. First, the buildup of heavy elements in galaxies like our own has not progressed much since several billion years before the formation of the Solar System. For example, stars 2 or 3 billion years older than the Sun have roughly the same composition as the Sun or stars currently forming. Second, stars older than about 8 to 10 billion years—the stars in the halos and nuclei of spiral galaxies, those in globular star clusters, and in elliptical galaxies—do have a measurably lower heavy element content. Third, stars that lie toward the inner part of the disks of spiral galaxies show a noticeably greater abundance of heavy elements than those lying further out.

Finally, heavy elements are manufactured only in the cores of stars much more massive than the Sun (to be discussed in some detail in chapter 3). This manufacturing process starts with the fusion of hydrogen into helium-4, much as it did during primordial nucleosynthesis (see chapter 1). Helium is then fused into carbon and oxygen, which subsequently undergo further fusion and are converted into still heavier elements. These nuclear reactions are the stars' source of energy. After some hundreds of thousands to millions of years of furious nuclear activity, the massive stars become structurally unstable and explode as supernovae, throwing much of the freshly synthesized elements back into space. There the elements intermingle with the primordial hydrogen and helium and gradually change the composition of interstellar matter.

Early Star Formation

These facts suggest that soon after the first fragments condensed out of the primordial clouds and began colliding and coalescing into protogalaxies, massive stars must have been forming in great numbers. They rapidly raced through their evolutions, like engines out of control, manufacturing heavy elements, exploding, and tossing their products back into space. The primordial gas thus became enriched with heavy elements. New generations of stars formed from the enriched gas and the cycle was repeated.

In protogalactic fragments that evolved into globular star clusters, elliptical galaxies, and the spherical components of spiral galaxies, the gas was used up rather quickly. Or, it was lost due to collisions between the galaxies. Very few stars were formed after the initial burst of star formation and nucleosynthesis ended after only very little hydrogen had become fused into heavier elements. In fragments that evolved into the disk components of spiral galaxies, much of the gas remained in uncondensed form and the cycle of stellar birth, nucleosynthesis, and supernova explosion continued for many generations. The heavy element content of the gas rose and within just a few billion years amounted to approximately 1 to 2% by mass. The rapid formation of massive stars then came to an end and the synthesis of heavy elements was reduced to a negligible level.

This theory of early star formation is further supported by the concentration of heavy elements toward the central regions of all galaxies. Any collapsing cloud of gas always develops the highest density near its center. The same must have been true during the collapse of the fragments and the growth of protogalaxies. Hence, the formation of massive stars, nucleosynthesis, and supernova activity may be expected to have been greatest there and to have produced the heavy element distribution observed in galaxies today.

To summarize, the buildup of heavy elements by massive stars was at its peak when the galaxies were forming. By the time the Universe was just a few billion years old, the rapid rate of star formation had come to an end. In spiral and irregular galaxies, star formation has continued to the present epoch, but at a much reduced rate. This suggests that, in comparison to today, the young galaxies must have been exceedingly bright, particularly toward their centers where the formation of massive stars and supernova explosions were concentrated. It may be that the early bursts of star formation and supernova activity created the giant black holes that are suspected to mark the centers of galaxies. We cannot be sure at present. But as new generations of telescopes are built that can extend our view into the Universe's distant past (in particular, space-based telescopes), we may be able to test these theories.

MILKY WAY GALAXY

When we look overhead on a clear night, we observe the faint glow of the Milky Way stretching from horizon to horizon. Away from the interfering lights and pollution of cities, observers with good eyesight can see approximately six thousand stars (3000 per hemisphere). The light of billions more, too faint to be seen individually, merges to form the silvery band of the Milky Way. This band is the disk of our galaxy and we are located right in the middle of it. In addition to stars, it consists of a patchwork of bright luminous clouds of gas interspersed with regions of dark, light-absorbing dust. We see it edge-on from within, just as we see some other galaxies edge-on from outside. A panoramic view of our galaxy is shown in figure 2.24. To appreciate this mosaic of photographs fully, remember that the photographs were taken from within the disk of the Galaxy, some 30,000 LY from its center, and constitute a full-circle view.

Overall Structure

Living right within the galactic disk and being surrounded by huge clouds of gas and dust has made it very difficult for astronomers to deduce the true structure of our galaxy. In the direction of the Milky Way disk, we literally cannot see the Galaxy for the clouds, particularly when we view it in the optical and UV parts of the spectrum. Only in directions perpendicular to the disk do we have a reasonably clear field of view.

The limitations on visibility are not nearly so severe in the radio spectrum. Infrared and radio waves, which are of longer wavelengths than optical and UV radiation, penetrate the interstellar gas and dust much more readily. Therefore, it became one of the tasks of radio astronomers to map the spiral structure of our galaxy. They focused their attention on clouds of neutral hydrogen gas that are concentrated along the arms and emit a characteristic radio signal at a wavelength of 21 cm. By observing this signal as it arrives from all directions and by noting its Doppler shifts due to the differential rotation of the gas about the galactic center, they deduced the size and the spiral layout of our galaxy (figure 2.25). This layout closely resembles the multiple-arm structures of some other large spiral galaxies, such as the Andromeda galaxy and M 81, and it marks the Milky Way as a galaxy of type Sb.

Figure 2.24. Panoramic view of the Milky Way. The coordinates shown are galactic longitude and latitude, where 0° longitude is the great semicircle in the direction of the Galaxy's poles and center (as viewed from Earth), and 0° latitude is the great circle that lies in the Galaxy's central plane (also called the galactic equator).

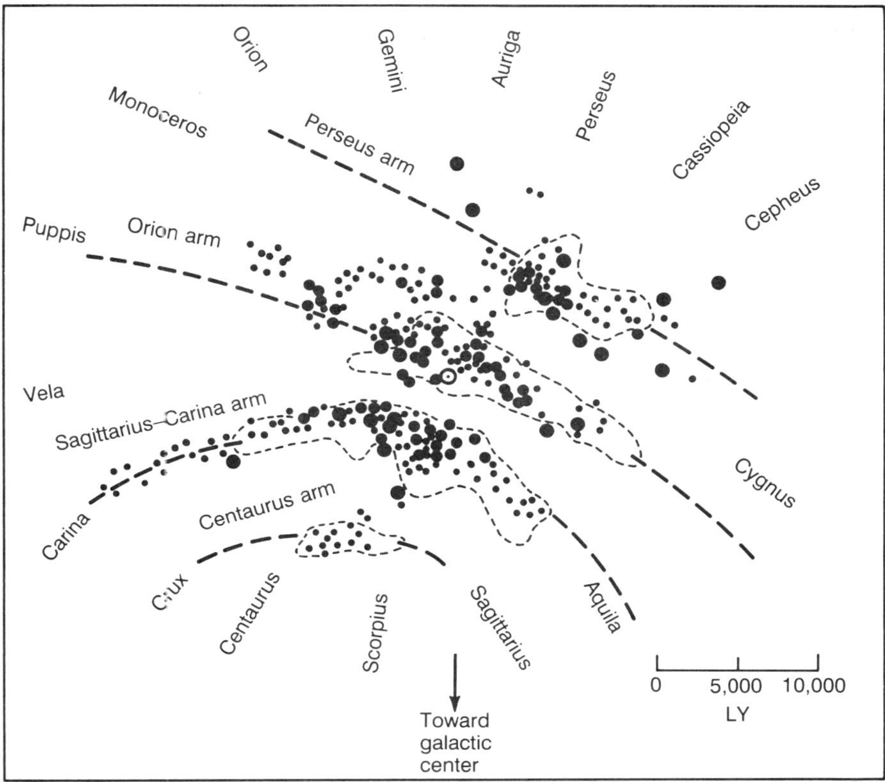

Figure 2.25. The spiral arm features of the Milky Way galaxy in the solar neighborhood. Four arms are shown—the Perseus, Orion, Sagittarius–Carina, and Centaurus arms. The Sun (⊙) is located on the inner edge of the Orion arm. The dots mark the distribution of open star clusters (•) and brightly illuminated clouds (●). The dashed lines enclose regions with the highest concentration of bright and recently formed stars. The directions of some of the constellations in neighboring regions in the sky are indicated along the periphery of the diagram.

Optical astronomers have also studied the size and the general distribution of matter in the Galaxy. Because they are so limited in seeing in the direction of the galactic plane, one of their most successful methods has been to determine the distances to the globular star clusters. Many of these clusters lie far above and below the galactic plane where there is little obscuring gas and dust. From the clusters' nearly spherical distribution about the galactic center and our off-center location relative to them, optical astronomers confirmed the results from the 21 cm observations.

According to both methods of observation, the visible part of the Milky Way (both in the optical and radio parts of the electromagnetic spectrum) is approximately 100,000 LY across, contains close to 150 billion stars, and has a mass of about 140 billion $M_{\odot}$. However, if the dark matter, which is suspected to be the dominant component of the Galaxy's halo, is included, the diameter of the Milky Way is at least 300,000 LY and its total mass is in excess of $10^{12}\ M_{\odot}$.

Figure 2.26. The Great Nebula in the constellation Orion, the brightest nebula visible from Earth. It lies at a distance of 1300 LY and is heated to incandescence by a number of young, hot, massive, and very luminous stars embedded in it. It is situated in Orion's Sword, roughly three degrees below the Belt, and is part of a much larger, dark cloud complex of gas and dust in which new stars are believed to be currently forming (see also figure 3.3).

Orion Arm

The Solar System is located some 30,000 LY from the galactic center at the inner edge of one of the spiral arms, known as the *Orion arm*. The two nearest neighboring arms are at distances of about 6000 LY each. They are the Sagittarius–Carina arm, which lies in the direction of the galactic center, and the Perseus arm, which lies in the opposite direction (see figure 2.25).

Because of its closeness, the Orion arm appears to wrap itself almost halfway around us in the night sky, from the constellations Monoceros and Orion to Gemini, Auriga, Perseus, Cassiopeia, Cepheus, and Cygnus. It contains most of the brightest stars that we see overhead—Sirius, Vega, Capella, Rigel, and Betelgeuse are just a few examples (see table 3.1 for data on these stars). It also contains many of the bright and dark nebulae that are visible through telescopes and that show up so spectacularly on photographs—such as the Orion (figure 2.26) and Eta Carinae nebulae

Nucleus

When the inhabitants of the middle to lower latitudes of the Northern Hemisphere look towards the southern horizon on a summer night, in the direction of the constellation Sagittarius, they see a gradual broadening in the band of the Milky Way. This broadening represents the outer regions of the central bulge of our Galaxy. The whole of it can only be seen from the Southern Hemisphere of

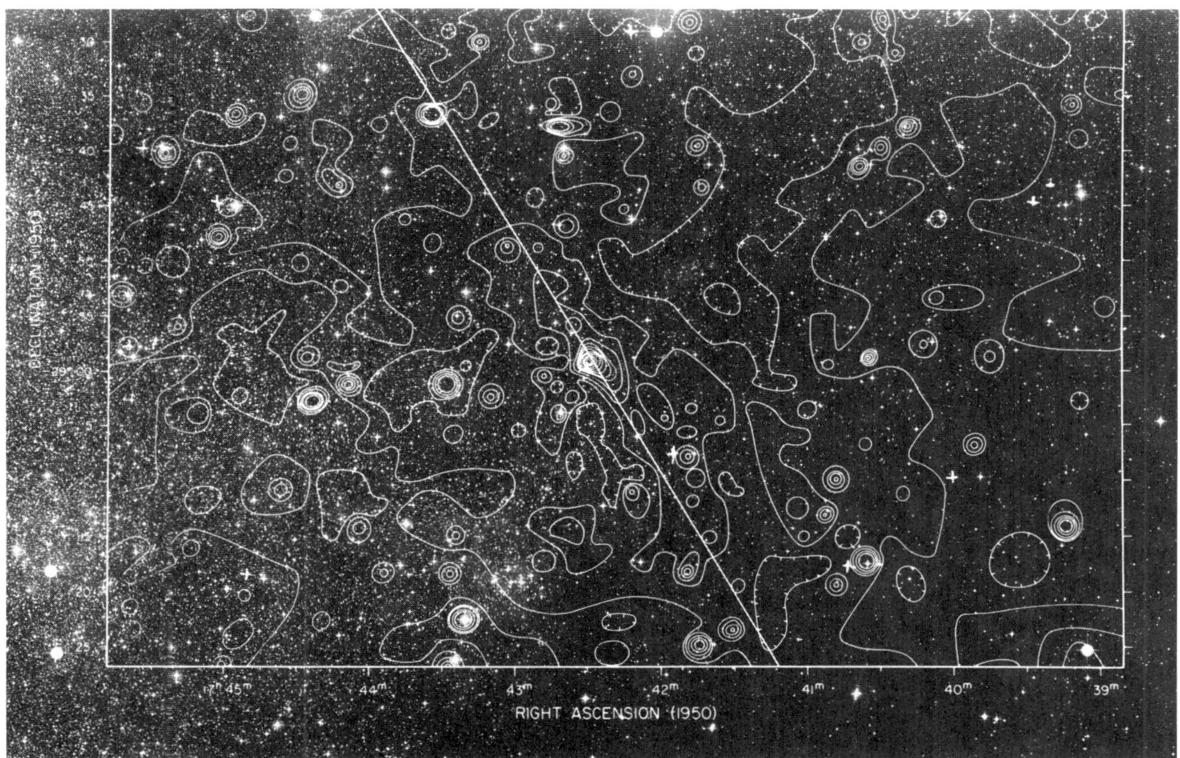

Figure 2.27. Photograph in the direction of the center of the Milky Way, with superimposed infrared map. The concentration of contour lines near the middle of the map signifies Sagittarius A West. The slanted line represents the central plane of the Galaxy's disk.

the globe and, with its high concentration of stars and nebulae, is easily the most striking visible feature of the Milky Way.

The clouds of gas and dust lie so thick in the direction of the Galaxy's nucleus that its center cannot be seen from Earth in optical light. Radio and infrared radiations do penetrate, however, and reveal a most interesting structure. As the contour map in figure 2.27 shows, the central region of the Galaxy contains several distinct infrared emission sources. The most intense of them is no bigger than Jupiter's orbit, yet the rapid motions of hydrogen gas in its vicinity indicate that it holds at least 4 million solar masses. It is called Sagittarius A West and signifies the center of the Galaxy.

The region surrounding Sagittarius A West shows ample evidence of a history of periodic, violent explosions. It is a strong source of synchrotron radiation, which indicates the presence of a mechanism capable of accelerating electrons to relativistic speeds in a magnetic field. It emits X rays and gamma rays, the most energetic kinds of radiation in the electromagnetic spectrum. The gamma rays have wavelengths that indicate they are produced by electron–positron pair annihilation, implying that there must be a source of positrons (the antiparticles of electrons, see the discussion of the birth of the Universe, age 10^{-12} second, chapter 1). The region also contains a number of IR sources. Furthermore, there are large streams of gas flowing rapidly outward from the center of the Galaxy. It is all reminiscent of quasars, Seyfert, and radio galaxies. Is Sagittarius A West a black hole, like those that are suspected to form the centers of the active galaxies, but on a smaller scale? The evidence is mounting for an affirmative answer. Matter passing through an accretion disk and falling into a black hole would explain the energetic activities taking place at the center of the Milky Way. Also, as far as we know, only a black hole can pack several million solar masses into a volume smaller than the Solar System.

Halo and Disk

The stars and globular star clusters of the halo of the Milky Way trace out large, elongated orbits about the galactic nucleus. These orbits are randomly oriented in space and give the halo its spherical symmetry. In contrast, the stars, open star clusters, and gas and dust of the disk trace out nearly circular orbits. They all lie very nearly in the same plane and give the disk its flat distribution.

As noted earlier, the matter of the disk does not rotate like a solid wheel. It rotates differentially, with the stars and the gas closer to the nucleus orbiting more rapidly than those farther out. The Solar System, which belongs to the disk and which lies at approximately 30,000 light-years from the galactic center, orbits with a speed of roughly 220 km/sec. This carries us around the Galaxy once every 250 million years. Despite this great orbital speed (equivalent to 800,000 km/hr) we have no sensation of motion. The distances to the stars are too great for any obvious positional changes to occur among the stars within the span of a human life. Still, careful observations through telescopes show that many of the nearby stars change their positions by up to a few seconds of arc every year.* These positional changes are mostly due to small random motions

*A second of arc is a measure of the spread of an angle. Sixty seconds of arc make up one minute of arc, and sixty minutes of arc make up one degree. There are 90 degrees in a right angle and 360

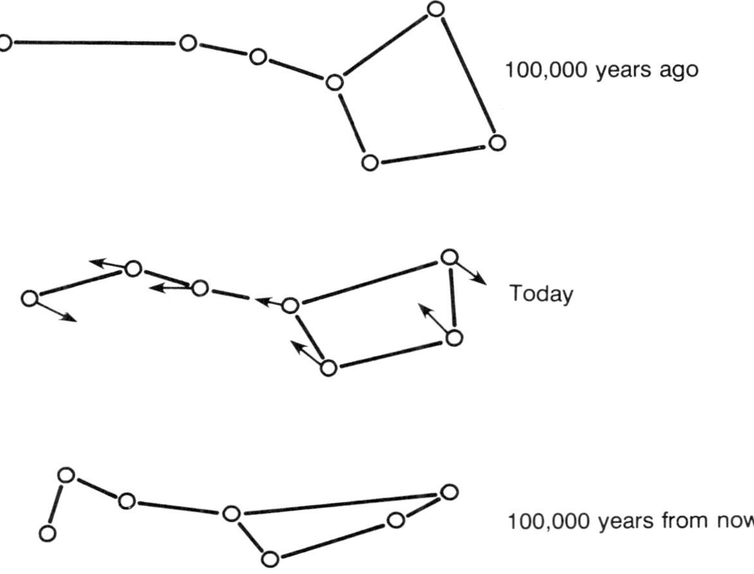

Figure 2.28. The Big Dipper through time. The arrows (center) indicate the motions of the stars. Adapted from Robert D. Chapman, *Discovering Astronomy* (San Francisco: W. H. Freeman & Co., 1978), 328.

superimposed on the circular orbital velocities. To a smaller extent they are due to the differential orbital motions. If we were to return to Earth 100,000 years from now, these small yearly changes in the positions of the stars would have added up to rather noticeable changes and, for example, most of the constellations would have become distorted beyond recognition (see figure 2.28).

We may regard the 250 million years of the Sun's orbital journey about the galactic center as one *galactic year*. By human standards this is an exceedingly long span of time. The last time we were at our present position in the Galaxy, the dinosaurs were just beginning to become the dominant land animals on Earth. Mammals did not yet exist. The continents were converging into one single supercontinent, *Pangaea*, and bore no resemblance to their present shapes. Two galactic years ago, the earliest vertebrates (jawless fishes) swam in the sea, and the land was still devoid of plant and animal life. Four galactic years ago, life was single-celled and microscopic. The origin of life itself dates back about 16 galactic years, and the age of the Earth and the rest of the Solar System is approximately 18 galactic years. Thus, in terms of our orbital motion around the Galaxy, the Earth and life on it are still very young.

degrees in a complete circle. As viewed from Earth, the disks of the Moon and the Sun are approximately 30 minutes, or 1800 seconds, of arc across. Hence, one second of arc is a very small angle indeed.

To recapitulate, the weak density fluctuations that characterized the Universe during its early evolution were the source of a new cosmic order. They led to the contraction of matter into superclusters of galaxies, which are the primary structural units of the cosmos. The superclusters measure, on the average, 100 million light-years across and are interspersed by voids of comparable size. Each supercluster contains millions of member galaxies and each galaxy contains millions to billions of stars. The Milky Way is a member of a small cluster, the Local Group, which is an outlying member of the Virgo supercluster of galaxies.

The creation of order did not stop with the formation of galaxies and stars. In the case of the Solar System, for example, it continued on progressively smaller scales with the making of planets and the origin and evolution of life on one of them. In Part Two (chapters 6–10) I will explore how this evolution began with random chemical reactions in the primordial soup and, in the course of some 4 billion years, produced the enormous diversity of life—from bacteria to protists, fungi, plants, and animals—that inhabits the Earth today.

Exercises

1 Name a few everyday examples supporting the statement that small random fluctuations in the physical properties of matter (such as in density, in temperature, and in the sizes of objects) cannot be avoided.

2 Waves of moderate energy (such as sound waves, waves induced in a taut rope, and water waves of moderate amplitude in deep water) pass through each other with little interaction. Confirm this by inducing waves at both ends of a rope or by throwing two rocks into a lake. Then watch the two waves approach and pass through each other. (*Note*: Waves of large energy, such as shock waves caused by jets flying faster than sound or breakers at a shallow beach, behave differently. They interact strongly when they collide, causing local turbulence and losing much of their energy through friction.)

3 (*a*) Turn on two garden hoses and point the streams of water at each other (the water should flow slowly). Slightly vary the directions and the flows of the two streams. Describe the interaction between the two streams.
(*b*) What does the interaction between the two streams of water tell you about the possible consequences of collisions of gas fragments in the early Universe? (*Note*: Liquid water is nearly incompressible, while gaseous matter can be compressed much more readily.)

4 Discuss the spatial distributions of globular star clusters, open star clusters, gas and dust clouds, and young stars in spiral galaxies.

5 Why does galactic rotation not "wind up" and destroy the spiral arms of our Galaxy and those of other spiral galaxies?

6 S0 galaxies resemble spirals except that they contain no, or very little, gas and dust. Explain this characteristic, using the fact that S0 galaxies are found predominantly near the central regions of rich clusters of galaxies.

7 In 1983 astronomers observed a supernova in Quasar QSO 1059 + 730. What can be deduced from this observation about the nature of this quasar?

8 Why do globular star clusters and elliptical galaxies contain only "old" stars, that is, stars with a much lower percentage of heavy elements than are found in the Sun and other disk stars of spiral galaxies?

9 On a clear night, and preferably on a mountain top and away from city lights, look at the starry sky overhead and follow the faint band of the Milky Way from horizon to horizon.
(*a*) Depending on the time of the year, hour of the night, and viewing conditions, locate the constellations Sagittarius, Aquila, Cygnus, Cepheus, Cassiopeia, Perseus, Auriga, Gemini, Orion, and Monoceros, all of which lie in or close to the disk of the Milky Way.
(*b*) Find the direction toward the Galactic center (ℓ, galactic longitude, $= 0°$), the direction in which the Solar System travels in its orbit around the Galactic center ($\ell = 90°$), and the direction opposite to the Galactic center ($\ell = 180°$).
(*c*) In analogy to our geographic coordinate system,

define and locate the two Galactic poles and the Galactic equator.
(This exercise requires a star chart, listed in the bibliography for this chapter. Star charts can also be found in most large dictionaries and in most introductory astronomy texts, several of which are listed in the bibliography for the Introduction to Part One.)

10 (*a*) Draw a face-on diagram of the Milky Way, including the Galactic nucleus, the Perseus, Orion, and Sagittarius–Carina arms, and the location of the Sun.
(*b*) By arrows indicate the motion of the Sun as well as the motions of disk stars closer to and more distant from the Galactic center than the Sun.
(*c*) How will the positions of the Sun and its neighboring stars change relative to each other in the course of time?

11 Locate as many of the following astronomical objects on the night sky as the season and observing conditions allow. Use a star chart and binoculars.

Object	Type	Constellation	Comments
Andromeda galaxy (M 31)	Spiral galaxy	Andromeda	Visible with the unaided eye as a faint, hazy spot of light.
h and Chi Persei	Open star clusters	Perseus	Two beautiful clusters, each about 45′ across and visible with the unaided eye; located midway between the stars Mirfak (α Persei) and Cih (γ Cassiopeiae).
M 34	Open star cluster	Perseus	Located midway between the stars Algol and Almach, just visible with the unaided eye.
Pleiades (Seven Sisters	Open star cluster	Taurus	Six or seven stars can be seen with the unaided eye, many more with binoculars or a small telescope.
Orion Nebula (M 42)	Nebula	Orion	Located in Orion's Sword and visible with the unaided eye.
Praesepe (Beehive, M 44)	Open star cluster	Cancer	A large, scattered cluster, almost resolved by the unaided eye.
"Great Cluster" in Hercules (M 13)	Globular star cluster	Hercules	A beautifully symmetric cluster, just visible with the unaided eye.
Lagoon Nebula (M 8)	Nebula	Sagittarius	A poorly defined nebulosity, with dark patches and stars, visible with the unaided eye.

Suggestions for Further Reading

Bahcall, J. N., and **L. Spitzer, Jr.** 1982. "The Space Telescope." *Scientific American*, July, 40–51.

Blandford, R. D., **M. C. Begelman**, and **M. J. Rees**. 1982. "Cosmic Jets." *Scientific American*, May, 124–42. A discussion of the violent activities at the center of many galaxies, manifested in the production of narrow, focused streams of ionized gas.

Blitz, L., **M. Fich**, and **S. Kulkarni**. 1983. "The New Milky Way." *Science* 220: 1233–40. A revised model of the Milky Way as deduced from the distribution of atomic hydrogen. The model shows that the disk of the Galaxy is about twice as extended as was previously thought and that beyond the Sun the gas is concentrated in large-scale, coherent spiral arms indicative of a regular four-armed spiral pattern.

Boer, K. S. de, and **B. D. Savage**. 1982. "The Coronas of Galaxies." *Scientific American*, August, 54–62. Satellite observations indicate that the disks of our and other spiral galaxies are surrounded by enormous envelopes of hot gas.

Bok, B. J. 1981. "The Milky Way Galaxy." *Scientific American*, March, 92–120. An article about the newly emerging picture of the structure and content of the Milky Way, by an eloquent lecturer and writer.

Burns, J. O., and **R. M. Price**. 1983. "Centaurus A: the Nearest Active Galaxy." *Scientific American*, Nov., 56–66.

Dressler, A. 1987. "The Large-Scale Streaming of Galaxies." *Scientific American*, September, 46–54. The Milky Way and other local galaxies are travelling through space as if attracted by a distant concentration of mass larger than any proposed by existing cosmological theories.

Geballe, T. R. 1979. "The Central Parsec of the Galaxy." *Scientific American*, July, 60–70.

Habing, H. J., and **G. Neugebauer**. 1984. "The Infrared Sky." *Scientific American*, November, 48–57. Comets, stars, interstellar gas and dust clouds, and galaxies as seen with the *Infrared Astronomical Satellite* (*IRAS*).

Hodge, P. W. 1981. "The Andromeda Galaxy." *Scientific American*, January, 92–101.

Hutchings, J. B. 1985. "Observational Evidence for Black Holes." *American Scientist*, January–February, 52–59. Recent observational research that tries to establish whether quasars, the cores of galaxies, and certain X-ray-emitting binary stars are powered by black holes.

Margon, B. 1983. "The Origin of the Cosmic X-Ray Background." *Scientific American*, January, 104–19. One source of the cosmic X-ray background may be a multitude of distant quasars.

Mathewson, D. 1985. "The Clouds of Magellan." *Scientific American*, April, 106–14. The turbulent history of our nearest galactic neighbors, including close encounters with each other and with the Milky Way.

Meier, D. L., and **R. A. Sunyaev**. 1979. "Primeval Galaxies." *Scientific American*, November, 130–44. The first galaxies to form after the Big Bang have not been seen, but the characteristics of older galaxies suggest what younger ones might have been like.

Mewaldt, R. A., **E. C. Stone**, and **M. E. Wiedenbeck**. 1982. "Samples of the Milky Way." *Scientific American*, December, 108–21. The authors explain that atomic nuclei found in cosmic rays are clues to the isotopic composition of other parts of the Galaxy.

Osmer, P. S. 1982. "Quasars as Probes of the Distant and Early Universe." *Scientific American*, Feb., 126–38.

Peebles, P. J. E. 1984. "The Origin of Galaxies and Clusters of Galaxies." *Science* 224: 1385–91. Some of the competing theories and unresolved problems concerning galaxy formation. For the advanced student.

Readhead, A. C. S. 1982. "Radio Astronomy by Very-Long-Baseline Interferometry." *Scientific American*, June, 52–61. Observations made simultaneously by radio telescopes thousands of miles apart can be combined with the aid of atomic clocks to yield the highest resolution ever achieved.

Rubin, V. C. 1983. "Dark Matter in Spiral Galaxies." *Scientific American*, June, 96–108. The newly emerging picture of the structure of spiral galaxies, indicating that much of their matter emits no light and that this matter is not concentrated near their centers.

Sandage, A. 1961. *The Hubble Atlas of Galaxies*. Washington, D.C.: Carnegie Institution of Washington. A beautiful and comprehensive set of photographs defining the Hubble classification of galaxies.

Strom, S. E., and **K. M. Strom**. 1979. "The Evolution of Disk Galaxies." *Scientific American*, April, 72–82. Through collisions or near encounters, spiral galaxies can evolve into S0 galaxies, consisting of smooth disks without spiral arms. Such evolutions occur most likely in rich clusters of galaxies.

Star Charts

Astronomy and *Sky and Telescope*, monthly popular astronomy magazines, have star and planetary charts in each issue, including discussions of what to look for in the night sky.

Chandler, D. 1977. *The Night Sky*. Cambridge, Mass.: Sky Publishing Corporation. A compact, easy-to-use star chart for the range 38°–50° north latitude.

Muirden J. 1985. *Astronomy with a Small Telescope*. London: George Philip and Son, Ltd. A very useful book for the amateur astronomer who wants to observe the night sky with the unaided eye, a pair of binoculars, or a small telescope.

Norton, A. P. 1978. *Norton's Star Atlas and Reference Handbook* (*Epoch 1950*). 17th ed. Edinburgh: Gall and Inglis. A comprehensive set of charts of stars, clusters, and nebulae; descriptive lists of objects observable through small telescopes; and definitions of astronomical terminology. For the serious amateur astronomer.

Philips' Planisphere (for Latitude 42° N). 1982. London: George Philip and Son, Ltd. A compact, easy-to-use chart.

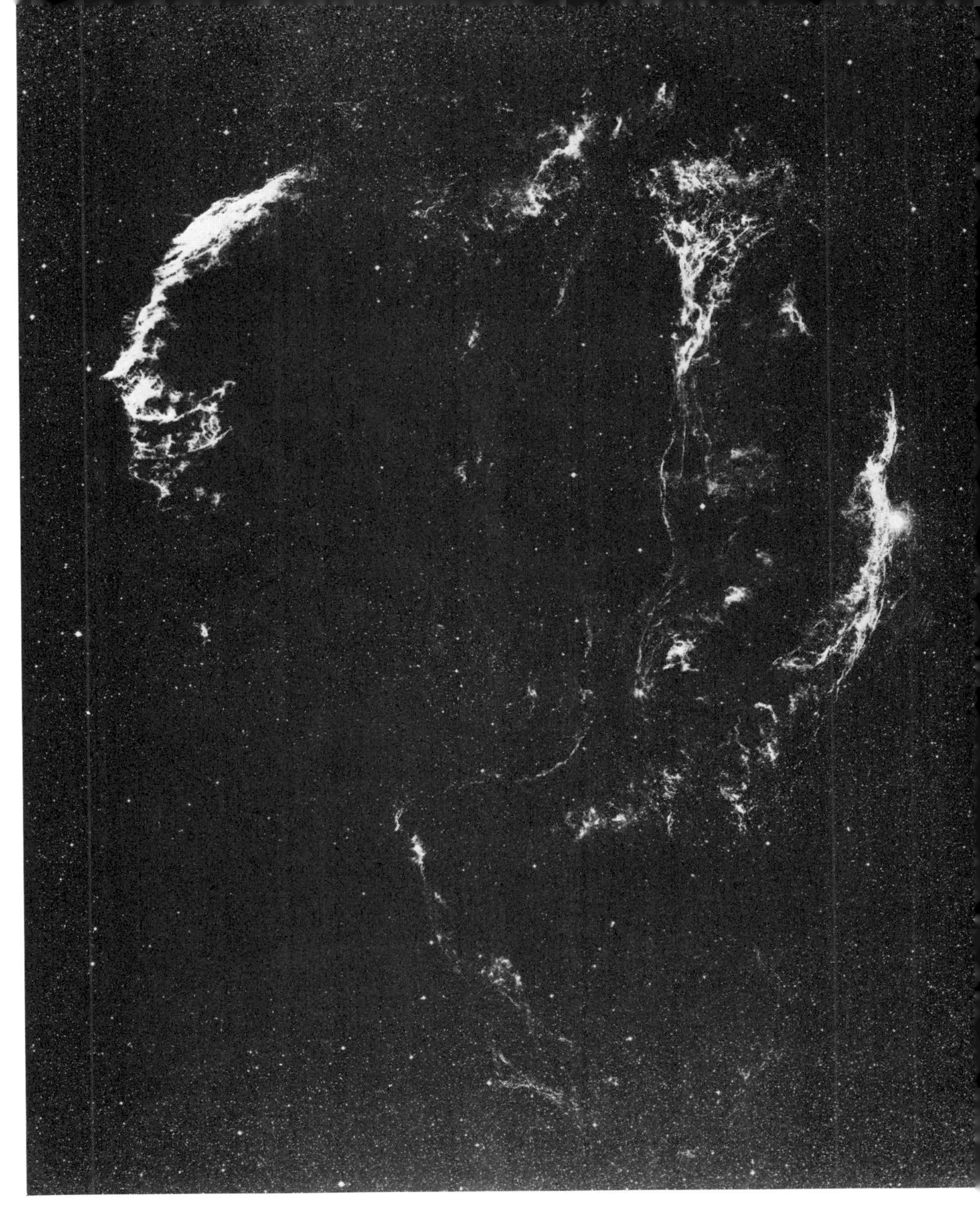

3 Stars and Heavy Element Synthesis

THIS STORY OF THE EVOLUTION of matter will now gradually progress from structures of cosmic dimension to ones closer in size to human experience. The previous chapter was devoted to the origin of galaxies and galaxy clusters, whose sizes range from thousands to roughly 100 million light-years and whose masses lie in the range from 10^6 to 10^{15} solar masses. In this chapter the discussion focuses on the structure and evolution of the stars, most of which have sizes of up to a few hundred or thousand Earth diameters, although some stars, namely the red supergiants, may be 10^5 times as large as the Earth, and some unusual stars, such as neutron stars, are smaller than the Earth.

The masses of stars are a few ten thousand to a few ten million times that of the Earth. The Sun is quite an ordinary star, with a radius and mass that are neither unusually large nor unusually small:

$$\begin{aligned} 1\,R_\odot &= 700{,}000 \text{ km} \\ &= 110 \text{ Earth radii} \\ 1\,M_\odot &= 2 \times 10^{30} \text{ kg} \\ &= 330{,}000 \text{ Earth masses} \end{aligned}$$

(As noted earlier, the symbol $\odot$ stands for the Sun. $R_\odot$ and $M_\odot$ refer to the Sun's radius and mass. $L_\odot$ refers to the Sun's luminosity. Radius of Earth [equatorial] = 6380 km, mass of Earth = 6.0×10^{24} kg.)

In general, stars distinguish themselves from all other objects in the Universe today by producing their own energy through nuclear reactions and pouring that energy forth into space as electromagnetic radiation. That is what makes them shine. The only exceptions are such stars as white dwarfs and neutron stars (see below) that have exhausted their nuclear energy supply. They shine by radiating stored-up heat energy. The Sun's rate of energy output is called *one solar luminosity* ($L_\odot$). The Earth intercepts only about 4.6×10^{-10} of this amount, which is the solar energy that heats our planet, provides the energy for photosynthesis, and gives us light (see box 3.1).

Most stars are very stable and neither expand nor contract perceptibly in the course of a human life span, although there are exceptions. Stars are not solid or liquid, like the bulk of the Earth—they are gaseous. However, unlike the gases of our planet's atmosphere, the gaseous matter of stars is ionized, with exceptions occurring only in the cooler, outermost layers (ionization is defined in the Introduction to Part One). The force that holds stars together is gravity, and it tends to shape them into nearly perfect spheres. We may, therefore, define stars as *giant, self-luminous, gaseous spheres.**

← ***Figure 3.1.*** The Cygnus Loop. This tenuous and rather delicate-looking shell of gas and dust is the ejecta from a supernova that exploded about 30,000 years ago. It lies at a distance of 2500 LY from Earth, measures roughly 130 LY across, and is expanding at an average rate of 116 km/sec. Initially, the rate of expansion was much greater, but the ejecta steadily decelerated as it plowed into the surrounding interstellar medium, compressing it and carrying it along.

*According to this definition, planets, such as Venus and Mars, are not stars because they do not have their own nuclear sources of energy. They shine by reflecting the light of the Sun. The same is true of the Moon, the asteroids, the comets, and all the other smaller bodies in the Solar System.

BOX 3.1.
Solar Luminosity Equivalents

IN our daily lives we commonly measure the rate of the production or consumption of energy in units of horsepower or watts. We use horsepower to express the rate at which engines convert chemical energy from fuel into mechanical energy; we use watts to rate the energy consumption of light bulbs and electrical appliances. Both horsepower and the watt are very small units and quite inconvenient for measuring the rate of energy output by stars. For example, it would take 4×10^{24} one-hundred-watt light bulbs to equal the brightness level of one solar luminosity.

One solar luminosity is also equal to 10^{23} calories of energy radiated per second [the same calorie (abbreviated Cal or kcal) as used to measure the energy content of food] or to about 10^{14} times the current rate of energy consumption by the United States. The last figure means that the energy generated by the Sun in one second would be sufficient to power the current United States economy for roughly 3 million years.

THE HERTZSPRUNG–RUSSELL DIAGRAM

When astronomers focus their telescopes on a given star, they intercept its light and obtain information about the star's surface temperature and luminosity. Stars that are green to bluish-white have surface temperatures hotter than the Sun's. Yellow stars like the Sun have surface temperatures of about 6000 K. Orange to red stars are 5000 K and cooler at their surfaces.

The amount of light that reaches Earth from a star depends both upon its luminosity and its distance: the nearer the star of a given luminosity, the brighter it is. The luminosity is the actual rate, independent of distance, by which a star pours forth radiant energy in all directions. This is the quantity to consider when studying stars. It allows meaningful comparisons to be made between them, even when they lie at different distances.*

In the 1910s, Ejnar Hertzsprung (1873–1967) of Denmark and Henry Norris Russell (1877–1957) of the United States recognized that a plot of the luminosities of stars versus their surface temperatures, as shown in figure 3.2, produces a very interesting pattern. Most stars fall along a fairly narrow band that stretches from the upper left corner of the diagram to the lower right. This band is called the *main sequence*. The stars at the upper left end of the main sequence have surface temperatures of up to 50,000 K and luminosities 10^6 times that of the Sun. Those at the lower right end are at about 3000 K and 1000 times fainter than the Sun. The Sun is located on the main sequence just below its middle.

*Observational astronomers commonly measure the brightness of stars in units of *magnitude*. This unit was introduced by antiquity's greatest astronomer, Hipparchus (born in Nicaea in Bithynia, fl. 146–127 B.C.). He divided the stars into six equally spaced brightness categories from one (brightest) to six (just barely visible with the unaided eye). With some modification, these categories have become the modern magnitude scale. It is a logarithmic scale because, very approximately, the human eye responds to physical stimuli logarithmically. More precisely, the difference in magnitude between two stars (m_1 and m_2) is related to the difference in their luminosities (L_1 and L_2) by the equation $m_2 - m_1 = 2.5 \log_{10} (L_1/L_2)$. *Absolute magnitude*, which is shown in some of the diagrams of this chapter, is the magnitude a star has if it is observed from a distance of 10 parsecs or 32.6 light-years. The *parsec* is the unit of interstellar and intergalactic distance measurement used by astronomers. It is the distance at which the radius of the Earth's orbit about the Sun subtends one second of arc. More specifically, 1 parsec = 3.26 LY.

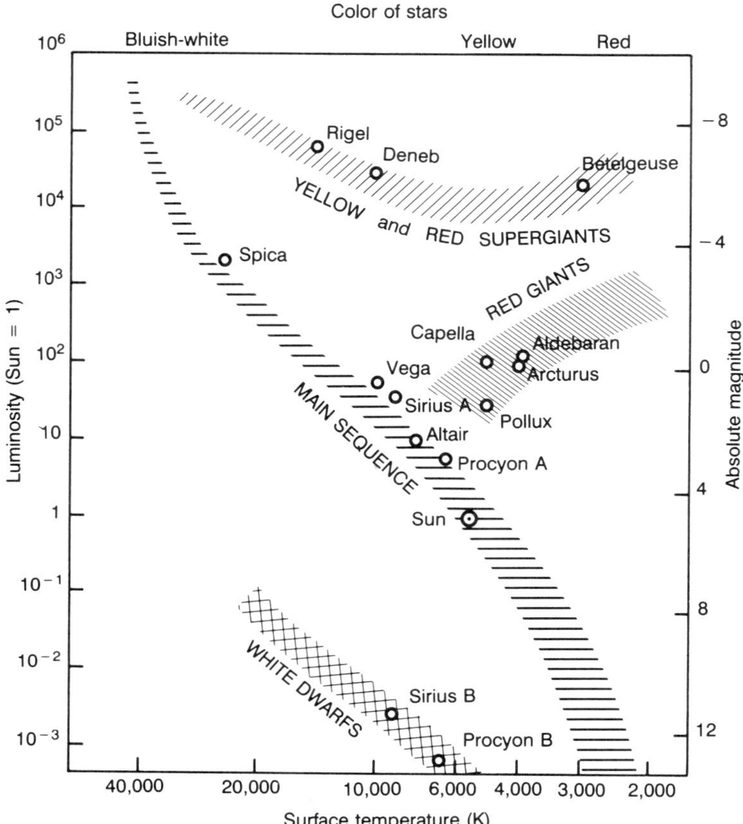

Figure 3.2. The Hertzsprung–Russell diagram, showing main sequence stars, red giants, yellow and red supergiants, and white dwarfs. The diagram also shows the twelve brightest stars visible from the Northern Hemisphere, the Sun, and the white dwarf companions of Sirius and Procyon.

In addition to main sequence stars, the diagram contains several other prominent groupings of stars. Along the right side and above the main sequence lie the *red giants*. As their name implies, they are very large stars and radiate predominantly in the red. Their surface temperatures range from about 2000 K to 6000 K and they are very luminous. To the right of the upper main sequence and extending clear across the diagram lie the *supergiants*. They are even larger and more luminous than the red giants. The lower part of the diagram is occupied by the *white dwarfs*. They are very compact in size and appear bluish-white to yellow. They are stars of low luminosity with surface temperatures somewhat hotter than the Sun's. Clearly, these names—red giant, supergiant, and white dwarf—were given to the stars during an age more

Table 3.1. The Twelve Brightest Stars Visible from the Northern Hemisphere (Listed in Order of Decreasing Brightness).

Star	Constellation	Type of star	Distance (LY)	Mass $M_\odot$	Radius ($R_\odot$)	Surface temperature (K)	Intrinsic luminosity ($L_\odot$)
Sirius A	Canis Major	Main sequence	9	2.3	2.5	9,000	35
Arcturus	Boötes	Red giant	36	4.5	20	4,000	100
Vega	Lyra	Main sequence	26	3	2.5	10,000	50
Capella	Auriga	Red giant	46	3.5	13	5,000	100
Rigel	Orion	Blue supergiant	800	20	36	15,000	60,000
Procyon A	Canis Minor	Main sequence	11	1.5	2	6,500	6
Betelgeuse	Orion	Red supergiant	600	20	500	3,000	20,000
Altair	Aquila	Main sequence	16	2	1.5	8,000	10
Aldebaran	Taurus	Red giant	70	5	20	4,000	100
Spica	Virgo	Main sequence	260	15	3	25,000	2,000
Pollux	Gemini	Red giant	35	4	8	5,000	30
Deneb	Cygnus	Yellow supergiant	1600	14	60	10,000	30,000

Note: The masses, radii, and luminosities are given in solar units. For example, Sirius A has 2.3 solar masses, is 2½ times the size of the Sun, and is intrinsically 35 times brighter than the Sun.

romantic than ours. Today contractions such as pulsar and quasar or simply letters and numbers are used to refer to new astronomical discoveries.

The diagram invented by Hertzsprung and Russell neatly separates stars of different types into distinct groupings according to just two easily measured parameters. It plays a fundamental role in astronomers' attempts to understand the physical nature of stars and is referred to as the *Hertzsprung–Russell diagram*, or *HR diagram*.

Of the twelve brightest stars (besides the Sun) visible from the Northern Hemisphere, five are main sequence stars, four are red giants, and three are supergiants. None of them is a lower main sequence star or a white dwarf, although both Sirius and Procyon have faint white dwarf companions, which are visible only through very large telescopes. Table 3.1 lists the twelve stars, including the constellations in which they are located, their distances, and a few of their physical characteristics.

There must be some underlying reason why stars fall into the distinct groupings that characterize the HR diagram. Nature does not produce this kind of order by accident. It turns out that the groupings are in part due to physical characteristics that the stars acquire at birth, such as mass and composition. Mostly, however, they are due to structural changes stars experience in the course of their lives. These structural changes affect their energy output and surface temperatures and, thereby, show up in the HR diagram.

By studying the pattern of the HR diagram, astronomers have learned a great deal about the evolution of the stars. This seems paradoxical, for humans have been observing stars with the purpose of understanding their physical structures for only a few hundred years—a time span far shorter than the millions and billions of years it takes stars to evolve perceptibly. How could we possibly have learned about events that take so long? Remember that the HR diagram contains information gathered from a great many stars. The collective nature of this information allows astronomers to deduce the evolution of individual stars. How this is done may be seen best by way of a simple analogy.

Suppose a visitor from outer space arrives on Earth and wishes to learn about us. For reasons beyond his control, he can stay for only one hour—far shorter than the average life span of humans. He lands his ship at a busy American shopping mall and, in the course of his one-hour stay, hurriedly videotapes the shoppers. After departure, he scrutinizes the tape. He notices that the majority of the people are about the same size, but some are quite small and a few are so tiny that they need to be carried or pushed around in carriages. Still others, although of normal size, walk bent over with the aid of canes. Being intelligent, our visitor soon realizes that he is observing an aging process. People are born very tiny; they grow up and spend most of their lives as active adults; and eventually they become old. Because the number of old people is relatively small, he concludes that in the end they die.

Astronomers are in the same position with regard to the stars as the visitor from outer space is with regard to people. In a relatively short span of time astronomers have observed hundreds of thousands of stars. They have measured their luminosities and surface temperatures, and they have taken detailed spectra of their light. By careful scrutiny of this information, and by relying on physical theory and computer modeling, they have deduced the following story about their evolution:

Stars are born within large interstellar clouds of gas and dust. They settle as young adults on the main sequence, where they spend most of their lives. They then evolve into red giants or supergiants, and eventually they die. They die either quietly or explosively. Those that die quietly end up as white dwarfs, growing dimmer and dimmer with time. Those that die explosively throw much of their matter back into space, leaving behind a collapsed object just a few kilometers across. The explosions are the supernovae mentioned in chapter 2 and the collapsed remnants are neutron stars or black holes.

Thus, stars do not occupy forever the same location in the HR diagram but follow evolutionary paths. The details of the paths depend on the stars' masses and compositions. In figure 3.14 such paths are shown for the Sun and for a star of 25 $M_\odot$, including the years they spend in their major evolutionary stages. The Sun will spend by far the greatest part of its adult life—10 billion years—on the main sequence, and it will die quietly as a white dwarf. In contrast, the 25 $M_\odot$ star will race through its entire evolution in just a few million years and end its life as a supernova. The remainder of this chapter is devoted to filling in details of the story of the origin and evolution of stars as observed today in the disk of our Galaxy.

THE BIRTH OF STARS

The birthplaces of stars in our and other spiral galaxies are the huge interstellar clouds of gas and dust that are mainly strung out along the spiral arms. Typically, these clouds range from a few thousand to several ten thousand solar masses and are roughly 10–50 light-years across, though a few are much larger. The most impressive one visible from Earth is the great cloud in the constellation Orion (figure 3.3). It is an enormous complex of gas and dust, measuring nearly 100 light-years across and filling most of the area outlined by the constellation. It contains at least 100,000 solar masses of gas and dust in various stages of compression. Much of the material is cold, dark, and impenetrable to optical and ultraviolet radiation. However, in a relatively small region on the side facing us and surrounding Orion's Sword, a group of very bright stars is heating the gas and dust to incandescence and creating the beautiful Orion nebula. The four brightest stars, known as the Trapezium, are not much older than a million years. Buried deep within the dark expanse of gas and dust are hundreds to thousands of compact, dense regions that astronomers call *clumps*. Some of these clumps are collapsing under their own gravity and 10,000 to a million years hence will have become stars. The brightest and most massive of the newly born stars (like the Trapezium stars) will race through their lives within a few million years and explode as supernovae, spewing freshly synthesized heavy elements back into space and sending shock waves through the cloud. Others will live for hundreds of millions to billions of years, traveling numerous times in nearly circular orbits around the center of the Galaxy.

Formation of Interstellar Clouds

Cloud complexes like the one in Orion form when powerful external forces act on interstellar gas and dust and compress them. The origins of these external forces are at present not well understood. It seems that they arise when interstellar gas and dust, in the course of their orbital motions about the galactic center, plow into a spiral arm. They probably also arise when the shock front from one or more nearby supernova explosions expands into the gas and dust. Both of these mechanisms of compression are rather complicated and involve a number of interlinked events.

While the gas and dust are still in the uncompressed, dispersed state, they are quite hot. Their average density amounts to about one atom per cubic centimeter and their temperature is maintained at roughly 10,000 K by steady bombardment from cosmic rays.* The high temperature creates a pressure large enough (see box 3.2) to resist successfully gravity's attempt at compression. This is analogous to the way the pressure of the air in an inflated tire or balloon offers resistance to compression, particularly on a hot day.

When interstellar gas and dust become compressed the situation changes drastically. The individual gas and dust particles start emitting infrared (IR) and

**Cosmic rays* are streams of atomic nuclei—mostly protons and nuclei of helium, plus a sprinkling of nuclei of heavier elements—that travel at nearly the speed of light through interstellar and intergalactic space. The Earth is continuously bombarded by them. They are believed to originate during supernova explosions and when matter falls onto a neutron star or into a black hole.

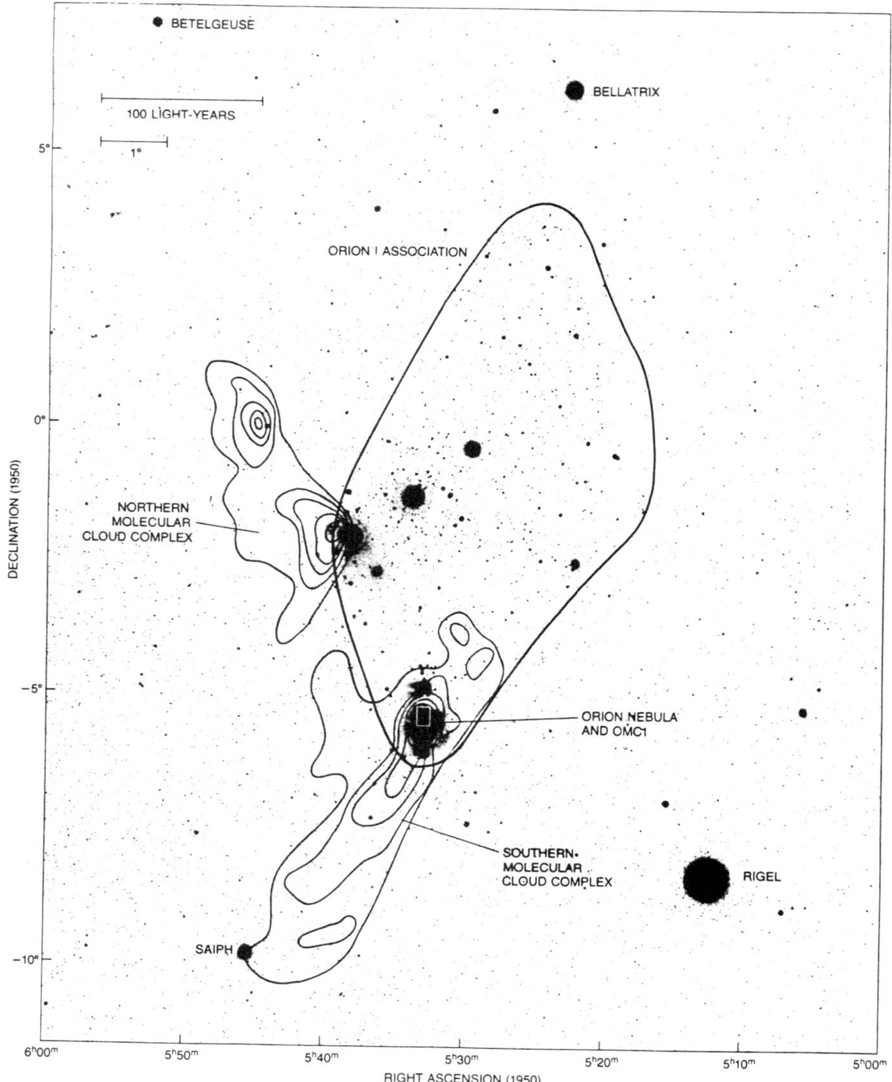

Figure 3.3. The constellation Orion shown in a negative print. In the area outlined by the constellation and at a distance of roughly 1300 LY from Earth lies an enormous cloud complex of interstellar molecules and dust grains. The contour lines superimposed on the photograph show the strength of the IR radiation emitted by the carbon monoxide of the cloud complex. The southern component of the cloud complex includes the Orion Nebula, outlined by the small rectangle (shown in greater detail in figure 2.26). The region also contains a very young open star cluster of about a thousand members, known as the Orion I Association (indicated by a heavy line).

radio waves. They do this because the compression squeezes them closer together, which makes them collide more often and, during each collision, some of their energy of motion is converted into IR and radio waves. The long wavelengths of this radiation allow it to escape from the cloud and to carry away

BOX 3.2.

The Concept of Pressure

IN physics, the concept of pressure refers to a force that is applied over an area. If the force (F) is applied at an angle to the area (A), only the component of the force perpendicular to the area ($F_\perp$) enters into the calculation of the pressure (P):

$$P = \frac{F_\perp}{A}$$

(see figure A). If the force is applied perpendicularly to the area, as in all of the examples below, then the full value of F contributes to the pressure because $F_\perp = F$.

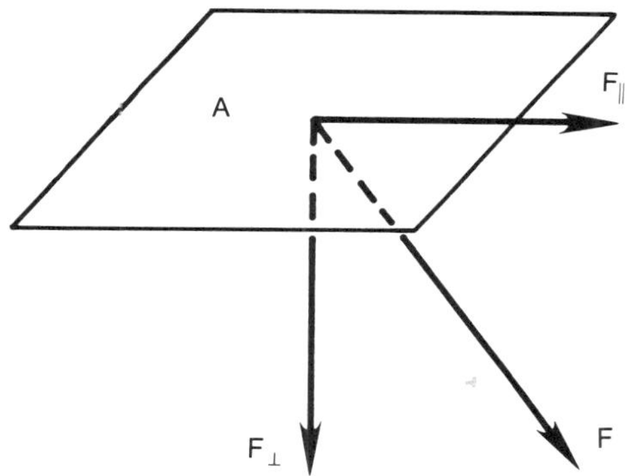

Figure A. Geometric decomposition of a force, F, acting at some angle over an area, A. The force may be thought of as consisting of two components: a component perpendicular to the area, $F_\perp$, and a component parallel to the area, $F_\parallel$.

A few examples will illustrate the concept of pressure. When we stand or walk on level ground, the pressure we create on the ground equals the force with which gravity pulls us downward divided by the surface area over which the force is distributed, in this case the surface area of the soles of the shoes in contact with the ground. Thus, a 160-pound man standing on both feet creates a pressure of 2.5 lb/in^2 on the ground, assuming that the surface of contact equals 64 square inches.

$$P = \frac{F_\perp}{A} = \frac{160 \text{ lb}}{64 \text{ in}^2} = 2.5 \text{ lb/in}^2.$$

If the man stands on just one leg, the pressure is doubled because the area over which the force is distributed is halved. If the man stands on skis, with a total surface area of, say, 350 in^2, the resulting pressure is slightly less than 0.5 lb/in^2. This pressure is low enough to allow him to walk through deep snow without sinking down very far. In contrast, a 120-pound woman wearing high-heeled shoes creates a pressure of nearly 500 lb/in^2 when she puts all of her weight on one heel (assuming the heel's surface area equals ¼ in^2), which is sufficient pressure to damage floors.

The last example illustrates that for a given force the pressure can be made very large by reducing the area over which the force is applied. People exploit this fact when they pierce an object with a needle or cut it with a knife. In these cases the area—the point of the needle or the cutting edge of the knife—is very small.

Pressure of Gases

The concept of pressure applies not only to solid objects pressing against an area, but also to gases. For example, the recommended air pressure in car tires today is 25–30 lb/in^2. And *the pressure of the Earth's atmosphere at sea level is approximately 14.7 lb/in^2 (= 1 atm*)*. This means that the column of air above every square inch of the Earth's sea level presses down with a force equal to 14.7 pounds. Interestingly, the atmosphere exerts this pressure not only on level surfaces, but on all surfaces no matter how they are oriented. For example, at sea level every square inch of our bodies is pressed upon by a force equal to 14.7 pounds. The reason is that air is a fluid that can flow in any direction.

*British units have been used in this discussion because few people in the United States are familiar with the metric units of pressure, dynes/cm^2 or Newtons/m^2, but most are aware of how many pounds per square inch of pressure are in a properly inflated car tire. In the rest of this box, and in the text, pressure is expressed in units of *atmospheres* (abbreviated atm). One atm is defined to equal the pressure of the Earth's atmosphere at sea level and a temperature of 15° C:

$$\begin{aligned} 1 \text{ atmosphere (atm)} &= 14.7 \text{ lb/in}^2 \\ &= 1.01 \times 10^6 \text{ dynes/cm}^2 \\ &= 1.01 \times 10^5 \text{ Newtons/m}^2 \end{aligned}$$

BOX 3.2 continued

The pressure of the Earth's atmosphere equals 14.7 atm only at sea level. At higher elevations, it diminishes because the greater the height, the less atmosphere there is above to press down. For example, on top of Mt. Mitchell, the highest point in the eastern United States with an elevation of 2037 m, the atmospheric pressure is 0.781 atm (11.5 lb/in^2). On top of Mt. McKinley, 6194 m, it is 0.453 atm (6.7 lb/in^2). On top of Mt. Everest, 8848 m, it is a mere 0.310 atm (4.6 lb/in^2). And at a height of 100 km it is 3.0×10^{-7} atm (4.4×10^{-6} lb/in^2).

The pressure of a gas exerts outwardly directed forces that expand the gas unless it is confined. In the cases of a balloon or car tire, the confinement is due to the enclosing rubber walls. The tension in the rubber walls creates an inward force per unit area that exactly balances and cancels the outwardly directed force of the pressure of the gas. If the walls are punctured, then the outwardly directed force becomes unopposed (assuming the puncture is large), the gas escapes, and the balloon or tire deflates.

In the case of the Earth's atmosphere, the confinement is due to gravity. To understand the meaning of this statement, consider the Earth's atmospheric layer between 2000 m and 2500 m, assuming that it is in equilibrium (that is, no winds are blowing). At 2000 m

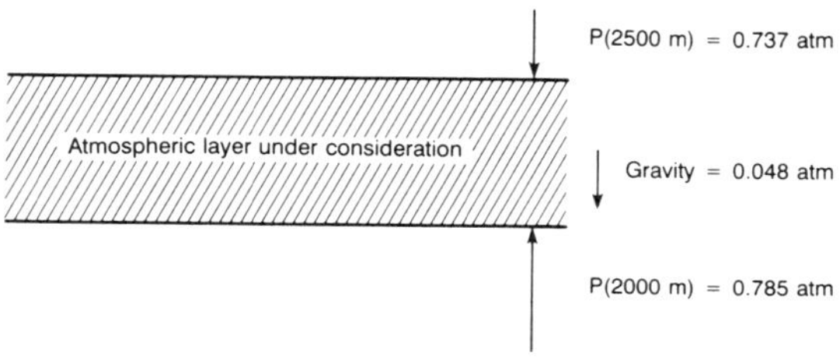

Figure B. Schematic illustration of the forces acting on each unit area of an atmospheric layer between 2000 m and 2500 m elevation: P(2000 m), P(2500 m), and gravity.

the atmospheric pressure is 0.785 atm and at 2500 m it is 0.737 atm, as shown in figure B. These numbers mean that the layer experiences a force per unit area of 0.785 atm from below and of 0.737 atm from above. The difference between these two pressures, namely 0.048 atm, represents a force per unit area that pushes on the layer in the upward direction (upward because the force from below is greater). However, the layer is not lifted off into space because gravity pulls down on it with an equal force per unit area. Thus, the upward force (due to pressure) and the downward force (due to gravity) cancel and the layer is moved neither upward nor downward.

There is another way to express these ideas. The difference in pressure between the bottom and top of the layer divided by its thickness is an estimate of the *pressure gradient* of the atmosphere between 2000 m and 2500 m.* This pressure gradient exerts an upward force on the layer. There also is a downward force, namely the force of the Earth's gravity that acts on the mass contained in the layer. The two forces are equal in magnitude, but oppositely directed. Hence, they cancel each other and the layer moves neither upward nor downward. We say that the layer is in *hydrostatic equilibrium*. The same considerations apply to the gaseous layers at all other heights in the Earth's atmosphere as well as to the layers in stellar interiors, provided that they are in hydrostatic equilibrium.

What happens when the gaseous layers of the Earth's atmosphere, or of stellar interiors, are not in hydrostatic equilibrium? The pressure gradients are then not balanced by opposing forces and, as a result, gas flows occur. In the case of our planet's atmosphere, we observe these gas flows as winds, storms, and hurricanes that, along with variations in temperature and precipitation, constitute our weather.

This raises another question: What brings about the unbalanced pressure gradients in our atmosphere? The chief cause is the heating of the atmosphere by the Sun and the variations of this heating from one location to another: more in the equatorial regions and above the continents than in the polar regions and above water, snow, and ice. Furthermore, there is a steady loss of heat from the atmosphere due

*The *pressure gradient is a measure of how rapidly the pressure changes with height*. In the case of the atmospheric layer in this example,

$$\frac{P(2500\text{ m}) - P(2000\text{ m})}{500\text{ m}}$$

is an estimate of the pressure gradient, where P stands for the pressure at a given height. The division by the thickness of the layer is necessary to cancel out the fact that if the layer were thicker (or narrower), there would be a proportionately greater (or smaller) pressure difference. The narrower the layer, the closer the estimate is to the true value of the pressure gradient.

BOX 3.2 continued

to the emission of infrared radiation, which also varies from one location to another. These differences in heating and cooling create differences in the pressure of the atmosphere from one location to another or, in meteorological terms, they create highs and lows. The highs and lows are responsible for the unbalanced pressure gradients in our atmosphere and, hence, for winds, storms, and hurricanes.

We may regard these atmospheric currents as nature's attempt to erase all pressure gradients except those that are balanced by gravity. However, unequal heating and cooling continually create new highs and lows. Therefore, equilibrium is never achieved globally in our planet's atmosphere and winds, storms, and hurricanes keep blowing.

Pressure, Temperature, and Volume

Now let us examine the relationship between the pressure of a gas and two other physical parameters—the temperature and volume of the gas. For gases such as air or those in the interior of the Sun and other similar stars, this relationship can be expressed by the *equation of state of ideal gases*:

$$P = \text{constant}\ \frac{T}{V},$$

where P, T, and V stand for pressure, temperature, and volume of the gas and the constant depends on the particular units used.

We all have some familiarity with the physics expressed by this equation. It states, for example, that if the temperature of a gas increases, then the pressure increases also (assuming for the time being that the volume remains unchanged). We experience this when driving on a hot day. The tires warm up and, as a result, the pressure of the air in the tires increases. Another example is the high pressure regions created by the heating of our planet's atmosphere. If the temperature of the gas decreases, then the pressure decreases also (again, assuming that the volume remains unchanged). This situation arises, for instance, when the temperature of interstellar gas clouds is lowered by the emission of IR and radio waves. The pressure is then lowered also, which aids in the compression of the gas and its collapse into stars.

The equation of state also tells how the pressure of a gas responds to a change in volume. If the volume of a gas decreases, the pressure increases (assuming that the temperature remains constant). This happens, for instance, when we pump up a bicycle tire with a hand pump. We compress the gas in the pump (that is, we decrease its volume). This increases the pressure, which, in turn, forces the gas through the valve into the tire. Likewise, if the volume of the gas increases, the pressure decreases (again, assuming that the temperature is kept constant).

In general, the temperature and volume of a gas may both change simultaneously. The equation of state still applies and tells the resulting pressure.

energy. This lowers the temperature and the pressure. The lowering of the pressure reduces the resistance to compression, and contraction continues. This, in turn, increases the rate of emission of IR and radio waves, and the whole process of cooling, lowering of pressure, and further contraction accelerates. The result is the formation of a large interstellar cloud with an average density of about 1000 atoms per cubic centimeter, a temperature of 100 K, and nearly total opaqueness (called *opacity* by astronomers) to all short-wavelength radiation. This large opacity makes the clouds appear so dark and ominous.

Interstellar Molecules

In addition to the steps just described, one other significant physical process occurs during the formation of interstellar clouds. The increasing rate with which atoms collide with each other and with the dust particles induces chemical reactions among them and they start combining into molecules.

Molecules are microscopic structures consisting of atoms held together by the electric force, as described in the Introduction to Part One and illustrated by a few examples in figures 3.4 and 3.5. The low temperature and the large opacity of the cloud assures that the molecules are not destroyed, as they otherwise would be, by collisions with each other or by bombardment from ultraviolet (UV) radiation and cosmic rays. The molecules formed most abundantly are those for which the atomic raw materials—hydrogen, carbon, nitrogen, oxygen, silicon, sulfur, etc.—are amply available. Most prominent among interstellar molecules are molecular hydrogen (H_2), the radicals hydroxyl (OH) and cyanogen (CN), water (H_2O), ammonia (NH_3), hydrogen sulfide (H_2S), carbon monoxide (CO), silicon oxide (SiO), and many organic molecules. The organic molecules found in interstellar space include methane (CH_4), the radical CH, hydrogen cyanide (HCN), formaldehyde (H_2CO), methanol (CH_3OH, popularly known as "wood alcohol"), formic acid (CHOOH, the acid that gives ant bites and nettles their sting), and many more. (A *radical* is a molecular fragment. For instance, the radical OH is a fragment of H_2O and the radical CH is a fragment of any of a number of organic molecules, such as CH_4 and CH_3OH. Organic molecules consist of carbon and hydrogen atoms, though other kinds of atoms may be present also.)

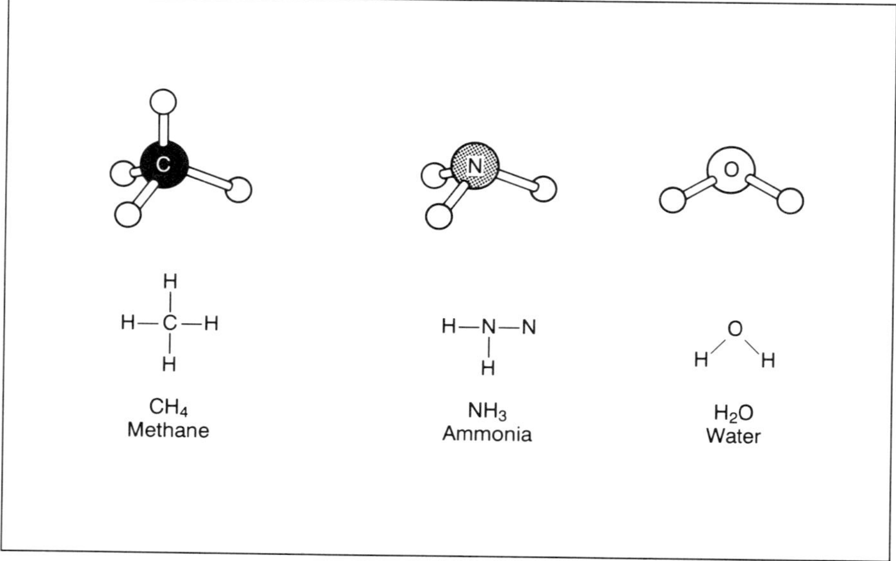

Figure 3.4. Three ways of representing the structures of the molecules methane, ammonia, and water. In the three-dimensional models (top) the spheres stand for the atoms carbon (C), nitrogen (N), oxygen (O), and hydrogen (H). The bars represent the chemical bonds that hold the atoms together in the molecular structures. (The bars are only simplistic representations of real bonds, which consist of pairs of electrons orbiting about pairs of atoms. For instance, methane has four such electron pairs bonding the central carbon atom with the four hydrogen atoms.) The two-dimensional models (middle) are simpler and more commonly used than the three-dimensional models. The symbolic notations (bottom) CH_4, NH_3, and H_2O are the simplest and the most commonly used representations.

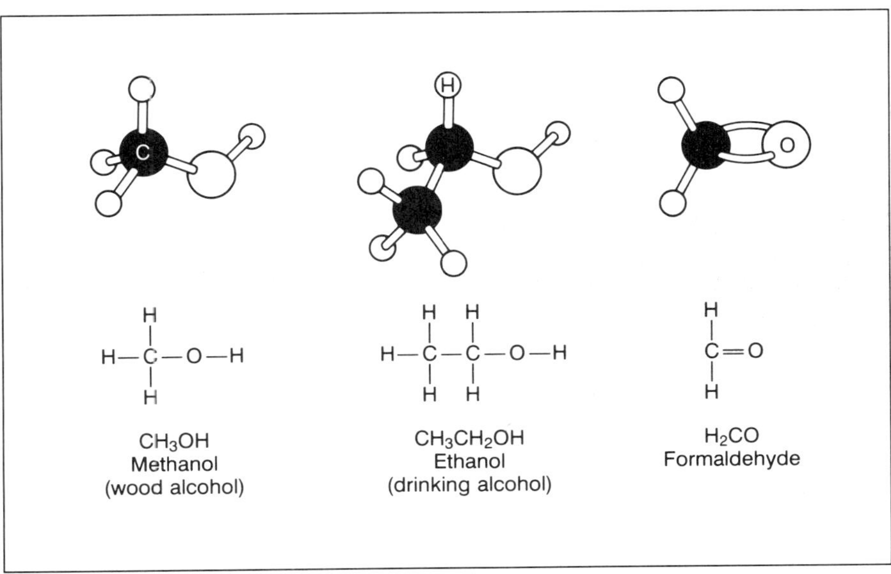

Figure 3.5. Three ways of representing the organic molecules methanol, ethanol, and formaldehyde, all three of which are produced both biologically on Earth and abiologically in interstellar clouds. (The spheres and bars have the same meaning as in figure 3.4.) In formaldehyde the carbon and oxygen atoms are joined by a double bond, indicating that two pairs of electrons orbit about these two atoms, holding them together.

The fact that the interstellar clouds harbor gigantic and very active chemical laboratories is a rather new discovery. Most of the IR and radio emissions that refrigerate the clouds originate from molecules, and every molecule radiates at characteristic wavelengths.* Up to the late 1960s, astronomers had been able to detect only the emission from the chemical radicals CH, CN, and OH. However, since then the technology of radio reception has become sufficiently sophisticated that the signals from many additional molecules can be distinguished. Interstellar ammonia and water were discovered in 1968, formaldehyde in 1969, and carbon monoxide and methanol in 1970. By the mid–1980s more than 60 molecular species had been found, most of which were organic molecules. Today the search continues for ever more complex interstellar molecules, in particular for such life-related ones as amino acids, sugars, and the nitrogenous bases.

New Stars

Consider again the compression of interstellar gas and dust into clouds. This compression is accompanied by much violence, rapid streaming of gas and dust, turbulent motions, and the creation of shock waves. As the compression

*When interstellar molecules collide with each other, with atoms, or with dust particles, some of their energy of motion goes into rotational and vibrational motions. Rotating and vibrating molecules possess energy due to these motions. They give up these energies and return to the quiet state (the lowest energy state) by emitting IR and radio waves of characteristic wavelengths.

Figure 3.6. The Trifid nebula. The name of this beautiful nebula means "cleft into three" and refers to the three dark, radial dust lanes that divide it into three sections. The luminous component of the nebula consists mostly of hydrogen gas that is heated to incandescence by a central bright, hot star. The star is a member of an open star cluster roughly 7 million years old.

continues, the force of gravity gradually becomes dominant and pulls the matter closer together. Finally, the cloud assumes a rather squashed and uneven shape, with some parts much more compressed than others. It all is reminiscent of an epoch billions of years earlier when the primordial clouds became unstable and broke up into protogalactic fragments—though those earlier events occurred on a much larger scale.

In most parts of the interstellar cloud, the compressive force of gravity sooner or later is balanced by the expansive forces of pressure and rotation. However, there usually exist small, clumpy regions in which the gas and dust have become much more compressed than elsewhere. In these regions, gravity remains dominant and contraction continues. The temperature drops to as low as 5 to 10 K, the pressure is reduced correspondingly, and the matter collapses into the very compact clumps mentioned earlier. These clumps are created along semiregular patterns throughout the cloud, first here, then there, and they can be seen as pitch-black splotches in photographs of luminous clouds, such as the Eta Carinae and Trifid nebulae (see figure 3.6). Sir William Herschel called them *Löcher im Himmel* ("holes in the heavens"), which quite accurately describes their visual appearance.

In the central regions of some of these clumps, the gravitational collapse eventually releases energy so rapidly that the emission of infrared and radio waves can no longer carry it all away. As a result, energy accumulates in the form of heat and the temperature begins to climb to 1000 K, 10,000 K, and still higher. The dust particles and molecules break up into individual atoms, the atoms become stripped of their electrons, and soon all the matter at the centers of the clumps is in fully ionized form. The resulting atomic nuclei and electrons fly around in a helter-skelter fashion and at very high speeds, continually colliding and bouncing off each other. Simultaneously, with the rise in temperature and the increase in the concentration of matter, the pressure rises. It is able to oppose gravity more and more effectively and gradually slows down the contraction.

The rise in temperature just described is greatest in the central regions of the clumps. This rise continues, but at a reduced rate because of the slowdown in contraction. When the central temperature of the clumps surpasses approximately 5 million degrees, the protons collide so violently with each other that they come within the range of each other's nuclear forces and begin to fuse into helium. Nuclear burning is now being induced and large amounts of energy are released. The centers of the clumps have contracted into self-luminous, gaseous spheres and a new generation of stars has been born.

A newly formed star is not immediately visible. For some time it remains shrouded in thick, opaque layers of uncondensed gas and dust. The first clue to the outside world that a star has actually been born comes in the form of strong infrared radiation. The radiation originates from the star as ordinary optical and UV light, but as it filters through the surrounding dust it becomes altered toward the infrared. Within a few thousand years, the most massive and luminous of the stars vaporize the cocoon-like envelopes of dust within which they were born and their intense radiation then escapes undiminished. It pushes the surrounding gas and dust of the cloud outward (by radiation pressure), heats the material to incandescence, and literally burns a hole into the cloud. That is how the bright interstellar nebulae, such as the one surrounding Orion's Sword or the Eta Carinae and Trifid nebulae, are created.

Depending on the conditions under which a clump forms—its mass, rate of rotation, disturbances from shock waves, exposure to radiation from nearby luminous stars, amount of dust compared to gas—it may give birth to a single star, a multiple star system (double, triple, or quadruple), or a star surrounded by planets. Altogether several hundred stars are usually formed from the clumps of a single interstellar cloud. (Actually, in most interstellar clouds of gas and dust, many more clumps than that start forming, but for reasons not well understood most do not complete the collapse into stars. Instead, they eventually become

Figure 3.7. The Pleiades, or Seven Sisters. These relatively young, hot stars are the brightest members of an open star cluster containing roughly 300 stars. Most of the stars occupy a nearly spherical volume 30 LY across and 400 LY from Earth, in the constellation Taurus.

dispersed.) For hundreds of millions to a few billions of years, the newly formed stars orbit around the center of the Galaxy as an open star cluster, such as the Pleiades (figure 3.7) and the double cluster h and Chi Persei (figure 1.2).

Presumably the Sun was also born as a member of an open star cluster, roughly 4.5 billion years ago. However, we do not know today which of our stellar neighbors are the Sun's brothers and sisters. These stars have long ago ceased to travel through space as an identifiable group. Open star clusters are loosely bound by gravity and within a dozen or so orbits about the Galaxy become dispersed beyond recognition by the tidal forces of the Galaxy.

THE MAIN SEQUENCE

A newly formed star adjusts its structure, by slight contractions and expansions, until at every point throughout its interior gravity is exactly balanced by the pressure gradient. At the same time it assumes a temperature profile, from its hot interior to its relatively cool surface, such that energy flows outward at precisely the rate at which it is generated in the core by the fusion of hydrogen (see box 3.3). Once a star has reached equilibrium in pressure and temperature, it is stable and changes its structure only very slowly.

BOX 3.3.
Heat Flow

HEAT is a form of energy. It is the energy a body possesses due to the kinetic motions of its molecules, atoms, or other constituent particles (such as the electrons and atomic nuclei in stellar interiors). The faster the kinetic motions, the greater is the heat energy of an object and the higher is its temperature, as discussed in box 1.2.

The heat energy of a body is not an intrinsic, unchanging quality of that body. It may increase or decrease. For example, the heat energy of a body is increased by the compression of that body or by nuclear and chemical reactions that release energy (see boxes 3.2 and 3.4). The heat energy of a body is decreased by expansion of that body or by nuclear and chemical reactions that absorb energy. Another way of changing the heat energy of a body is by the flow of heat from one place to another. This flow of heat is the topic discussed here.

It is a common experience of daily life that heat energy flows from places of high temperature to places of low temperature. The flow of heat generally occurs by three mechanisms: radiation, convection, and conduction. In *radiative heat flow*, heat energy is carried by electromagnetic radiation. If the radiating body is at a low temperature, such as a human body, the radiation lies mainly in the IR part of the spectrum. If the radiating body is very hot, such as the surface of the Sun or other stars, the radiation lies mainly in the visible or in a still shorter wavelength part of the spectrum. In *convective heat flow*, heat energy is carried by a fluid, such as a gas or a liquid. Finally, in *conductive heat flow*, heat energy is passed from molecule to molecule, atom to atom, or electron to electron of the conducting body. Conduction by electrons occurs mainly in metals, which contain large numbers of free electrons capable of picking up heat energy at one place and depositing it at another place.*

*One of the characteristic features of metals is that they contain electrons that are not firmly attached to atoms. These free electrons are able to move about with little resistance, which makes metals good conductors of heat. It also makes them good conductors of electricity. In contrast, the electrons of most other substances in our terrestrial environment—for example, wood, plastic, glass—are not free. They are firmly attached to atoms, making such substances poor conductors of heat and electricity.

We continually experience the flow of heat by one or more of the three mechanisms just described, as the following list illustrates.

Radiative Heat Flow

Radiant energy given off by the Sun and other stars.

Infrared radiation given off by heaters, lamps, fires, our own bodies.

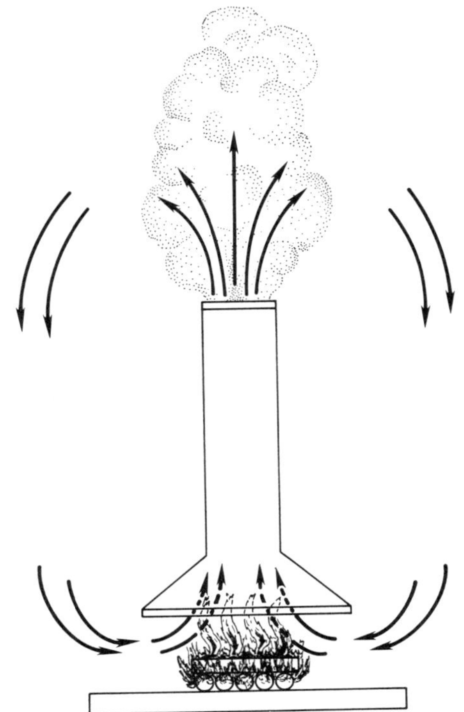

Figure A. The cycling of air through a fireplace is an example of convection. The convective cycle is driven by forces that result from the large temperature difference between the bottom and the top of the chimney. Air heated by the fire expands, becomes less dense, and rises by the force of buoyancy. Rising hot air coming out of the chimney cools by radiating off its heat and mixing with the surrounding cold air. As a consequence, it contracts, becomes denser, and sinks. The same mechanism drives convective currents in stellar layers whose temperature decreases rapidly with height.

BOX 3.3 continued

Convective Heat Flow

Hot air coming from a blower (such as a hair drier) or rising from a heater or a fire (see figure A).

Transport of heat or cold by ocean and air currents.

Transport of heat by the blood stream from the cores of our bodies to the tissues below the skin.

Conductive Heat Flow

The heating (or cooling) of the handle of a metal spoon in a hot (or iced) cup of tea.

The loss of heat through the walls of a stone house.

The flow of heat through the tissues of our bodies from the core to the skin (a process that takes place simultaneously with convective heat transport by the blood).

As stated above, heat always flows from hot to cold. Because temperature differences exist everywhere in our environment—in our bodies, in homes, car engines, the ground and atmosphere—the flow of heat is a common phenomenon.

The same phenomenon occurs in stars. They possess large temperature differences between their cores and their surfaces—from many millions of degrees to thousands of degrees. Hence, heat energy flows continuously from a star's core, where it is generated by nuclear reactions, to its surface, where it is radiated off into space.

As in our immediate environment, the flow of heat through a star's interior may occur by radiative, convective, and conductive transport. Radiative heat transport operates in all stars. It does so because in the hot interior the photons (that is, electromagnetic radiation) that are inevitably present are more energetic than the photons in the outer, cooler layers. Hence, the photons that diffuse from the interior outward carry more energy than those that diffuse inward. The result is a net radiative transport of heat energy toward the star's surface.

Convective and conductive heat transports are of significance in stars only under special circumstances. Conduction of heat is important in stars that are very compact, such as white dwarfs and the cores of red supergiants. Their matter contains free electrons, as in metals, that can travel over great distances and, thereby, transport heat energy. Convective heat transport occurs only if the temperature drops very rapidly from the interior to the surface. In that case, hot gases from the interior rise and deposit their heat in cooler layers further out. The cooled-off gases then sink, heat up, and rise again (see figure A).

In the solar interior, the heat flow from the center outward occurs by radiation. However, in approximately the outer one-third of the Sun, where the temperature drops very rapidly, convective transport takes over. The presence of convection in the Sun's surface layers can be seen in photographs of the Sun taken through special filters. They show a granular pattern, which is caused by convection (see figure 3.11). The centers of the granular cells are brighter and their boundaries are dimmer. In the bright areas, hot gases from the interior arrive at the surface. From there they stream horizontally toward the boundaries of the cells, radiating their heat content off into space and cooling. At the boundaries, the cooled-off and, hence, dimmed gases sink toward the interior and the cycle is repeated.

In many stars, yet another mechanism of heat transport occurs: heat loss by neutrinos. Some of the neutrinos are the by-product of certain nuclear reactions, such as those of hydrogen burning. Others come into existence spontaneously under extreme temperature and density conditions, such as are found in white dwarfs and the cores of red supergiants. All such neutrinos carry energy. Because neutrinos are very elusive particles and tend not to interact with matter, they carry their energy directly off into space. In the case of main sequence stars, this energy loss does not amount to very much compared to the energy loss by radiation and convection. However, in the case of white dwarfs and red supergiants, it is rather substantial and greatly speeds up the evolution of those stars.

The flow of heat energy from stellar interiors to the surfaces is crucial for the evolution of the stars. It is as crucial for them as it is, for example, for the proper running of car engines or the maintenance of our own bodies. Without the efficient expulsion of heat, car engines and human bodies would quickly heat up past tolerable levels and destroy themselves. Similarly, if stars were unable to get rid of the energies released by their nuclear reactions, they, too, would heat up, expand, and destroy themselves.

There is, however, a difference in the time scales involved over which human bodies and car engines on

BOX 3.3 continued

one hand and stars on the other rid themselves of their heat. Human bodies and car engines expel their heat energies rapidly (that is why we feel cold so quickly when we step outside without adequate clothing in cold weather). In contrast, stars require considerably more time to transport their energy to the surface. In the case of the Sun it takes roughly 10 million years. That means the energy of the radiation we receive today from the Sun was produced by nuclear reactions at the center of our star approximately 10 million years ago, long before our species walked on Earth.

In the HR diagram, the star is now approaching the main sequence from the right (that is, from the low-temperature region). During this approach it still experiences periodic instabilities in its surface layers, which evidence themselves as surface eruptions, accompanied by mass ejection and intermittent bursts in its light output. Stars passing through this phase are called *T Tauri stars*, after the prototype variable star T in the constellation Taurus.

The violent surface eruptions cease once the star reaches the main sequence. It stays in this extremely stable evolutionary phase as long as hydrogen remains in its core for fusion into helium and release of energy. How long that takes depends on the star's mass. A 50 $M_\odot$ uses up its hydrogen so swiftly that in just 1 million years its main sequence life is over. The Sun, with its smaller mass, burns its hydrogen more conservatively and stays in this evolutionary stage for roughly 10 billion years. A star less massive than the Sun lasts still longer. This large difference in the rate of nuclear fuel consumption and energy release is what makes massive stars so luminous and low-mass stars so relatively faint.

In table 3.2, the lifetimes and other physical characteristics of main sequence stars of different masses are compared. The masses listed range from 0.1 to 50 $M_\odot$, the approximate mass range of most stars observed in nature. Stars much more massive than 50 $M_\odot$ are not stable and rather rapidly shed their excess mass in the form of strong stellar winds. Objects below approximately 0.1 $M_\odot$ never reach sufficiently high temperatures for the ignition of hydrogen. In the absence of a nuclear energy source these objects are not self-luminous and, hence, are not stars. The planet Jupiter, with just under 0.001 $M_\odot$ is an example of such a low-mass object. Had Jupiter been born with one hundred times more mass, it would be a faint star (in the bottom right-hand corner of the main sequence) and, together with the Sun, would form a double star system.

The Sun

The variation of temperature and pressure of a main sequence star, from the center to the surface, depends only on its mass and composition. Figure 3.8 shows the distribution of these and a few other parameters for the Sun, which is approximately halfway through its main sequence life. The Sun's center has a temperature of 16 million degrees and its density is 160 times that of water. Furthermore, its central pressure equals about 300 billion earth atmospheres. These are large values and they testify to the severity of conditions in stellar interiors. Temperature, pressure, and density all decrease from the Sun's center

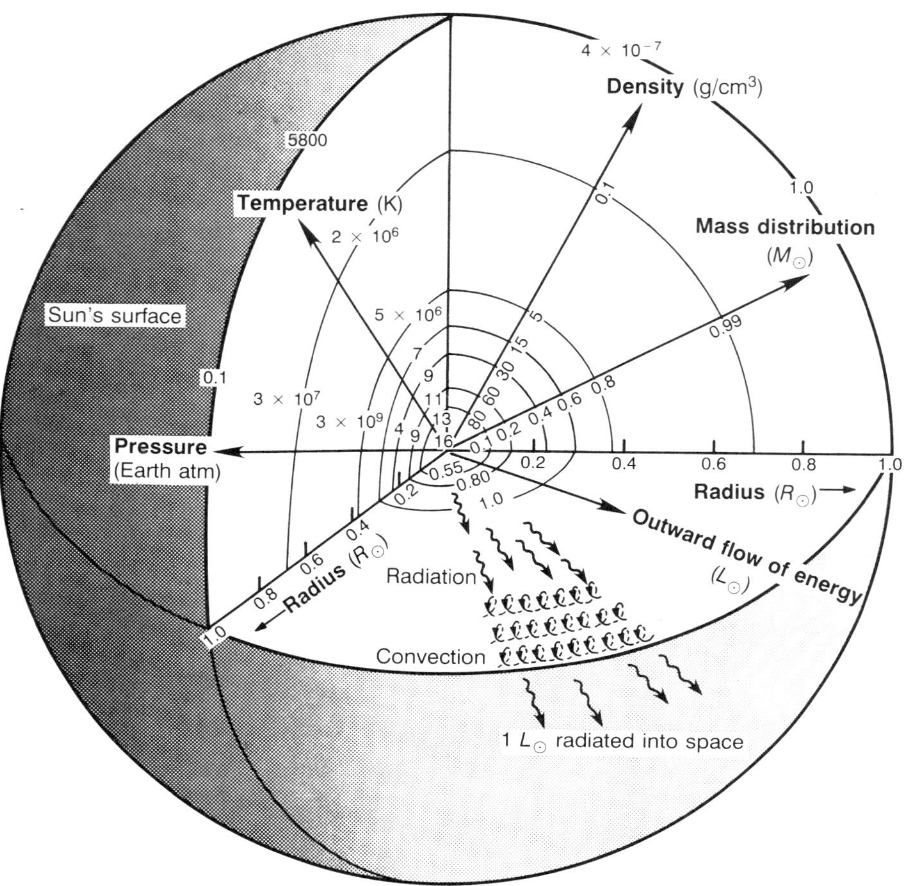

Figure 3.8. Diagram illustrating the internal structure of the Sun, as derived by computer calculations.

outward. The rate of this decrease is very gradual in the interior and very sharp in the outer layers. For instance, in the outer one percent of its total mass, the Sun's temperature drops from 2×10^6 K to 5800 K, the pressure decreases from 3×10^7 earth atmospheres to about one-tenth of an earth atmosphere, and its radius goes from 0.69 to 1.0 $R_\odot$. The enormous spatial extent of the Sun's outermost layers is responsible for its low surface density of a mere 4×10^{-7} g/cm^3, which is approximately 3000 times less than the sea level density of the Earth's atmosphere.

Note the strong concentration of nuclear energy generation toward the Sun's center, as indicated by the increase in the outward flow of energy in figure 3.8: the central 10% of the Sun's mass produces 55% of its luminosity; the next 10% produces another 25%; and beyond 60% in mass from the center, the luminosity increases no further. The outflowing energy is generated by nuclear reactions. The flow itself is due to the steep drop of temperature from 16 million degrees at the Sun's center to a mere 5800 K at the surface. Throughout roughly the inner two-thirds of the Sun this energy flux is carried by photons, which slowly filter outward much as light filters through dense fog or through a cloud here on Earth. In the outer third of the solar surface the energy is transported mainly by convective motions of the gases, much as heat energy is carried by gases up the chimney of a fireplace (see box 3.3). As the energy arrives at the solar surface, it is radiated off into space, making the Sun shine.

The surface of the Sun, as well as of other stars, is not well defined, for stars do not possess sharp outer edges. Gases of extremely low density expand for hundreds of thousands of kilometers into space. In the case of the Sun, these outermost layers contain the chromosphere and the corona. When astronomers speak of the surface of a star, they do not mean those tenuous outermost gases. They are referring to the layer that emits the light that we see. This surface layer

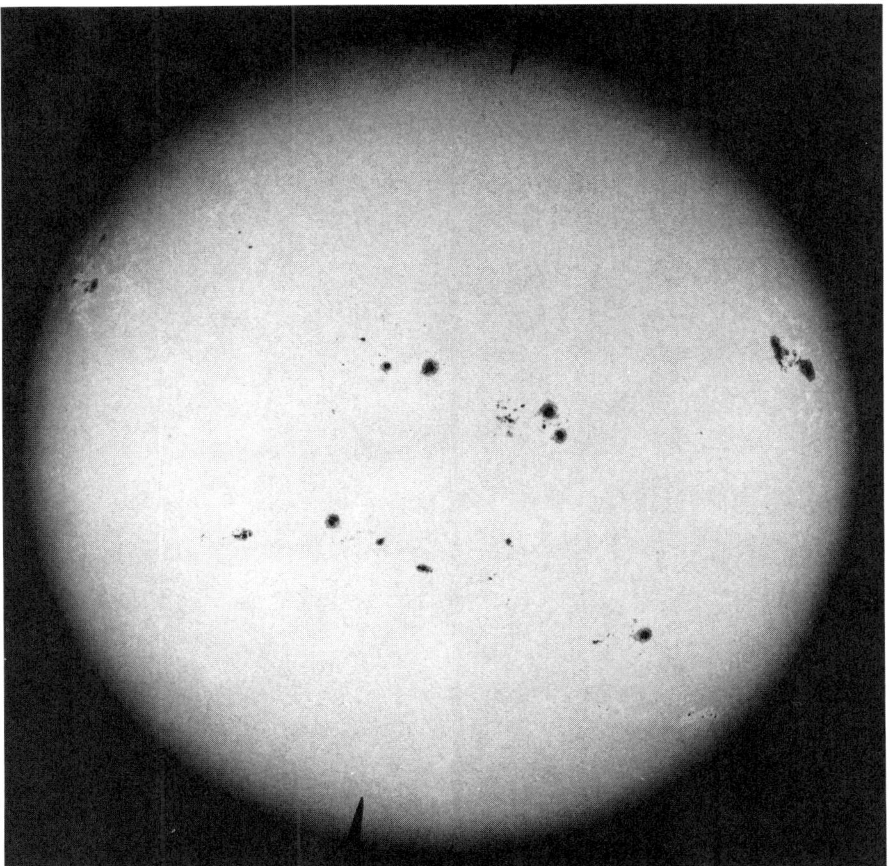

Figure 3.9. The Sun, photographed near sunspot maximum on July 13, 1937. Sunspots are the most conspicuous features of the Sun's photosphere. They are regions in which the gases are up to 1500 K cooler than those of the surroundings. The number of spots varies over an 11-year cycle (next maximum is expected to occur in 1991).

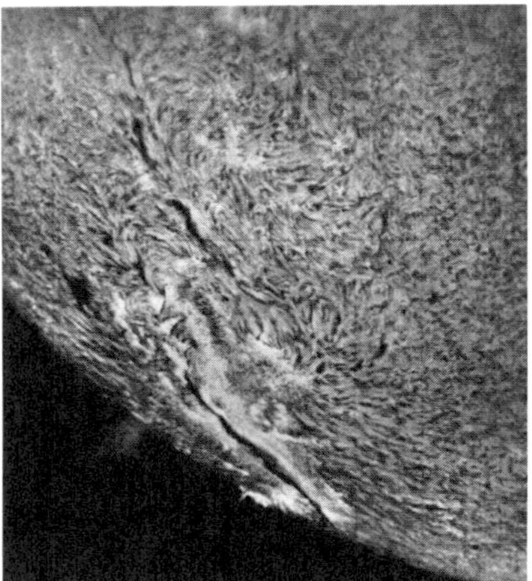

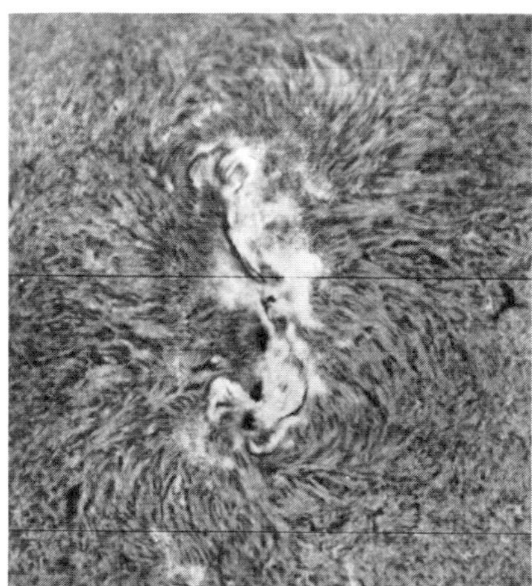

Figure 3.10. A group of *sunspots*, photographed in the light of the Hα line of hydrogen.

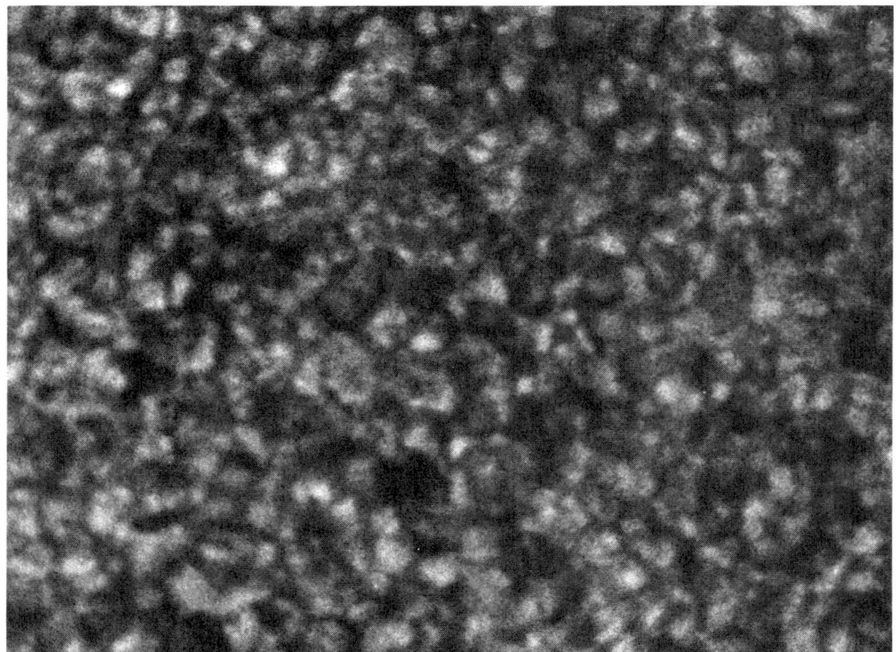

Figure 3.11. The granular pattern of the solar photosphere. The *solar photosphere* is the Sun's visible surface. Its granular appearance arises from the convective motions of gases by which energy is transported to the Sun's surface. Regions in which convective currents arrive at the surface from below are hotter and brighter, while regions in which convective currents move downward are cooler and darker.

Figure 3.12. The total eclipse of the Sun of June 8, 1918, photographed at Green River, Wyoming. As the Moon moves in front of the Sun, it blocks out the light from the solar photosphere and makes it possible to photograph the Sun's extended halo of very tenuous and hot gas (2 million K and higher), known as the *corona*.

is known as the *photosphere* (figures 3.9–3.11) and is spatially rather narrow and well defined. If we were to go below it, we would find the matter becoming opaque rapidly, like dense fog. None of the radiation present there can reach us directly because it is absorbed by the overlying gases. In contrast, the layers above the photosphere—the *chromosphere* and *corona* (figure 3.12) in the case of the Sun—are transparent and the radiation emitted from the photosphere readily passes through them and into space. Very little radiation is emitted from the chromosphere and corona, and without special instruments they cannot be seen.

Open Star Clusters

Next let us consider the HR diagram in figure 3.13. It is constructed from the data of several open star clusters. This diagram demonstrates the effect of stellar aging. When the clusters were born, all of their stars lay along the main sequence. Today, each of the graphs of the different clusters has its own distinct

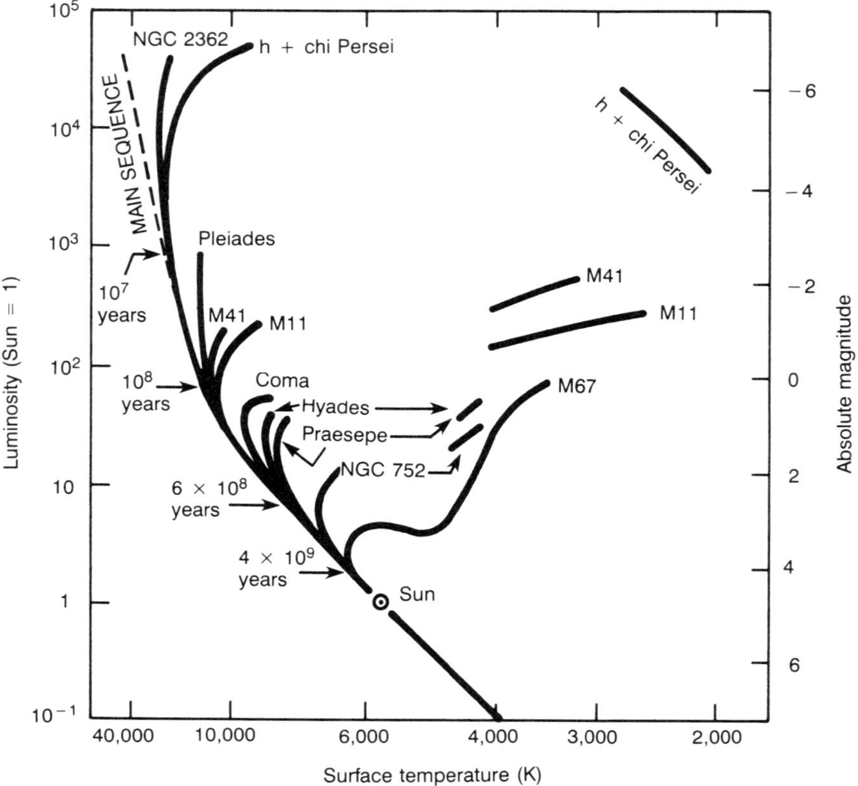

Figure 3.13. Hertzsprung–Russell diagram constructed from the data of ten open star clusters and showing the effect of stellar aging. The ages of all ten star clusters can be deduced from the age scale given to the left of the main sequence.

termination point on the main sequence. In the case of the two youngest clusters, NGC 2362 and h and Chi Persei, this point lies far to the left. However, even they no longer include any of the most massive stars that evolve through their main sequence lives in fewer than 10 million years. These stars have already burned all of the hydrogen in their cores and evolved into yellow and red supergiants or exploded as supernovae. In contrast, less massive stars, which stay on the main sequence longer, are still there; and thus we may conclude that these two clusters are roughly 10 million years old. In the case of the Hyades cluster, only stars with main sequence lives of 600 million years and longer are still on the main sequence; the others have evolved off it. Hence, this cluster has an age of about 600 million years. Cluster M 67 is one of the oldest known open star clusters in the Galaxy. Its main sequence terminates at the point corresponding to an age of 4 billion years, which is nearly as old as the Solar System. M 67 must be an extremely tightly bound system to have lasted so long, for most open star clusters disperse within 1 to 2 billion years.

Core Hydrogen Burning

We may regard the main sequence as a way station in the evolution of stars. The fusion of hydrogen into helium in the stars' cores temporarily arrests or slows down the evolutionary changes of the stars' surface features and makes them bunch up along the main sequence. The burning of helium into heavier elements affects the evolution of the stars similarly (see the next section). It produces surface conditions that concentrate those stars into the red giant and supergiant regions in the HR diagram. Likewise, the cooling of stars after they have exhausted their nuclear fuel creates yet another way station, namely the white dwarf region.

Why is the slowing down of the evolution along the main sequence not similar for all stars? Why do those with 10, 25, or 50 times more mass than the Sun complete their main sequence lives 1000, 2000, and 10,000 times faster than solar mass stars? The reason is that the core temperatures of the massive stars are higher than those of the less massive stars (see table 3.2) and that hydrogen burning is extremely temperature sensitive. It is much more rapid at high temperatures than at low temperatures. Consequently, the massive stars convert their hydrogen into helium much more rapidly than the smaller-mass stars. This means that the massive stars are not only brighter, but they also run out of hydrogen fuel in their cores much sooner and have shorter main sequence lifetimes than the less massive stars.

Interestingly, the mechanisms by which hydrogen is converted into helium tends to be different for the high- and low-mass stars. Main sequence stars more massive than about 2 $M_\odot$ burn their hydrogen mainly by the *carbon cycle*, in which carbon-12 (^{12}C) is used as a catalyst. Main sequence stars less massive than about 2 $M_\odot$ burn their hydrogen mainly by the *proton–proton chain*. The details of these two and other nuclear reaction mechanisms of importance in stellar evolution are discussed in box 3.4.

The Sun has been on the main sequence for roughly 4.5 billion years and will remain in this evolutionary phase for about another 5.5 billion years. It burns its hydrogen relatively slowly and mainly by the proton–proton chain. If the Sun

Table 3.2. Comparison of Physical Characteristics of Main Sequence Stars.

Mass ($M_\odot$)	Radius ($R_\odot$)	Luminosity ($L_\odot$)	Surface temperature (K)	Central temperature (10^6 K)	Central density (g/cm^3)	Main sequence lifetime (years)
0.1	0.15	0.001	2,500	5	200	10^{12}
1	1	1	5,800	16	160	10^{10}
5	3	500	20,000	25	25	10^8
10	5	10,000	25,000	30	10	10^7
25	10	100,000	35,000	40	5	5×10^6
50	20	500,000	45,000	45	3	10^6

used up its hydrogen more rapidly, say as rapidly as a 10 $M_\odot$ star, it would have passed through its entire life long ago and as a main sequence star would have been much brighter and hotter than it actually is. In that case it would be very difficult to imagine how life on Earth could have come into existence or, if it had, how it could have survived and evolved. This is particularly true of land life, which is directly exposed to the solar radiation.

During the Sun's main sequence life up to now, its surface temperature has remained nearly constant and its luminosity has increased by about 30%. Considering the length of time involved, this is a remarkably small change and attests to the stability of the Sun's structure.

POST–MAIN SEQUENCE EVOLUTION

While stars are on the main sequence, only their cores experience major evolutionary changes—the conversion of hydrogen into helium. The outer layers remain largely unaffected. This situation ends rather abruptly when all of the hydrogen in the central 10–15% of a star's mass has been converted into helium. The star's main sequence life comes to a close and, in the absence of central nuclear energy generation, the stellar interior does what it did earlier. It contracts and releases gravitational energy, bringing about a number of significant physical changes:

1. About half of the energy released by the contraction heats the star's interior and the temperature rises. As a consequence, the hydrogen just exterior to the helium core burns in a shell.
2. The rest of the energy released by contraction plus the energy generated by the newly created hydrogen shell source diffuse slowly outward and inflate the star's outermost layers. Thus, even though the core of the star contracts, its outer layers expand.
3. The expansion lowers the star's surface temperature and the color of its radiation shifts toward the red.
4. Finally, as the temperature at the center of the contracting core approaches 100 million degrees, helium ignites and fuses into carbon and oxygen. In the case of stars with masses similar to the Sun's, the ignition of helium occurs rather suddenly and violently, and it is known as the *helium flash*. In the case of more massive stars, the ignition of helium occurs more gradually. The star now has two nuclear energy sources: helium burning in the core and hydrogen burning in a thin, adjacent shell. Together, the energy released by these two sources brings the contraction of the core almost to a halt and the outer layers stop expanding. Equilibrium is reestablished between the rate of nuclear energy release in the interior and the rate of energy loss (mainly by radiation) from the star's surface.

Red Giants and Supergiants

The physical events just described are the beginning of the star's post–main sequence evolution. This evolution is characterized by tracks in the HR diagram that may be rather complex depending on the star's mass, as illustrated in figure 3.14 for a 1 $M_\odot$ and a 25 $M_\odot$ star. A fuller description is in the following pages.

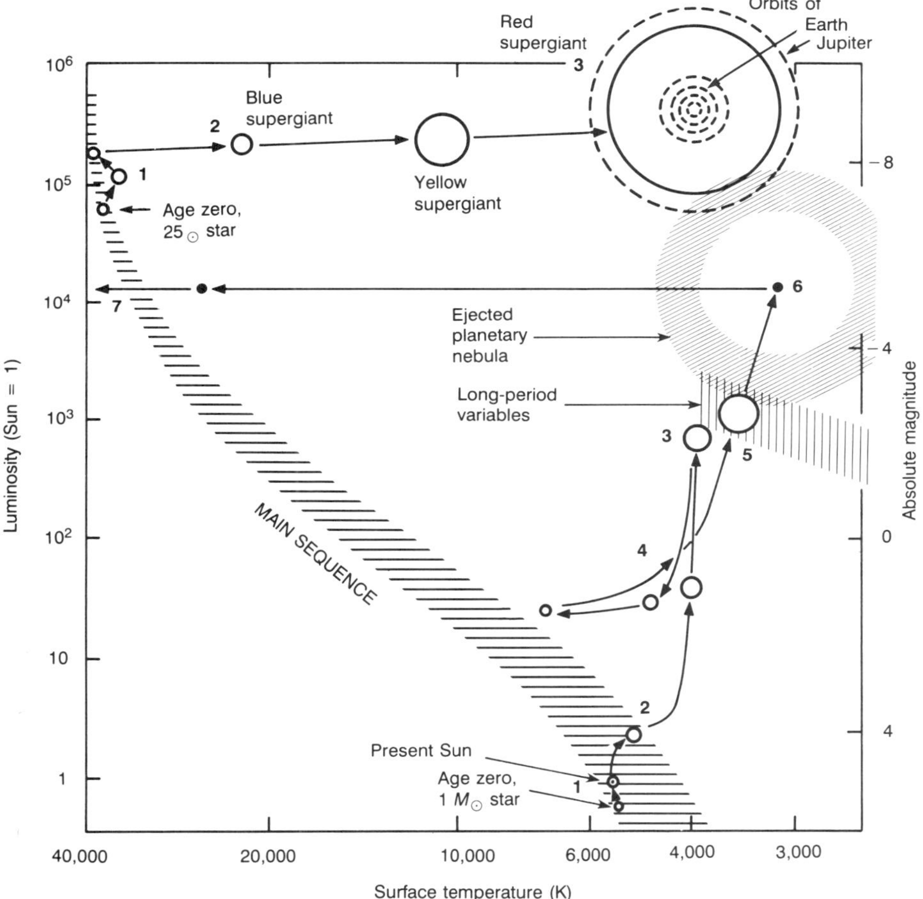

Figure 3.14. Evolutionary tracks of 1 $M_\odot$ (Sun) and 25 $M_\odot$ stars in the Hertz-sprung–Russell diagram. The major stages, indicated by numbers in the figure, are:

1 $M_\odot$ star

1. Main sequence (present Sun). H burns in core. Age 4.5 billion years.
2. End of main sequence. Core H burning ceases, H burning shifts to shell, He core contracts, H-rich envelope expands. Age 10 billion years.
3. Helium flash. He ignites explosively in core.
4. Red giant. He burns in core, H burns in surrounding shell.
5. Evolved red giant, long-period variable. Very compact C–O core, He and H burn in shells, highly extended H-rich envelope.

Figure 3.14 continued.

6. Ejection of H-rich envelope as a planetary nebula. Nuclear reactions are extinguished. Age 14 billion years.
7. Upon ejection of planetary nebula, stellar remnant moves to the left in the HR diagram and its surface temperature exceeds 100,000 K. Then the star cools and fades; within approximately 100 million years of the ejection of the planetary nebula it becomes a white dwarf (not shown in this figure).

25 $M_\odot$ star

1. End of main sequence. Core H burning ceases. Age 5 million years.
2. Beginning of supergiant phase. He, C, O, and Si burn in successive stages in the core and surrounding shells. Outer H-rich envelope expands. Star becomes a blue, yellow, and then a red supergiant.
3. Red supergiant, with enormously expanded H-rich envelope. Star explodes as a supernova about 500,000 years after the end of its main sequence of life.

Note that the stars are not drawn to scale. Structural details of the evolution of the 1 $M_\odot$ star are found in figure 3.16. (The track of the 25 $M_\odot$ star is taken from Weaver 1980.)

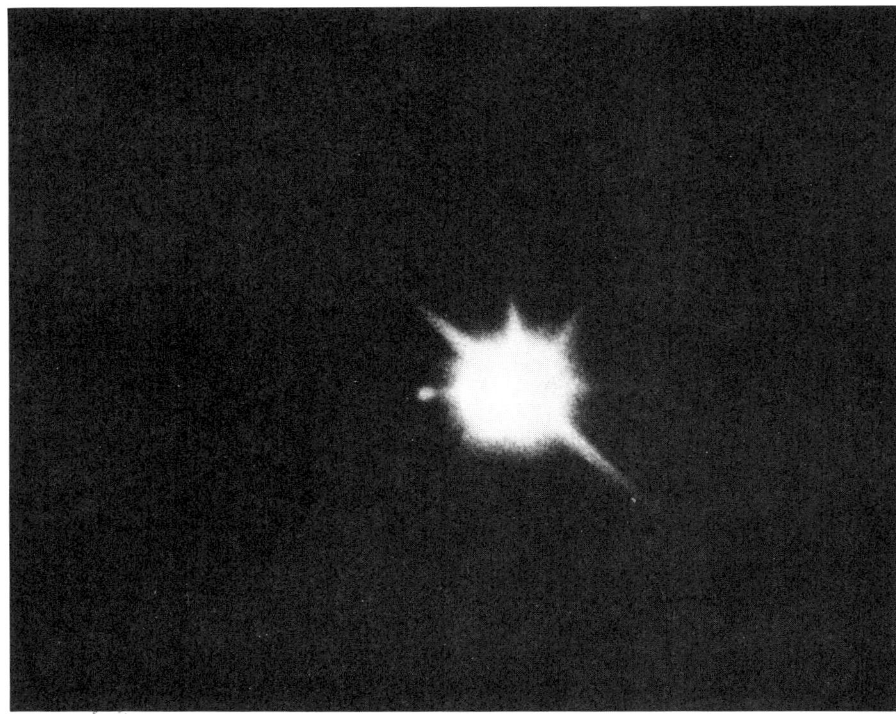

Figure 3.15. Sirius, also known as Alpha Canis Majoris or the Dog Star. This star is a double star system, consisting of Sirius A and B. Sirius A is by far the brighter of the two stars and usually the only one seen. It is, in fact, the brightest star in the sky and, along with its companion, constitutes the sixth nearest stellar system to Earth, at a distance of 8.7 LY. Sirius A is a main sequence star of about 2.3 $M_\odot$, while Sirius B is a white dwarf of about 1 $M_\odot$.

Stars below approximately 8 $M_\odot$ become *red giants*. During this evolutionary phase they usually shed enough mass, by a steady stellar wind and by sudden ejection, that eventually they become less massive than 1.4 $M_\odot$ and end their lives as *white dwarfs* (figure 3.15). Stars above approximately 8 $M_\odot$ become *supergiants*. They also lose mass, but generally not enough to become reduced below 1.4 $M_\odot$, which is the upper limit for white dwarfs (more about white dwarfs and their upper mass limit below). These more massive stars undergo complicated nuclear evolutions and end their lives as *supernovae*.

Long-period Variables and Planetary Nebulae

The final evolution of a star like the Sun, which is typical of stars that become white dwarfs, is illustrated in figure 3.16. When the helium in the core of such a star has been depleted by conversion into carbon and oxygen, the central parts of the star contract again. Gravitational energy is released and the temperature rises. Helium starts burning in a thin shell adjacent to the carbon–oxygen core, and hydrogen continues to burn (in a shell) further out. The

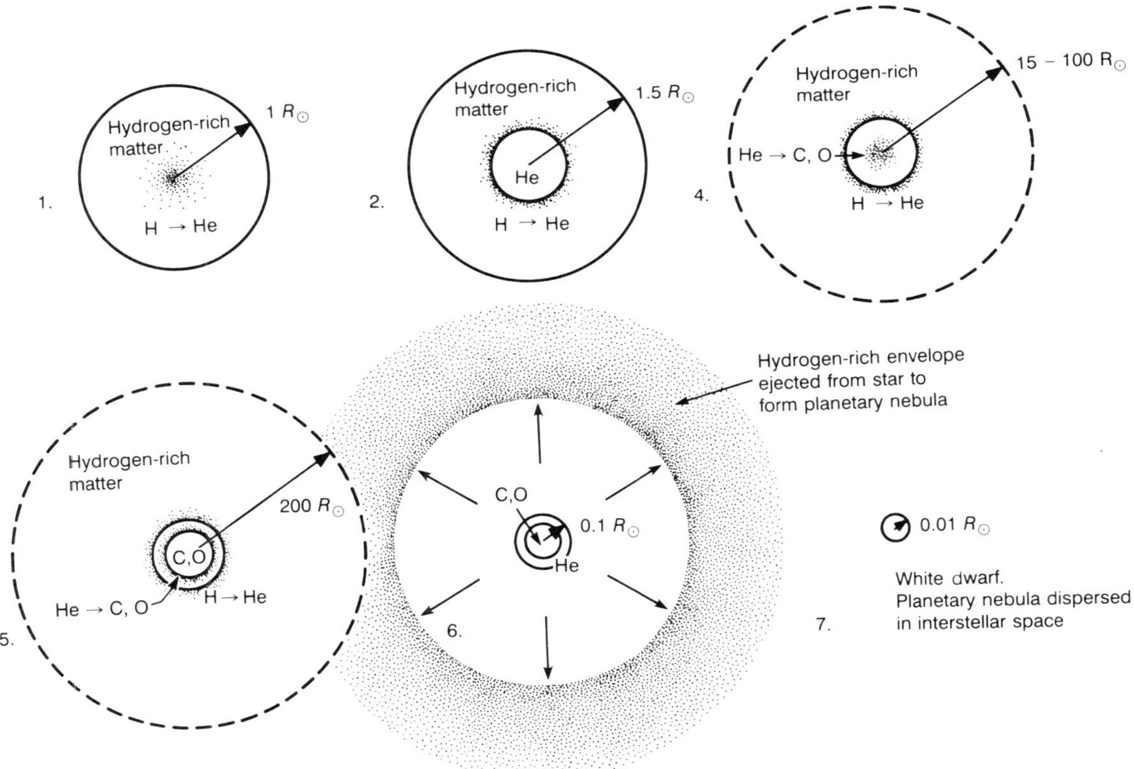

Figure 3.16. Schematic illustration of the evolution of a 1 $M_\odot$ star from the main sequence to the white dwarf stage. The numbers refer to the stages indicated and defined in figure 3.14. Dots represent regions of nuclear burning. Stellar radii are not drawn to scale; therefore, the outer surfaces of the two red giants are indicated as dashed circles.

hydrogen shell source produces helium, and the helium shell source below it converts the helium into carbon and oxygen. Thus, the two shell sources create an ever-growing carbon–oxygen core (note the core grows in mass, though in size it slowly contracts). The temperature never becomes high enough for the nuclear ignition of carbon or oxygen.

As the helium and hydrogen shell sources continue to burn, and the carbon–oxygen core continues to grow in mass and to contract, the rate of nuclear energy generation speeds up and the flow of radiation outward increases. Gradually the radiation flow becomes so intense that it creates instabilities in the remaining hydrogen-rich surface layers. Periodically the pressure gradient exceeds gravity and then falls below the value of gravity. As a result, the star begins to pulsate. Over a period of months to a year, the surface layers slowly expand and then contract again in repeating cycles. Such a pulsating star is known as a *long-period variable*. It is rather cool and luminous, as indicated by its location in the HR diagram. It is also characterized by a steady loss of mass from its surface in the form of an outflowing wind.*

In the course of time, the pulsations of a long-period variable grow stronger until eventually expansion is no longer followed by contraction. The pressure gradient created by the outward flowing radiation becomes sufficiently strong to push the entire hydrogen-rich envelope off into space. The ejection is helped along by the fact that the envelope is already quite extended and, hence, relatively weakly bound by gravity. Furthermore, the ejected matter cools to the point where the ions and electrons combine into atoms and molecules, which releases energy and contributes to the outward push. Such an ejected stellar envelope is known as a *planetary nebula*. This name was coined by early astronomers who, through their low-power telescopes, found that planetary nebulae resemble both interstellar gaseous nebulae and the Sun's outer planets (although planetary nebulae bear no relationship to planets, and they have no appreciable motions in the sky as do planets). Subsequent analysis of the spectra of their light confirmed that they are indeed gaseous. There are more than one thousand planetary nebulae in our part of the Milky Way. Approximately 4 billion years have elapsed between the end of the star's main sequence life and the ejection of the planetary nebula.

A planetary nebula typically contains about 0.2 $M_\odot$ of hydrogen-rich matter, with some enrichment of helium and other products of the parent star's nuclear history, and it expands at 10–20 km/sec (about 50,000 km/hr or 30,000 mi/hr). Initially it is brightly illuminated by the intense radiation from the central star as, for instance, in the case of the Ring nebula shown in figure 3.17. However, as the

*All stars with surface temperatures like the Sun's and cooler lose mass from a hot corona surrounding them. The corona consists of a very sparse expanse of gas and is created by energy released from convective and magnetic activities in the surface layers of these stars. The wind emanating from the Sun is extremely weak and carries away only about 0.0001 $M_\odot$ in the course of the Sun's entire main sequence lifetime. Still, the solar wind creates noticeable effects in the Earth's environment. It passes Earth with a speed of about 450 km/sec. By injecting protons, electrons, and other charged particles into the Earth's magnetic field, it causes auroras (northern and southern lights); and when it strikes the skin of a spacecraft, X rays are created which can harm the astronauts. In red giants and supergiants, which are much more luminous and have more extended coronae than the Sun, the stellar winds are much stronger and may carry away up to 1 $M_\odot$ in just a few million years.

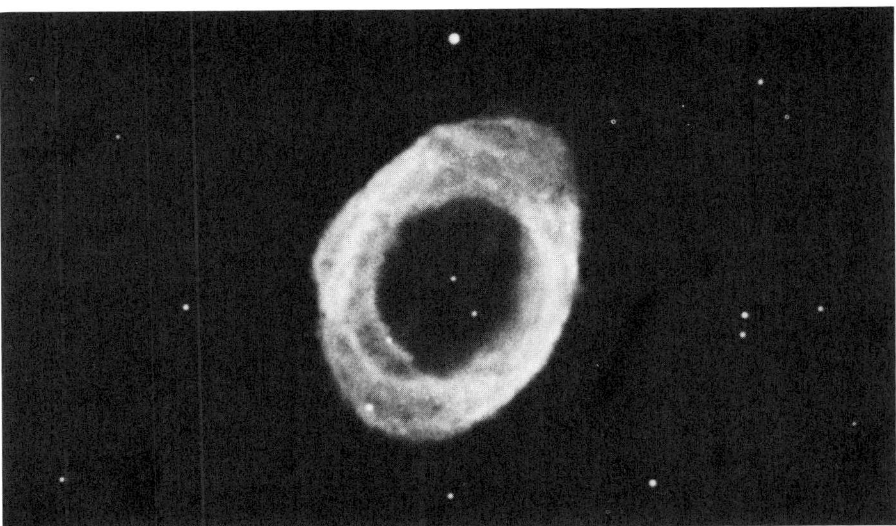

Figure 3.17. The Ring nebula, a planetary nebula in the constellation Lyra. This nebular shell of gas is expanding at roughly 20 km/sec radially away from the star seen near its center and from which it was ejected approximately 5000 years ago. The star is exceptionally compact and hot and, like all central stars of planetary nebulae, is evolving toward the white dwarf stage. The nebula lies at approximately 2000 LY from us and has a radius of about one-third of a LY.

star fades (see below), the nebula fades also, and within a few ten thousand years it can no longer be seen. The expansion continues until the ejected matter is completely dispersed and intermingled with the gas and dust of interstellar space.

The star that remains after the ejection is compact and luminous. As the exposed interior of a former red giant, it has a very high surface temperature, which initially may exceed 100,000 K. In the HR diagram it lies far to the left. Despite the high temperature at the surface, conditions in the interior are no longer favorable for nuclear reactions. The burning of hydrogen has become extinguished during the ejection, and the burning of helium does not last much longer either because of its proximity to the star's surface. For some time the star continues to shine very brightly by radiating away its stored-up heat energy. Gradually the star cools and grows fainter and fainter. At the same time it contracts, but only slightly because it is already quite compact. Little else happens to the star and over a period of about 100 million years it evolves into a white dwarf.

White Dwarfs and Black Dwarfs

Even though white dwarfs do not possess any nuclear energy sources and shine by radiating stored-up heat energy, they are rather interesting objects. They typically have a mass of about 0.5 to 1 $M_\odot$ and a size comparable to that of the Earth, making them very compact. Their average densities range from about 10^5 to 10^7 g/cm^3, which means that a volume the size of a fingertip contains from

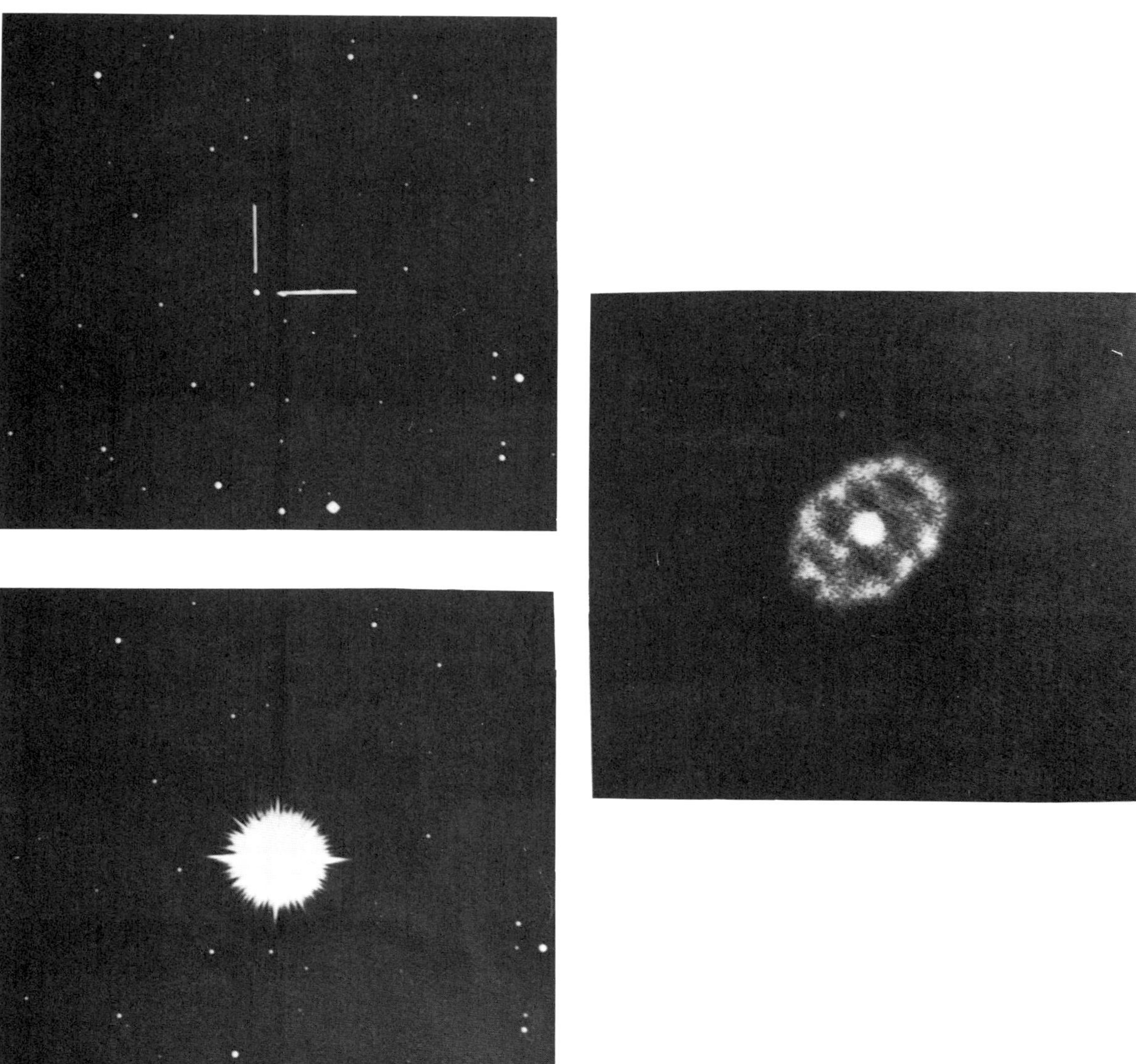

Figure 3.18. Three photographs of Nova Herculis 1934. The first two show the star before and during the outburst in 1934. The third was taken in 1972 and shows a magnified view of the star and the expanding ejecta.

0.1 to 10 tons of matter. At such high densities the electrons (which are part of the matter along with nuclei of helium, carbon, oxygen, and a few other heavier elements) touch each other and cannot be compressed further. Hence, gravity is no longer able to contract such stars. In the course of billions of years the white dwarfs cool and fade to a point where they can no longer be seen, even through telescopes. They have then become *black dwarfs*—dead cinders of once living stars.

A white dwarf is incompressible, however, only if its mass is less than approximately 1.4 $M_\odot$, the *Chandrasekhar mass limit* (named after Subrahmanyan Chandrasekhar, an astrophysicist from India who since 1937 has been at the University of Chicago and who in 1984, along with William A. Fowler of the California Institute of Technology, received the Nobel Prize in Physics for the fundamental contributions he has made to our understanding of stellar evolution). If the mass of the white dwarf is greater than 1.4 $M_\odot$, gravity is able to overwhelm the pressure and collapse the star further, which leads to a supernova explosion (see discussion of the evolution of a 25 $M_\odot$ star, next section).

Novae

Occasionally exceptions arise to the rule that white dwarfs merely cool and fade. If a white dwarf has a close stellar companion that loses hydrogen-rich matter, the white dwarf may become a *nova* meaning "new star" (figure 3.18). A nova outburst is illustrated in figure 3.19 in four stages:

1. A white dwarf (carbon–oxygen core, surrounded by a helium layer) and a red star orbit each other in close proximity. The red star has an extended hydrogen-rich envelope, which keeps expanding.
2. Much of the hydrogen-rich matter the red star loses is attracted gravitationally by the white dwarf. The matter spirals through an accretion disk and falls

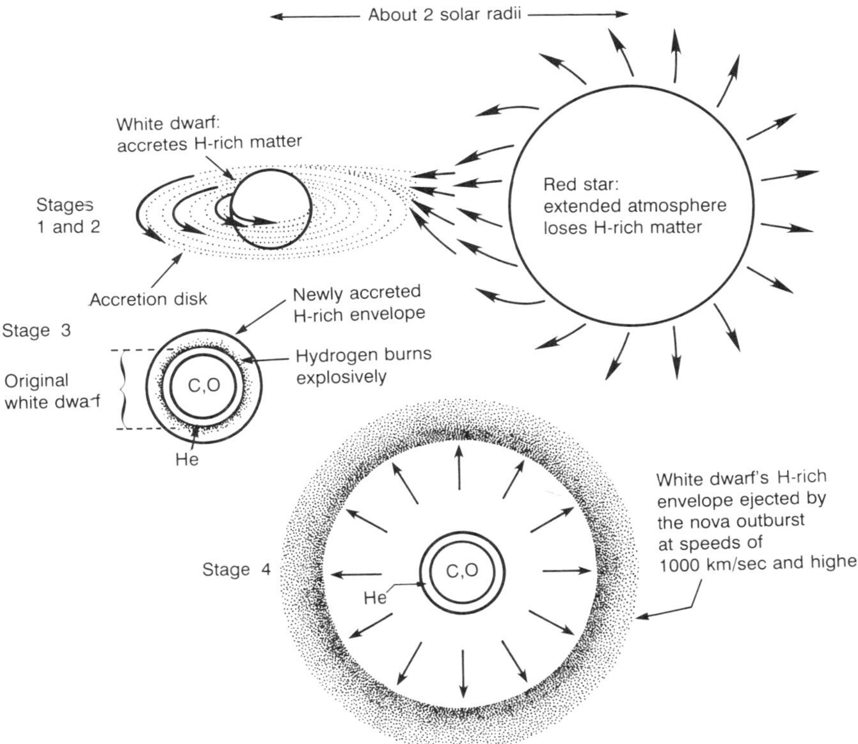

Figure 3.19. Schematic illustration of a nova outburst.

onto the white dwarf. A hydrogen-rich envelope builds up on the white dwarf.

3. When the envelope has grown to approximately $10^{-4} M_{\odot}$, the temperature and density of its base become severe enough for the fusion of hydrogen into helium. Because of the extreme densities existing in the white dwarf, the nuclear burning occurs explosively and a great deal of energy is released in a short time.
4. The nuclear energy released is sufficient to eject most of the accreted matter at speeds of 1000 km/sec and higher (3.6 million km/hr and higher) and to cause the star's brightness to increase suddenly by a factor of several million. This sudden brightening makes the star visible to great distances, which is why such an explosion is called a nova.

Nova outbursts are repeated periodically as long as the companion star loses hydrogen-rich matter and the white dwarf accretes it. Once this mass exchange ceases, the nova outbursts come to an end and the white dwarf resumes its evolution of cooling and fading. (Note that novae are the result of explosions in a white dwarf's surface layers. Supernovae are much more energetic and involve the entire star.)

FINAL EVOLUTION OF MASSIVE STARS

The more massive stars do not end their lives quietly as white dwarfs nor does their nuclear history terminate with the burning of helium. Considerable excitement awaits them during their final years, culminating in a superenergetic and catastrophic explosion that destroys the normal structure of the star and expels much of its matter into interstellar space at very high speeds.

Consider a computer simulation of the final evolution of a $25 M_{\odot}$ star to see how these events come about. When the star has completed its main sequence life, it starts burning helium in the core and hydrogen in a surrounding shell. Up to now the evolution has been much like that of the less massive star discussed in the previous section, except that the burning of hydrogen in the $25 M_{\odot}$ star took place mainly by the carbon cycle and the luminosity was higher, as shown by the star's evolutionary track in the HR diagram of figure 3.14. Upon depletion of helium in the core, gravitational contraction releases energy once again, raises the temperature, and greatly expands the star's outer layers, making it a supergiant.

Heavy Element Synthesis

Now the evolution begins to deviate sharply from that of the less massive star. Helium starts burning in a shell next to the carbon–oxygen core, hydrogen continues to burn in a shell further out, and soon carbon ignites at the center. The synthesis of the heavy elements has now begun in earnest. When the carbon in the core becomes exhausted, contraction raises the temperature to the point where oxygen burns in the core and carbon in an adjacent shell. Simultaneously, helium and hydrogen continue to burn in shells further out.

BOX 3.4.

Nuclear Reactions: The Energy Source of Stars

NUCLEAR reactions convert atomic nuclei of one kind into nuclei of another kind. The forces participating in these conversions are the nuclear and the weak forces, with the latter coming into play only when neutrinos are present.

Nuclear reactions are to be distinguished from chemical reactions. Chemical reactions never affect the nuclei of atoms. They affect the electrons surrounding the nuclei and either create or break the chemical bonds that hold atoms together in molecules, as discussed further in the Introduction to part 2. The only force participating in chemical reactions is the electromagnetic force. Examples of chemical reactions are the formation of molecules in interstellar space, the burning of coal and other fossil fuel, and the digestion of food. Chemical reactions liberate considerably less energy than most nuclear reactions. For example, if the Sun derived its energy from the burning of coal, it would shine at its present rate for only about 5000 years (assuming it were made entirely of coal and burned all of it). In reality, the Sun will shine for well over 10 billion years by fusion reactions.

Nuclear reactions may be divided into two major classes: fusion and fission reactions. In *fusion reactions*, nuclei of lighter elements are fused into nuclei of heavier elements. *Fission reactions* are the reverse of fusion reactions. In them, nuclei of heavy elements are split into nuclei of lighter elements.

Stars obtain their energy by fusion reactions. These reactions liberate enormous quantities of energy that comes from a slight lowering in mass of the nuclei undergoing fusion. For example, in the fusion of hydrogen into helium-4, the mass of the initial hydrogen is lowered by roughly 0.7% compared with the mass of the resulting helium-4. If we call the reduction in mass Δmass (read "delta mass"; the symbol Δ is the letter D in the Greek alphabet and stands for "difference"), the energy liberated is given by Einstein's equation

$$\text{energy} = \Delta\text{mass} \times c^2,$$

where c stands for the speed of light. The amount of mass that is converted into energy in fusion reactions is very little (a mere 0.7% in this example), but because of the factor c^2 in Einstein's equation, the energy liberated is very large (note that $c^2 = 9 \times 10^{16}$ m^2/sec^2). To appreciate just how much energy is released, note that the fusion of 1 kg of hydrogen into helium-4 yields as much energy as the chemical burning of 25,000 tons of coal.

Fusion reactions that build up nuclei of elements beyond iron do not liberate energy; they require energy.* Hence, in general the elements beyond iron are formed only during such violent events as supernova explosions, which are able to provide the required energy. The energy that thereby becomes locked up in the elements beyond iron may be released by fission reactions as, for example, by the fission of uranium into simpler products (such as lanthanum, barium, krypton, and bromine) in atomic power generators.

Fusion of ^1H into ^4He

By far the most energetic of all the fusion reactions occurring in stars is the conversion of hydrogen into helium. There are two major routes by which this conversion may proceed. In stars of about 2 $M_\odot$ and less, including the Sun, whose core temperatures are below 20 million K, it proceeds mainly by the *proton–proton chain*. In stars more massive than about 2 $M_\odot$, whose core temperatures are above 20 million K, it proceeds mainly by the *carbon cycle*:

The Proton–Proton Chain
(5 million K and above)

$$p + p \rightarrow {}^2H + e^+ + \nu$$
$$p + {}^2H \rightarrow {}^3He + \gamma$$
$${}^3He + {}^3He \rightarrow {}^4He + 2p + \gamma + \text{energy}$$

(*Note*: The first two reactions must occur twice in order to yield the two ^3He nuclei required by the third reaction.)

The Carbon Cycle
(20 million K and above)

$${}^{12}C + p \rightarrow {}^{13}N + \gamma$$
$${}^{13}N \rightarrow {}^{13}C + e^+ + \nu$$
$${}^{13}C + p \rightarrow {}^{14}N + \gamma$$
$${}^{14}N + p \rightarrow {}^{15}O + \gamma$$
$${}^{15}O \rightarrow {}^{15}N + e^+ + \nu$$
$${}^{15}N + p \rightarrow {}^{12}C + {}^4He + \text{energy}$$

*The Periodic Table of the Elements, page 23, shows which elements are lighter and which are heavier than iron.

BOX 3.4 continued

The net result of the reactions of both the p–p chain and the carbon cycle is the conversion of nuclei of hydrogen into nuclei of helium-4 along with the release of large amounts of energy: 4 p → ^4He + energy.

The positrons (e^+) produced by the reactions eventually collide with electrons (e^-) and are annihilated, giving off additional energy. The neutrinos escape from the star and carry away some energy. The photons become rapidly absorbed by the stellar matter and deposit whatever energy they carry as heat.

The unusual feature of the carbon cycle is the way nuclei of ^{12}C participate in it. In the course of the cycle, these nuclei change into nuclei of ^{13}N, ^{13}C, ^{14}N, ^{15}O, ^{15}N, and, in the final reaction step, back into ^{12}C. Thus, nuclei of ^{12}C act as catalysts.* They cycle through the reactions, temporarily changing into nuclei of other elements and facilitating the reactions.

During the late stages of a massive star's evolution, when the carbon cycle (which occurs then in a shell beyond the star's core; see text) is interrupted by a supernova explosion, some of the nuclei are in the form ^{14}N and, along with the rest of the envelope material, are ejected into space. This constitutes the main source of nitrogen in the Universe, an element that is very abundant in our atmosphere and is essential for all earth-bound life.

Temperature dependence. The next question is, "Why do the p–p chain and the carbon cycle, as well as other fusion reactions, take place only at the high temperatures found in stellar interiors?" The answer is that the nuclear force, which plays the key role in all fusion reactions, has an extremely short range, namely about 10^{-13} cm (see tables I.1 and I.2). Unless the collisions among the nuclei bring them within that distance of each other, they will not feel each other's nuclear forces and nuclear reactions will not occur. It is not easy, however, to get nuclei to within 10^{-13} cm of each other because of the repulsive force exerted by their positive electric charges. This force increases the closer two nuclei come to each other. It also increases the more charge the nuclei carry. Only at temperatures of several million K and higher do nuclei fly around sufficiently rapidly and collide sufficiently violently to overcome their mutual repulsions and come within the range of each other's nuclear forces.

Because protons carry only one positive electric charge, the minimum temperature required for the p–p chain is the relatively low temperature of about 5 million K. In the case of the carbon cycle, in which protons react with nuclei that carry six or seven positive charges, the temperature must be at least 20 million K before the reactions occur at appreciable rates. And for most other fusion reactions taking place in stars, some of which are discussed below, the temperatures required are higher still.

Except for scale and the forces involved, the requirement of high temperatures to induce nuclear reactions is analogous to the requirement of a minimum temperature for burning wood. At room temperature, wood is quite stable, just as nuclei are quite stable at sufficiently low temperatures. However, if the temperature rises above approximately 500° C, then the molecules of wood oscillate and knock into each other so violently that chemical reactions are induced, unlocking the molecules' stored-up energies and creating the phenomenon we call fire.

Reaction rates. There is one final point to consider regarding the p–p chain and the carbon cycle, namely the rates at which these reactions take place. The rapidity with which fusion reactions occur depends in part on the violence and frequency with which the nuclei collide with each other. The higher the temperature, the more violent and frequent are the collisions and, hence, the more rapid is the nuclear burning.

Temperature is, however, not the only factor influencing the reaction rates. Another factor is the probability that a nuclear reaction occurs once two nuclei have come within the range of each other's nuclear forces. It turns out that the presence of the weak force tends to reduce the reaction probability considerably. This effect is particularly pronounced in the first reaction of the p–p chain (the emission of a neutrino

**Catalysts* are particles that participate in nuclear or chemical reactions and speed them up without themselves undergoing any permanent changes. The catalysts of nuclear reactions are aways atomic nuclei as, for example, the nuclei of ^{12}C in the carbon cycle. The catalysts of chemical reactions are usually molecules or chemical compounds. As will be discussed in Part Two of this book, all of life's chemical processes depend on the participation of catalysts, such as the enzymes.

BOX 3.4 continued

indicates the presence of the weak force). This reaction is a kind of bottleneck in the p–p chain. Protons must collide many times before two of them react and fuse into a ^{2}H nucleus. Once this has happened, the rest of the reaction sequence to ^{4}He goes relatively quickly. In contrast to the p–p chain, the carbon cycle is not much slowed down by the weak force, even though this force participates in the second and fifth reaction steps.

Fusion of ^{4}He into ^{12}C and ^{16}O

The fusion of hydrogen into helium in stellar evolution is followed by the fusion of helium into carbon and oxygen. The fusion into carbon occurs by the *triple-alpha reaction*. This reaction is so called because in it three nuclei of ^{4}He—which are also known as *alpha particles*—collide and fuse into a nucleus of ^{12}C:

$$^{4}He + {}^{4}He + {}^{4}He \rightarrow {}^{12}C + \text{energy}$$

(100 million K and above)

The fusion into oxygen occurs by the addition of alpha particles to nuclei of ^{12}C:

$$^{12}C + {}^{4}He \rightarrow {}^{16}O + \text{energy}$$

(100 million K and above)

Because this reaction requires the presence of nuclei of ^{12}C, it takes place only after the triple-alpha reactions have built up a supply of carbon nuclei.

The production of carbon and oxygen by these reactions requires much higher temperatures than the earlier fusion of hydrogen into helium. The reason is simple. Each alpha particle carries two positive electric charges and the nuclei of ^{12}C carry six charges. Furthermore, in the triple-alpha reactions, three alpha particles must be brought almost simultaneously within the range of each other's nuclear forces. Consequently, the electric forces of repulsion that need to be overcome are considerably greater than those in the reactions of the p–p chain and the carbon cycle, and carbon and oxygen are manufactured only at temperatures of 100 million K or higher. That means that these reactions do not occur while stars are on the main sequence. Only after their cores have contracted further and their central temperatures have risen to 100 million K or higher, do those reactions take place.

Fusion of ^{12}C and ^{16}O into Heavier Elements

The next major phase of nuclear burning consists of the fusion of ^{12}C and ^{16}O into still heavier nuclei.

Carbon Fusion
(600 million K and above)

$$\begin{aligned} {}^{12}C + {}^{12}C &\rightarrow {}^{24}Mg + \gamma \\ &\rightarrow {}^{23}Na + p \\ &\rightarrow {}^{20}Ne + {}^{4}He \\ &\qquad + \text{energy} \end{aligned}$$

Oxygen Fusion
(1 billion K and above)

$$\begin{aligned} {}^{16}O + {}^{16}O &\rightarrow {}^{32}S + \gamma \\ &\rightarrow {}^{31}P + p \\ &\rightarrow {}^{31}S + n \\ &\rightarrow {}^{28}Si + {}^{4}He \\ &\qquad + \text{energy} \end{aligned}$$

Because the electric repulsive forces are even greater in these reactions than they were in the earlier reactions, the minimum temperatures required are greater also. They are approximately 600 million and 1 billion K for the carbon and the oxygen reactions, respectively. This means that they can occur only after the stellar core has contracted and heated up still further than during the helium-burning phase.

As the equations indicate, several different products result during carbon and oxygen burning: ^{20}Ne, ^{23}Na, ^{24}Mg, ^{28}Si, ^{31}P, ^{31}S, and ^{32}S, as well as protons, neutrons, and alpha particles (^{4}He). The protons, neutrons, and alpha particles do not last long. As rapidly as they form, they react either with each other or with the heavier nuclei and produce still additional nuclear species and isotopes. The result is a broad spectrum of nuclear products.

The nuclear reactions discussed so far were induced by violent collisions between nuclei. Upon completion of oxygen burning, further nuclear reactions are induced by photons, which are inevitably present in stellar interiors. The temperatures are now in excess of 2 billion degrees, and the photons are so

BOX 3.4 continued

energetic that they knock protons, neutrons,and alpha particles out of the heavier nuclei. These photon-induced fission reactions are rapidly followed by fusion reactions in which the protons, neutrons, and alpha particles react with each other and with heavier nuclei, much as they did during the burning of carbon and oxygen earlier. However, now, because of the higher temperatures, they push the buildup of the heavier nuclei all the way to the formation of iron, nickel, and other elements of similar atomic weight. These *iron peak nuclei* (see figure 3.22) are more stable than the lighter nuclei and are, therefore, relatively immune to the breakdown by photons. They accumulate, which explains their relatively high abundance in nature.

With the formation of the iron peak elements, fusion reactions and the release of nuclear energy come to a close in stellar evolution. The buildup of still heavier elements requires energy and occurs only during such violent events as supernova explosions.

Summary

A summary of the most dominant nuclear reactions in stars follows.

hydrogen burning	^{1}H	→	^{4}He, ^{14}N (byproduct of the carbon cycle)
helium burning	^{4}He	→	^{12}C, ^{16}O
carbon burning	^{12}C	→	^{20}Ne, ^{24}Mg
oxygen burning	^{16}O	→	^{28}Si, ^{32}S
photon-induced burning	^{28}Si, ^{32}S	→	^{56}Fe

Because these reactions are dominant, their nuclear products—^{4}He, ^{12}C, ^{14}N, ^{16}O, ^{20}Ne, ^{24}Mg, ^{28}Si, ^{32}S, and ^{56}Fe—are the most abundant elements in the matter ejected during supernova explosions.

The matter ejected during supernova explosions intermingles with and enriches the gas and dust of interstellar space and, thereby, becomes part of the raw material for the birth of new generations of stars. Some 4.5 billion years ago, the Sun was formed from such enriched matter and that is why the elements listed above are so abundant in the solar matter, as indicated by the graph of the relative abundance of the elements in the Sun (figure 3.22). It is also the reason why, with the exception of the chemically inert gases helium and neon, these elements are among the most abundant ones on Earth.

Step by step the star synthesizes the heavy elements. The ashes from the burning of one element become the fuel for the successive nuclear reactions: hydrogen fuses into helium, helium into carbon and oxygen, and these two elements fuse into neon, sodium, magnesium, silicon, phosphorus, and sulfur (see box 3.4 for details). The energy liberated by these reactions keeps the star shining and retards the gravitational contraction of its interior which (see below) would otherwise occur very rapidly.

Throughout the carbon and oxygen burning phases, neutrinos are created in great abundance in the core of the star. They come into existence spontaneously because of the extreme temperatures and densities that exist there—approximately 0.6 to 1.0 × 10^9 K and 10^5 to 10^7 g/cm^3. As noted before, neutrinos are very elusive and as quickly as they are created they dart off into space, hardly ever interacting with the intervening layers of the star. With them they carry energy. This loss of energy by neutrino emission greatly increases the speed with which the star evolves from here on. It causes the star's core to contract much more quickly than it would without neutrino emission. The contraction, in turn, raises the density and temperature of the core, and carbon as well as oxygen burn faster and faster. Carbon burning in the core is so intense that it is completed in 600 years. Oxygen burning is even more extreme and is over in just one-half year. By stellar standards, these are very fast evolution times indeed.

In the next phase of nuclear burning, the elements produced by the fusion of oxygen — mainly ^{28}Si and ^{32}S — are ignited, and in a series of reactions they are converted into iron and other elements of similar weight (again, see box 3.4). The structure of the star now consists of a rapidly growing iron core (growing in mass, but shrinking in size), surrounded by successive shells in which the burning of sulfur, silicon, oxygen, carbon, helium, and hydrogen are the major sources of nuclear energy, as shown in figure 3.20.

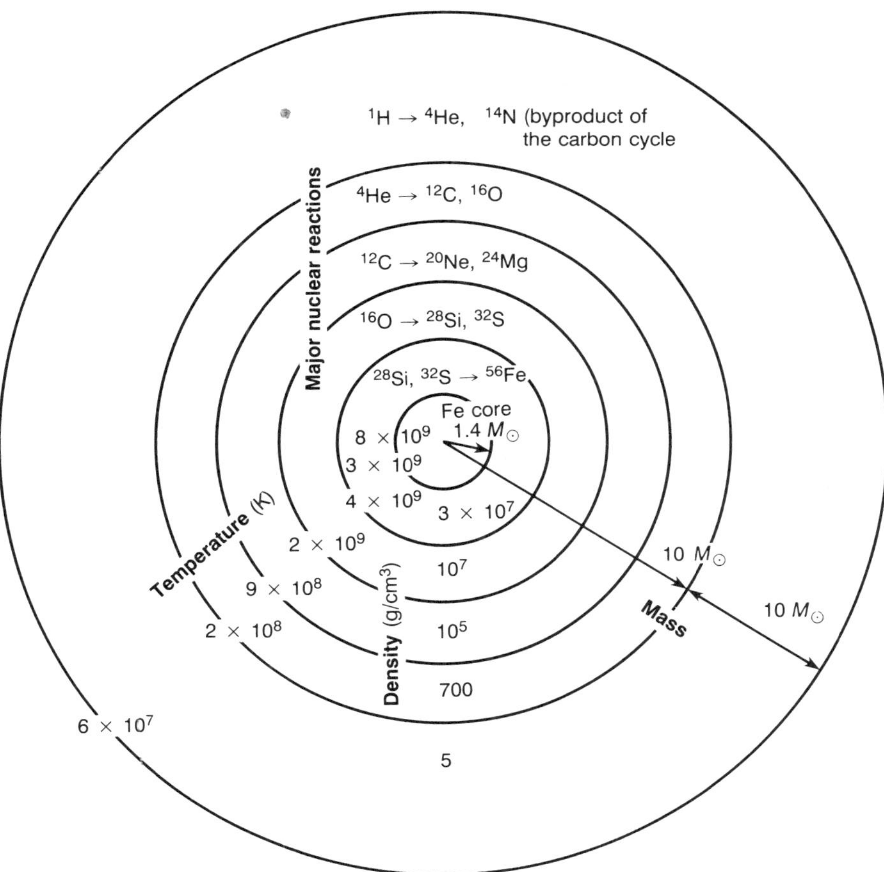

Figure 3.20. Structure of a 20 $M_\odot$ star just before implosion of its iron core and explosion of its outer parts as a supernova. The diagram shows the major nuclear reactions and the distribution of temperature, density, and mass. (The shells are not drawn to scale.)

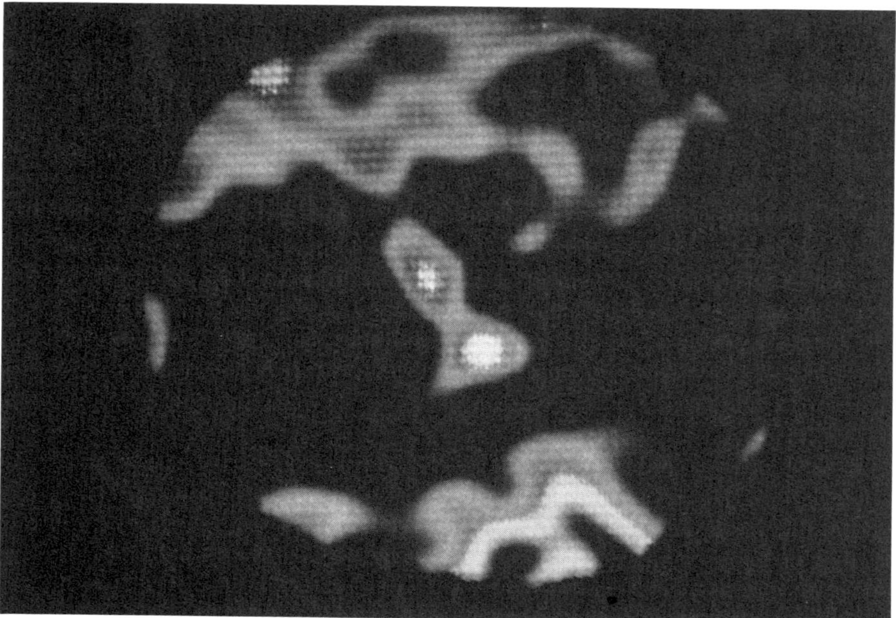

Figure 3.21. Betelgeuse, a red supergiant and the brightest star in the constellation Orion. This photograph was obtained by a special technique (called *speckle interferometry*) designed to overcome the problem of twinkling, which is caused by the Earth's atmosphere. It shows several enormous starspots, each roughly as large as the Earth's orbit around the Sun. The star's total size is approximately 1.5 times greater than the orbit of Mars.

The ongoing gravitational contraction of the star's central regions raises their density to approximately 10^9 g/cm^3. This means that the central one-solar mass of the star is compressed into a sphere ten times smaller than the Earth. Along with the contraction of the core, the steady outflow of large amounts of radiant energy toward the surface greatly expands the hydrogen-rich outer layers. If the star were at the location of the Sun, the outer layers would easily engulf the inner part of the Solar System to beyond the planet Mars. Simultaneously, steady loss of mass in the form of a strong stellar wind reduces the star's mass to perhaps as low as 20 $M_\odot$. Because of its enormous size, the surface layers of the star are now quite cool and it radiates largely in the red. It has become a red supergiant (figure 3.21) and in the HR diagram it lies in the upper right-hand corner. Approximately 500,000 years have elapsed since the star left the main sequence.

Core Collapse

With the formation of iron in the star's core, the traditional sequence of successive stages of nuclear burning comes to a halt. The reason is that iron cannot undergo nuclear reactions *and* release energy; it requires energy for fusion into still heavier elements. Hence, the iron core of the star resembles a white dwarf: its density is high, the electrons touch each other and resist contraction, and it possesses no nuclear energy source. However, the white dwarf configuration of the iron core does not last very long. The neutrino emission has speeded up the evolution to such a feverish pace that within just one day of silicon burning the iron core grows beyond the Chandrasekhar limit of 1.4 $M_{\odot}$ and the structural stability that has sustained the star so far breaks down. The star is coming face to face with its final destiny.

As soon as the iron core grows larger than 1.4 $M_{\odot}$, gravity becomes so strong that the electrons are no longer able to resist contraction and the core begins to collapse. In just a few milliseconds the following sequence of events runs its course:

1. The collapse crunches the iron nuclei together so violently that they break up into their constituent elementary particles—protons and neutrons.
2. The electrons present in the collapsing core are forced into the protons and produce neutrons. The collapsing core now consists entirely of neutrons. The equation of the reaction between electrons and protons is

$$e^- + p^+ \rightarrow n + \nu.$$

3. With the disappearance of the electrons, virtually all resistance to gravity vanishes and the collapse becomes "free-fall," reaching speeds of up to 10% the speed of light.
4. As the density of the collapsing matter approaches 10^{14} g/cm^3, the neutrons begin to touch each other and strenuously resist further compression (just as the electrons did earlier at a lower density). As suddenly as it began, the collapse stops. It stops with a bounce, followed by a brief period of sharp oscillations, somewhat akin to the ringing of a bell when it is struck by its clapper. After the oscillations have damped out, the collapsed core reaches a new equilibrium configuration about 20 km across, corresponding to the size of an average American town. The core has become a neutron star.

The collapse of the star's core from white dwarf dimensions into a neutron star takes just a few milliseconds because of the enormously strong gravitational forces that exist when 1.4 $M_{\odot}$ of matter are compressed into a volume smaller than the Earth. Furthermore, with the removal of the electrons, the pressure virtually vanishes and gravity becomes nearly unopposed, so it is able to collapse the core at an extremely fast rate.

Supernova Explosion

The collapse of the stellar core goes far too quickly for the envelope to be immediately affected. It is simply left behind. However, that period of grace does not last long. The collapse releases colossal amounts of energy, easily exceeding the energy emitted by 100 stars like the Sun during their entire main sequence lives. That energy has to go somewhere. A lot of it goes into breaking up the iron nuclei into their constituent protons and neutrons and into squeezing the electrons into the protons. Some of it is carried away by neutrinos, mainly during the bounce and the ensuing oscillations. The remainder of

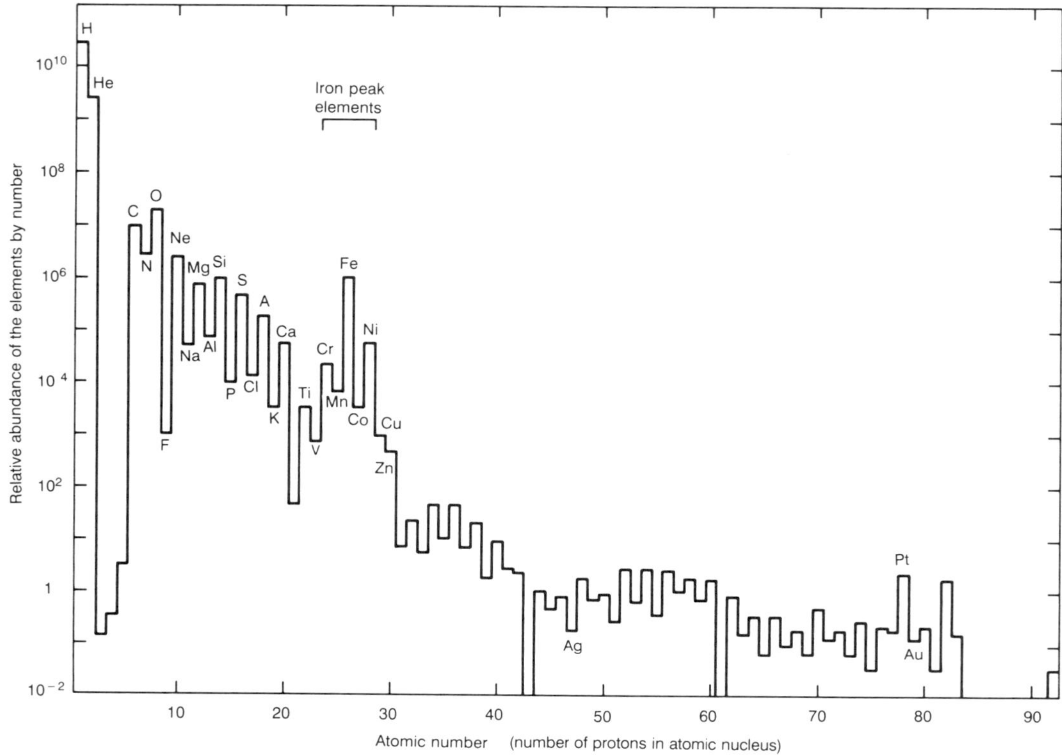

Figure 3.22. Relative abundance of the elements in the Sun. The abundance is by number, with silicon arbitrarily set at 10^6. For instance, for every 10^6 silicon atoms, the solar material contains 3.0×10^{10} hydrogen atoms, 2.5×10^9 helium atoms, and 10^7 carbon atoms. The 26 most abundant elements are indicated by their chemical symbols (see the Periodic Table of the Elements, table I.3). Silver (Ag), platinum (Pt), and gold (Au) are shown as well. Gold is an uncommon element: for every 10^6 silicon atoms, there exist only 0.13 gold atoms. Converting the abundance by number into abundance by mass gives 73.6% hydrogen, 24.8% helium, and 1.6% other elements.

Data: Robert W. Noyes, *The Sun, Our Star* (Cambridge and London: Harvard University Press, 1982), Table 2.1.

the energy goes into the creation of a superstrong shock wave, which runs radially outward toward the star's surface layers and affects them in a number of irreversible and far-reaching ways:

1. All along its way toward the surface, the shock wave loses energy by friction to the matter of the envelope, thus heating it and causing brief bursts of explosive nuclear burning in the silicon and oxygen layers. Both of these elements fuse into iron. The iron, in turn, undergoes further nuclear reactions and is converted into still heavier elements, such as copper, zinc, silver, and gold (see figure 3.22 and box 3.4). The energy required by these fusion reactions is supplied by the shock wave.

2. When the shock wave reaches the surface of the star, it has a whiplash effect on the outermost layers. The nuclei and electrons that make up these layers

are accelerated to nearly the speed of light and sent flying off into space. For millions and, possibly, billions of years they keep speeding through interstellar and intergalactic space until they are absorbed by gaseous clouds and other objects. They are one source of the cosmic rays that bombard the Earth continuously.

3. The heating produced by the shock wave plus the energy released by the brief bursts of nuclear reactions raise the pressure throughout the envelope to explosive conditions. The star now resembles a hydrogen bomb that has just gone off, except that it is bigger by far and more energetic. It is a gaseous sphere almost as large as Jupiter's orbit and contains approximately twenty times the mass of the Sun. At first slowly and then faster and faster the star's envelope gathers momentum. Layer after layer begins to roll and thunder outward, each driven by its own internal pressure and the push from the layers below. Within a few weeks, speeds exceeding 5000 km/sec (corresponding to 20 million km/hr or 10 million mi/hr) are reached and all the original, unburned hydrogen plus the freshly synthesized elements beyond the collapsed core—from helium to carbon, oxygen, neon, sodium, magnesium, silicon, phosphorus, sulfur, iron, and beyond—are ejected into space. The star is exploding as a supernova.

Outside observers, such as astronomers on Earth, who watch a supernova go off are much more limited in what they see. They cannot see the implosion of the core, for it is hidden deep within the star's envelope. Nor can they see the shock wave running through the envelope. They first become aware that something unusual and violent is happening when the shock wave arrives at the surface and the whiplash effect tosses the star's outermost layers into space at relativistic speeds. At that instant the star suddenly begins to brighten and within a day or two it outshines the rest of the galaxy in which it is situated (figure 3.23). Almost simultaneously with the brightening, the Doppler shift of the star's light reveals that the envelope is gathering speed and expanding. Soon it becomes apparent that the expansion is explosive and will end in the expulsion of enormous amounts of matter at speeds of thousands of kilometers per second. The large luminosity lasts for a few weeks and then gradually fades.

Figure 3.23. A supernova in galaxy NGC 7331. The left-hand photograph was taken before the supernova explosion, and the right-hand one was taken during the supernova's maximum brightness (1959). The other stars seen in these photographs (the sharp, circular images) are all foreground stars within our own galaxy, but despite their relative nearness they are much fainter than the supernova.

However, the ejected envelope can be seen for hundreds of thousands of years. It keeps expanding into the surrounding interstellar gas and dust, pushing them out of the way and compressing them (figure 3.1), and, perhaps, inducing the formation of yet another generation of stars. Gradually the expansion slows and the ejected matter mixes with the interstellar medium. Eventually the mixing becomes so thorough that the supernova ejecta can no longer be recognized, except for the presence of freshly synthesized heavy elements.

Black Holes

This description of supernovae focused on a star with an original mass of 25 $M_\odot$. If the star is much more massive, the mass of the imploding core may exceed 3 $M_\odot$. In that case the neutrons are unable to bring the collapse to a halt. They are squeezed into each other like the electrons and protons were earlier, and the collapse of the core continues undiminished. When its size becomes less than about 10 km across, the gravitational force at its surface is so strong that nothing can escape from it—neither material particles nor electromagnetic radiation—and anything that falls into it is irretrievably lost. The core has become a *black hole*. However, gravitational energy is still released during the collapse and leads to a supernova explosion and the ejection of the envelope, as in the case of the 25 $M_\odot$ star.

Black holes do not interact with the outside world except by gravity. Should some future human astronauts ever travel among the stars and suddenly feel that their spaceship is strongly attracted towards a point in space that appears to contain nothing, they are being pulled into a black hole. If they have the misfortune of insufficient engine power to climb back out, they will be doomed. The enormous gravitational forces suck them at ever-increasing speeds into the bottomless well. Tidal forces rip the spaceship apart and disintegrate everything into atomic form. The atoms are broken up into electrons and nuclei and, with the speed of light, these particles cross the boundary of the black hole. Except for a slight increase in the mass of the hole, the spaceship and occupants have vanished from the Universe. For the astronauts and their ship the event is over in seconds.

Observers on a sister ship, at a safe distance from the black hole, see the catastrophe as if it were happening in slow motion. For many hours they continue to receive electromagnetic signals of the descent and destruction of the ship and its unfortunate occupants. That radiation arrives more and more redshifted, because it has to climb out of an ever-deepening gravitational well. At first it is shifted from the optical to the infrared and then to the radio part of the spectrum. Eventually the wavelengths become stretched so much that the observers' receivers are inadequate for reception. For them, too, the event is now over. (See chapter 1 for additional discussion of the gravitational redshift and of the difference in time perception between observers in gravitational fields of differing strengths.)

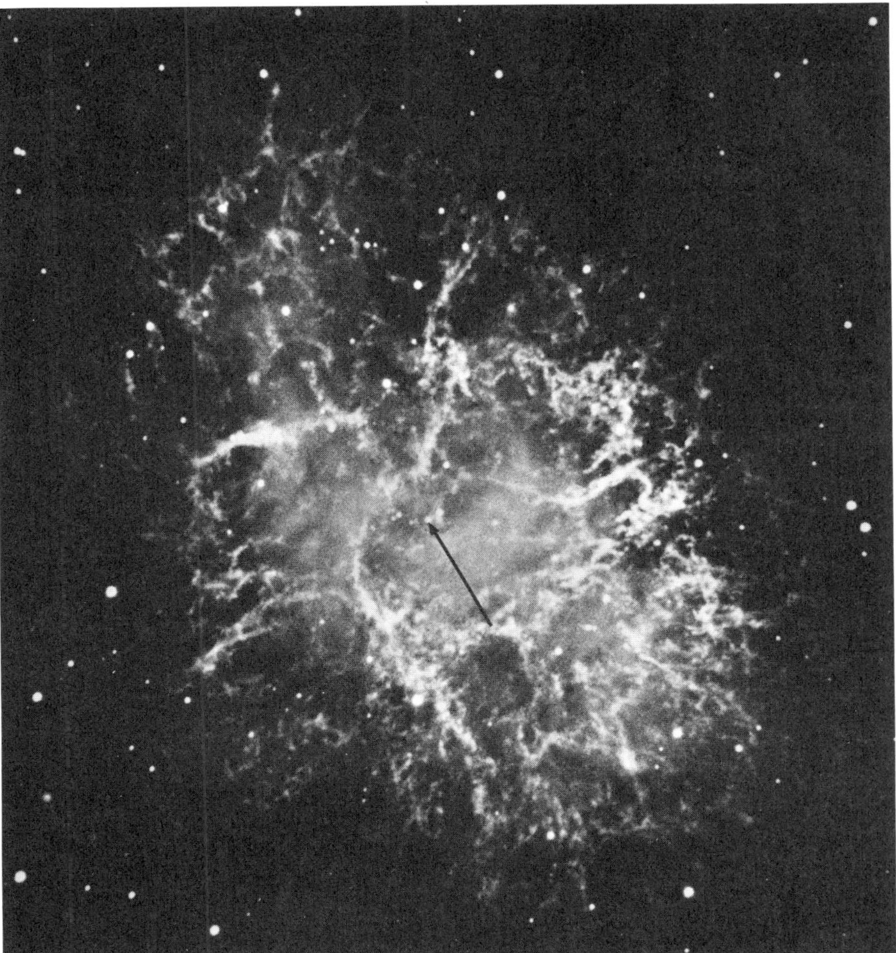

Figure 3.24. The Crab nebula. This nebula lies at a distance of about 6000 LY and consists of the gaseous matter ejected by the supernova observed in 1054 A.D. The energy of these emissions is believed to come from the stellar remnant of the explosion, the Crab pulsar (indicated by an arrow).

Historical Supernovae

At the present epoch, supernovae are rather rare events. Only seven are known to have occurred in our Galaxy during recorded history. Of these only the supernovae of 1006, 1572, and 1604 A.D. were recorded by European scholars. References to the others (the earliest dating from 185 A.D.) are found in Chinese, Japanese, and Korean chronicles.* The most famous historical supernova is one

*Observations of galaxies like our own have shown that on the average one supernova explodes in them every 100 years. The last one observed in the Milky Way was almost 400 years ago. Hence, we may be tempted to think that soon we should be treated to the spectacle of another one. This may

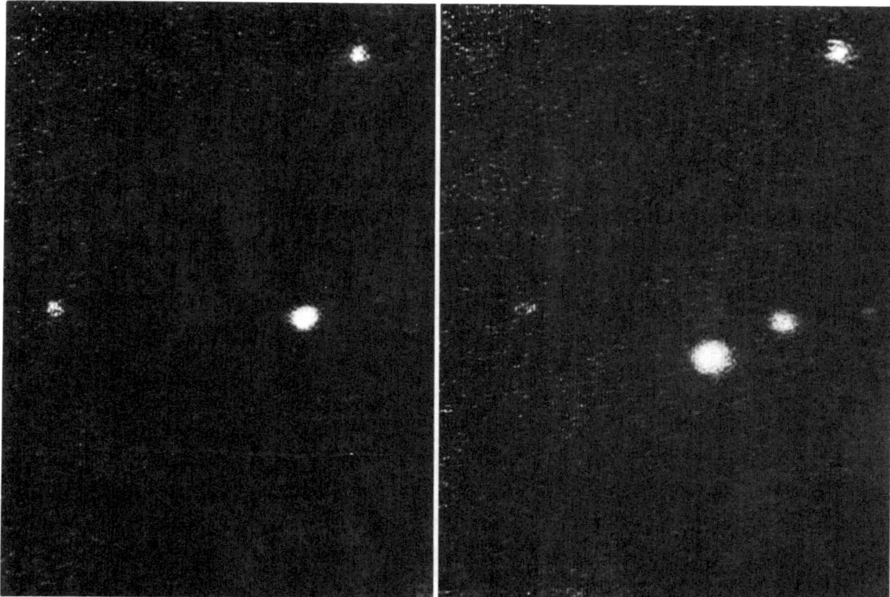

Figure 3.25. The Crab pulsar. This pair of photographs illustrates the on–off optical pulsation of the pulsar that lies embedded in the Crab nebula and that is the stellar remnant of the Crab supernova explosion.

that occurred in 1054 A.D. in the constellation Taurus and was recorded by the Chinese and Japanese.

Despite a distance of about 6500 LY, the supernova of 1054 A.D. was brighter than Venus for weeks before fading from view. At the location in the sky where the supernova exploded, modern astronomers have found a highly contorted and agitated nebula, the Crab nebula (figure 3.24), and a very hot and compact star. The Crab nebula is the stellar envelope ejected by the explosion, and it still is expanding at about 1500 km/sec. It radiates in all wavelengths from γ and X rays to UV rays, optical light, and IR and radio waves, attesting to the violence and turbulence still present today, more than 900 years after the outburst. The compact star is a neutron star—the collapsed remnant of the parent star's core.

Interestingly, the neutron star in the center of the Crab nebula sends out pulses of light and radio waves with startling regularity 30 times every second. Many of the neutron stars resulting from supernova explosions do that and, hence, they are known as *pulsars* (figure 3.25). The pulses are believed to be caused by a lighthouse effect. The neutron star rotates and, at the same time, strong magnetic fields on its surface focus the light and radio waves into a narrow beam. During every rotation, the beam sweeps out an arc of 360°, much like the rotating light beam from a lighthouse. As the beam crosses our line of

be the case, but we cannot be sure. The explosion of supernovae is a random phenomenon and there is no certain way of predicting when the next one will take place. It may happen in one year or in one or more centuries.

sight we see a pulse. Because the pulsing of neutron stars is observed in the frequency range from about once every few seconds to several hundred times per second, we may assume that neutron stars rotate at those rates.* The most rapidly pulsing pulsar discovered to date is Pulsar PSR 1937 + 214, which pulses 642 times per second.

Our galaxy, like all others, may well be populated by millions of neutron stars and black holes. We don't know for sure how many there are because after some time neutron stars stop pulsing and can no longer be detected. Furthermore, black holes cannot be seen at all, except when matter (as from a nearby companion star) falls into one and emits X rays and other electromagnetic radiation. Most of the neutron stars and black holes are believed to date back to the formative years of our galaxy, when nearly all of the heavy elements present today were manufactured and when supernova activity must have been at its peak.

On February 23, 1987, a supernova was sighted in the Large Magellanic Cloud. For a discussion of this event see page xi, Note to the Student.

The birth, evolution, and death of the stars illustrate perhaps better than any other phenomena taking place in the Universe today the interplay between the four fundamental forces of nature. The gravitational force pulls the interstellar gas and dust together into clouds. Emission of IR and radio waves (electromagnetic force) contributes to the contraction by removing heat, and it allows the further collapse of the gas and dust into clumps. Nuclear reactions (nuclear force) supply the energy necessary to bring the collapse to a halt and to transform the clumps into stars. Emission of light (again, the electromagnetic force) carries away the energy liberated by the nuclear reactions and makes the stars shine. Neutrino emission (weak force) speeds up the final evolution of the massive stars. In the end, gravity asserts itself and the stars, or parts of them, collapse into white dwarfs, neutron stars, or black holes. None of the four forces acts independently of the others. Like the different sections of an orchestra, they perform together, influencing each other and taking turns in controlling the evolution.

While this evolution runs its course, conditions arise that stand in stark contrast to each other. For instance, in following the life cycle of stars we have encountered cold, dark clouds with temperatures as low as ten degrees and hot stellar interiors with temperatures of tens of millions and billions of degrees. In interstellar space densities may be as low as 10^{-24} g/cm^3, while those of neutron stars are 10^{14} g/cm^3 and those of black holes are higher still. Nuclear reactions require that nuclei come to within 10^{-13} cm of each other; the size of a star may be as large as 10^{13} cm. The low-mass stars stay on the main sequence for 10 billion years and longer, while the final stages of nuclear burning in massive stars are completed in years or days and the collapse of their cores is over in a fraction of a second.

*Normal stars rotate much more slowly than neutron stars. For instance, the Sun and other stars like it rotate approximately once every 20 to 40 days. More massive stars generally rotate more rapidly, but never faster than about once every two or three days. When the core of such a star collapses, its rate of rotation greatly increases, for the same reason that skaters spin much more quickly when they bring their extended arms close to their bodies. A physicist would say that this speeding up of the rotation is evidence of the conservation of angular momentum.

Ultimately, these contrasting conditions result from the properties of five elementary particles—the electron, proton, neutron, neutrino, and photon—and the four fundamental forces by which they interact. Stars may be regarded as laboratories that allow the study of these particles and their interaction properties under conditions far broader than is possible on Earth. In particular, stars permit scientists to study the synthesis of heavy elements as well as the triggering of interstellar cloud collapse, both of which were preconditions for the formation of the Solar System and life on Earth.

Exercises

1 Locate on the night sky as many of the stars listed in table 3.1 as the season and observing conditions allow. Estimate the stars' relative brightness and order them accordingly, from bright to faint. Compare your order with that of table 3.1. (This exercise requires a star chart, listed in the bibliography in chapter 2.)

2 Fill a glass with water, place an index card over it, invert the filled glass while pressing the card against the glass, and let go of the card. Unless the card is warped or bent or the glass is uncommonly tall, the water will stay in the inverted glass. Why? (*Hint*: Compare the two kinds of pressures pushing on the card from above and below.)

3 (*a*) The boiling point of water diminishes with increasing elevation. For example, on the tops of Mt. Mitchell (2037 m), Mt. McKinley (6194 m), and Mt. Everest (8848 m), it is 93°, 79°, and 70° C, respectively. Why is this so?
(*b*) On what principle does a pressure cooker work?

4 (*a*) Describe the physical conditions under which molecules form in interstellar space.
(*b*) How do such conditions arise?
(*c*) Why are H_2, CO, OH, NH_3, CN, H_2O, H_2S, SiO, HCN, CH_3OH, and H_2CO among the most abundant molecules found in interstellar space?

5 Using one of the molecule kits listed in the bibliography in chapter 6, build three-dimensional models of the molecules listed in figures 3.4 and 3.5.

6 Distinguish between fusion and fission reactions, and give a few examples of each.

7 Discuss the importance of nucleosynthesis for the origin and evolution of life on Earth. Make reference both to the nucleosynthesis occurring presently in the Sun and to the nucleosynthesis that occurred in the Galaxy's early generations of massive stars.

8 (*a*) Describe the three mechanisms of energy transport in stellar interiors.
(*b*) Name a few examples other than those listed in box 3.3 illustrating the application of these mechanisms in everyday life.

9 Why do stars, in the course of their evolution, bunch up along the main sequence?

10 Why are massive stars much more short-lived than low-mass stars?

11 What would happen if:
(*a*) the temperature of the Sun's core were doubled?
(*b*) the temperature of the Sun's core were halved?
(*c*) all of the hydrogen in the central 15% of the Sun's interior were helium?
(*d*) all of the matter of the Sun were iron?
(*e*) the mass of the Sun were increased by 50% and all of its matter were iron?

12 Discuss the major differences among:
(*a*) main sequence stars, red giants, supergiants, and white dwarfs.
(*b*) stellar wind, long-period variable stars, planetary nebulae, novae, and supernovae.
(*c*) neutron stars and black holes.
(*d*) pulsars and quasars.

Suggestions for Further Reading

Bethe, H. A., and **G. Brown**. 1985. "How a Supernova Explodes." *Scientific American*, May, 60–68. The article focuses on the supernova explosions of massive stars (8 $M_\odot$ and greater), which give rise to Type II supernovae. It briefly touches upon Type I supernovae, which occur in less massive stars. (Also see the articles below by Kirshner and by Wheeler and Nomoto.)

Blitz, L. 1982. "Giant Molecular-Cloud Complexes in the Galaxy." *Scientific American*, April, 84–94. The sizes, content, physical conditions, evolution, and galactic distribution of interstellar cloud complexes, including star formation that takes place in them.

Boss, A. P. 1985. "Collapse and Formation of Stars." *Scientific American*, January, 40–45. Computer modeling of star formation.

Dyson, F. J. 1971. "Energy in the Universe." *Scientific American*, September, 50–59. An interesting article about the genesis and flow of energy both here on Earth and in stars, galaxies, and the Universe at large.

Greenberg, J. M. 1984. "The Structure and Evolution of Interstellar Grains." *Scientific American*, June, 124–35.

Kafatos, M., and **A. G. Michalitsianos**. 1984. "Symbiotic Stars." *Scientific American*, July, 84–94. Binary star systems that consist of a red giant and a nearby compact companion star. Mass transfer from the red giant to the compact star may give rise to an accretion disk, nova outbursts, and ejection of highly directional jets of matter.

King, I. R. 1985. "Globular Clusters." *Scientific American*, June, 78–88.

Kirshner, R. P. 1976. "Supernovas in Other Galaxies." *Scientific American*, December, 88–101.

Lada, C. J. 1982. "Energetic Outflows from Young Stars." *Scientific American*, July, 82–93.

Leibacher, J. W., **R. W. Noyes**, **J. Toomre**, and **R. K. Ulrich**. 1985. "Helioseismology." *Scientific American*, September, 48–57. Oscillations of the Sun, observed on its surface, hold clues to the structure, composition, and dynamics of the Sun's interior.

Mathis, J. S., **B. D. Savage**, and **J. P. Cassinelli**. 1984. "A Superluminous Object in the Large Cloud of Magellan." *Scientific American*, August, 52–60. A giant nebula in this small galaxy holds an object that is 50 million times brighter than the Sun. If it is one body, it is far more massive than any known star.

Noyes, R. W. 1982. *The Sun, Our Star*. Cambridge, Mass.: Harvard University Press. A nonmathematical, up-to-date account of current understanding of the physical structure of the Sun, including chapters on solar observations, the Sun's impact on the Earth's climate, and solar energy.

Penrose, R. 1972. "Black Holes." *Scientific American*, May, 38–46.

Philip, A. G. D., and **L. C. Green**. 1978. "The H-R Diagram as an Astronomical Tool." *Sky and Telescope*, May, 395–98. Highlights from a symposium of the International Astronomical Union on "The HR Diagram," held in memory of Henry Norris Russell in 1977 in Washington, D.C.

Scoville, N., and **J. S. Young**. 1984. "Molecular Clouds, Star Formation, and Galactic Structure." *Scientific American*, April, 42–53. Radio observations show that the giant molecular cloud complexes where stars are born are distributed in various ways in spiral galaxies, perhaps accounting for the variation in the clouds' optical appearance.

Seward, F. D., **P. Gorenstein**, and **W. H. Tucker**. 1985. "Young Supernova Remnants." *Scientific American*, August, 88–96. Space observations of the expanding shells from recent supernova explosions reveal X-ray and radio emissions, shock fronts, and clumping of the ejected matter.

Stephenson, F. R., and **D. H. Clark**. 1976. "Historical Supernovas." *Scientific American*, June, 100–107. The seven supernovae that were seen between 185 A.D. and 1604 A.D.

Thorne, K. S. 1967. "Gravitational Collapse." *Scientific American*, November, 88–98. In the case of burnt-out stars of critical mass, star clusters, and perhaps galaxies, gravity may overwhelm all other forces and crush matter out of existence to form black holes.

Waldrop, M. M. 1987. "Sighting of a Supernova." *Science* 235: 1143 (6 March). The discovery of the first nearby supernova since the invention of the telescope. (Also see the article by S. E. Woosley and M. M. Phillips referenced in this chapter.)

——. 1987. "Supernova 1987A on Center Stage." *Science* 238: 1038–1040 (20 November). A readable account of the interpretations of and questions raised by Supernova 1987A, based on the proceedings of The Fourth George Mason Fall Workshop in Astrophysics held 8 months after the supernova exploded (George Mason University, 12–14 October, 1987). (Also see the article by S. E. Woosley and M. M. Phillips referenced in this chapter.)

Weisberg, J. M., J. H. Taylor, and **L. A. Fowler.** 1981. "Gravitational Waves from an Orbiting Pulsar." *Scientific American*, October, 74–82. Pulsar PSR 1913 + 16 is a member of a binary system whose second member is an ordinary star. The orbital radius of the pulsar is slowly shrinking, which is interpreted as being due to the emission of gravitational waves.

Wheeler, J. C., and **K. Nomoto.** 1985. "How Stars Explode." *American Scientist*, May–June, 240–47. Two classes of supernovae are observed: Type I and Type II. Relying on observations of their light curves and remnants and on computer modeling, the authors describe the physical mechanisms that appear to give rise to each class.

Wheeler, J. C., and **R. P. Harkness.** 1987. "Helium-Rich Supernovas." *Scientific American*, November, 50–58. Computer modeling and observations suggest a new class of supernovas (besides the traditionally recognized classes types I and II), which result from the collapse of a massive star that previously lost its hydrogen-rich outer envelope. (Also see the article by J. C. Wheeler and K. Nomoto referenced in this chapter).

Williams, R. E. 1981. "The Shells of Novas." *Scientific American*, April, 120–31.

Wilson, O. C., A. H. Vaughan, and **D. Mihalas.** 1981. "The Activity Cycles of Stars." *Scientific American*, February, 104–19. Variations in activity similar to the Sun's 11-year sunspot cycle have been followed and studied in 91 nearby stars.

Wolfson, R. 1983. "The Active Solar Corona." *Scientific American*, February, 104–19.

Woosley, S. E., and **M. M. Phillips.** 1988. "Supernova 1987A." *Science* 240: 750–759 (6 May). An update of the dramatic and surprising developments of this supernova, some 14 months after it exploded. (Also see the two articles by Waldrop referenced in this chapter.)

Wynn-Williams, G. 1981. "The Newest Stars in Orion." *Scientific American*, August, 46–55.

Additional References

Weaver, T. A. 1980. "The Evolution of Massive Stars and the Origin of the Elements." *Energy and Technology Review*, February, 1.

Figure 4.1. Saturn and its ring system, photographed by *Voyager 2* from a distance of 21 million km. Four of Saturn's satellites are visible. Below the planet, from left to right, are Tethys (which casts a dark shadow on Saturn's disk), Dione, and Rhea. Mimas is barely visible against the planet's disk, below the rings and above and to the left of Tethys.

4 The Solar System: Orbital Layout

NEAR THE INNER EDGE of the Orion Arm and approximately 30,000 LY from the center of the Milky Way, the Sun speeds silently through interstellar space at nearly 1 million kilometers per hour. Opposing gravitational and centrifugal forces keep it firmly locked into a nearly circular orbit around the massive, central bulge of the Galaxy, as noted in chapter 2. Every 250 million years the Sun completes one circuit. It has done so since its birth some 4.5 billion years ago and, barring some near encounter or collision with another star, will do so in the future.

From a distance of a few light-years or more the Sun appears to be just another star among the 150 billion populating the Milky Way. However, when viewed from nearby it is a very special star. It is the gravitational hub of a planetary system that includes Earth—our home and shelter in the vastness of space—and, through its steady outpouring of radiation, provides the energy that sustains terrestrial life.

It is, therefore, not surprising that since the dawn of history humans have been fascinated by the Solar System. Our ancestors observed the daily rising and setting of the Sun, the waxing and waning of the Moon, and the wandering of the planets across the firmament. Periodically they were startled and frightened by the appearance of a comet or the falling of meteorites, and they ascribed plagues and other calamities to these events. In their myths and religions they deified the Sun and the planets and they explained their motions by ingenious and, sometimes, fanciful theories.

Today we are still fascinated by the Solar System and employ science to probe its secrets. We use telescopes to observe the Sun, the planets, and the lesser bodies; we bounce radar waves off their surfaces; we send space probes to them; and in 1969, we first set foot on the Moon. As a result of the space missions, we have extensive photographic records of the surface features of the planets and their satellites, excepting only Neptune and Pluto. We know about their magnetic fields. We possess Moon rocks. We have on-site analyses of the surface compositions of Venus and Mars. We are beginning to understand the structures of the planets and their satellites. And, with the aid of high-speed computers, we are developing new theories of the origin and evolution of the Solar System. The goal of this chapter and the next is to describe the orbital distribution and structures of the member bodies of the Solar System, and to make you familiar with some of the current theoretical efforts concerning its origin and evolution. We shall begin with a brief summary of the historical background to our contemporary perspective.

EMERGENCE OF THE HELIOCENTRIC VIEW OF THE SOLAR SYSTEM

Most ancient scholars thought that the Earth was the center of the Universe, with all other celestial objects revolving around it. However, an early exception was Aristarchus of Samos (*c.* 310–230 B.C.), who advocated a heliocentric Solar System. The ancients clearly distinguished between the *wandering stars* or *planets* and the *fixed stars*. The planets included the Sun, the Moon, Mercury, Venus, Mars, Jupiter, and Saturn, all of which, in the course of days to weeks, can be seen to wander across the firmament.* The fixed stars were the objects still called stars today and that do not appear to partake in the wandering motions. Today we know that the planets and fixed stars differ not only in their apparent motions across the firmament, but are fundamentally different objects. The planets belong to the Solar System and shine by reflecting sunlight. In contrast, the fixed stars (that is, those in our galaxy) lie at distances from a few to tens of thousands of light-years, and they shine by radiating energy released by nuclear reactions in their interiors.

The commonly held view among the ancients (figure 4.2) was that the seven planets as well as the fixed stars are carried on crystalline spheres around the Earth once every day and that the friction between the spheres gives rise to the "music of the spheres." This *geocentric* or *Earth-centered* view was formalized in the second century A.D. by the great Egyptian astronomer and mathematician, Ptolemy, in his book the *Almagest* (derived from the Arabic, meaning "The Great Treatise"). He stated that all the planets move along small circular paths, called epicycles, whose centers in turn move along larger circles around the stationary Earth. This system allowed him to approximate the motions of the planets across the sky, including the retrograde motions of the outer three.

Copernicus and Kepler

Ptolemy's geocentric view prevailed for more than 1300 years, until the early part of the sixteenth century. By then a new intellectual spirit had taken hold in Europe and many traditional teachings came under critical scrutiny. The mood was one of adventure and of looking beyond what was known or had been done before. Explorers sailed the seven seas and merchants expanded their trade to the far corners of the globe. This new adventurism also spread to astronomy. Growing discrepancies were noticed between the Ptolemaic predictions of the positions of the planets and their true positions in the sky, and a few astronomers dared to acknowledge that something fundamental was wrong with the geocentric model. The first major step in resolving the problem was taken by the Polish astronomer Nicolaus Copernicus, who in his celebrated work *De revolutionibus orbium coelestium* ("On the Revolutions of the Celestial Spheres") proposed a new and much simpler model of the Solar System. He placed the Sun at the center of the system and let the six known planets, including Earth, move in circular orbits around it. The Moon lost its status as a planet and became a satellite of Earth. This is the *Sun-centered* or *heliocentric* model of the Solar System.

Although Copernicus' heliocentric model came closer to reality than the geocentric model, it still did not quite correctly predict the motions of the planets. This became particularly evident in the case of the planet Mars, whose positions were very accurately measured by the Danish astronomer Tycho

*The ancients attributed mystical meanings to the planets and named most of them after their gods:

Mercury	Roman god of commerce and messenger of the gods
Venus	Roman goddess of love
Mars	Roman god of war
Jupiter	chief Roman god
Saturn	Roman god of agriculture

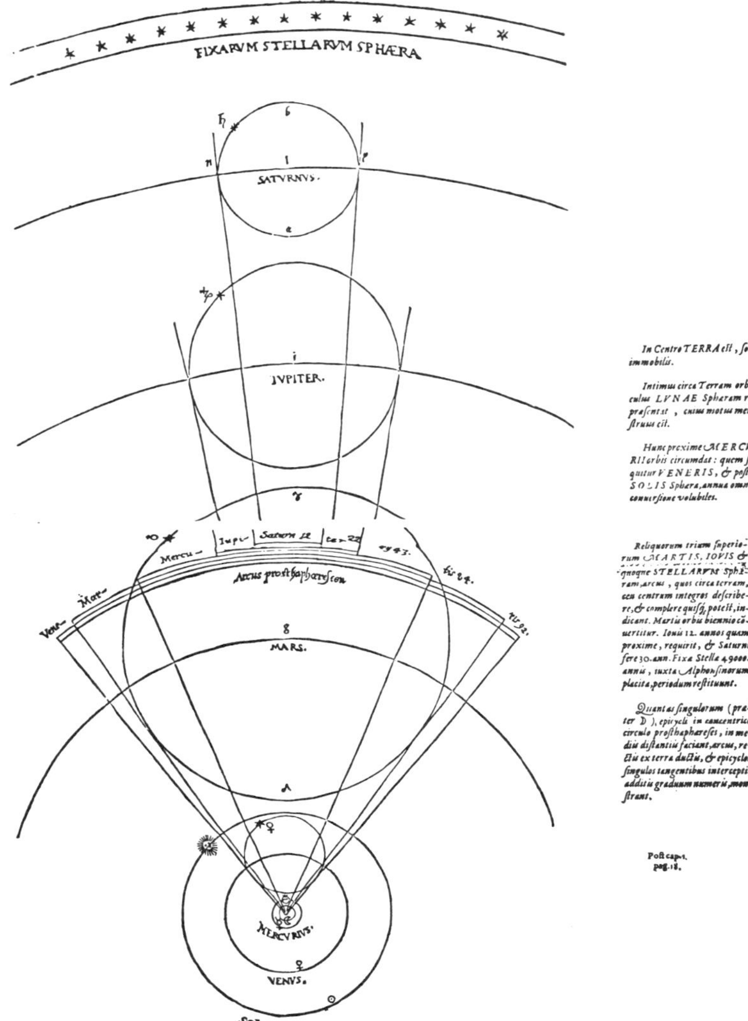

Figure 4.2. The orbital layout of the Solar System according to the theory of the ancients in the western world, showing the order of the planets, the relative sizes of their orbits, and epicycles. The Latin text is translated (Kepler 1981):

In the center is the Earth, which alone is motionless.

The innermost small orbit round the Earth represents the Sphere of the Moon, of which the motion is monthly.

The next round that is the orbit of Mercury; it is followed by that of Venus and after it is the Sphere of the Sun, which all go round in an annual revolution.

The orbits of the other three, the superior planets, Mars, Jupiter, and Saturn, as well as the Sphere of the Fixed Stars, are indicated by arcs, which anyone can complete by describing the whole of them about the Earth as the center. The orbit of Mars makes a turn in two years. That of Jupiter requires 12 years, as nearly as possible, and that of Saturn about 30 years. The Fixed Stars complete a period in 49,000 years, according to the tenets of the Alfonsine tables.

The amounts of the equations for each of them (except the Moon) produced by the epicycles on the concentric circle at their mean distances are shown by the arcs intercepted by straight lines drawn from the Earth and touching each of the epicycles, that number of degrees being added.

Brahe (1546–1601) over a period of some 20 years. Brahe attempted to resolve the disagreement between theory and observations, but without success.

Brahe's assistant, the German mathematician Johannes Kepler (1571–1630), succeeded with the task. Kepler found that he could achieve agreement if he made two assumptions: One, the planets do not move in circular orbits, as proposed by Copernicus, but in elliptical orbits; and, two, the Sun lies not at the center of the orbits, but at one of the foci of the ellipses. He also noted that the planets travel faster when they are near the Sun and slower when they are more distant. Finally, he discovered a simple relationship between the planets' distances from the Sun and their orbital periods. Kepler summarized these results in his famous *three laws of planetary motion*, which to this day form the basis of our understanding of celestial orbital motions (they are described in more detail in box 4.1). They apply to the motions of planets around the Sun, as well as to the motions of satellites, including artificial ones, around planets. They also apply to the motions of double stars and double galaxies around each other.

These accomplishments by Copernicus, Brahe, and Kepler contributed greatly to the intellectual optimism of their times. They had proved that it was possible to gain valuable insights into natural phenomena by making careful observations and measurements and by using them in the construction of new theoretical models. No longer was it necessary to accept ancient dogmas on faith. We must rank this development—reliance on observation and experimentation combined with theoretical reasoning—among the great revolutionary breakthroughs in the intellectual history of humankind. It was, in fact, the beginning of modern science.

Galileo and Newton

The heliocentric model of the Solar System was met initially with much opposition, as Galileo's trial by the Inquisition amply demonstrates. Only in the succeeding centuries, after a series of remarkable discoveries confirmed the model's validity, did it become universally accepted. Many scientists contributed to this success, but two of them stand out particularly. They are the two giants of seventeenth century science, Galileo Galilei and Isaac Newton.

Galileo was the first to observe celestial bodies through the then newly invented telescope. He noticed that Venus has phases just like the Moon and correctly explained them as being due to that planet's motion around the Sun, during which increasing and decreasing portions of its illuminated side face toward us. When he trained his telescope toward Jupiter, he saw four small bright objects orbiting around that planet with periods of between two and seventeen days. They are Jupiter's four largest satellites, which today we call the Galilean satellites—Io, Europa, Ganymede, and Callisto.* Galileo fully realized

*It is possible that Galileo was not the first to observe the satellites of Jupiter. This honor may go to Gan De (about 365 B.C.), one of the earliest Chinese astronomers. He is quoted in *The Kaiyuan Treatise on Astrology*, which was compiled between 718 and 726 A.D., as having said (Hughes 1982): "In the year of chan yan . . . , Jupiter was in Zi, it rose in the morning and went under in the evening together with the lunar mansions Xunü, Xü and Wei. It was very large and bright. Apparently, there was a small reddish star appended to its side. This is called 'an alliance'," The term *alliance* probably indicates that Gan De thought the reddish star to be a subsidiary or satellite of Jupiter. We do not know for certain that what Gan De saw was one of Jupiter's satellites. But if it was, it may have been Ganymede, Jupiter's largest and brightest satellite.

BOX 4.1.

Kepler's Laws of Planetary Motion

KEPLER formulated his three laws of planetary motion by carefully analyzing Tycho Brahe's data of the planets' positions in the sky. He began with this work in 1600 when he was hired by Brahe, who was then the imperial mathematician at the court of Emperor Rudolf II in Prague, and was asked to find a theoretical explanation for the orbit of Mars. Upon Brahe's death one year later, Kepler succeeded him in his court appointment and continued for some twenty years with the studies of planetary orbits. At first he experimented with traditional epicycles and tried to fit various combinations of circles to the orbit of Mars, but without obtaining satisfactory agreement. He then used "oval" curves and, finally, ellipses, which he soon realized can be made to agree quite accurately with the observations. By 1609 he had worked out his first two laws of planetary motion, which he published under the title *Astronomia nova* ("New Astronomy"). He published the third law in 1619 in his *De harmonices mundi* ("Harmonies of the World"). The three laws are:

Law I: Each planet revolves around the Sun in an elliptical path, with the Sun occupying one of the foci of the ellipse.

Law II: The straight line joining the Sun and a planet sweeps out equal areas in equal intervals of time.

Law III: The squares of the planets' orbital periods are proportional to the cubes of the semimajor axes of their orbits.

Kepler's *first law* states that the planets' orbits are ellipses. An *ellipse* may be constructed on paper by taking a piece of string of a certain length, attaching it firmly to two fixed points, and tracing out a line, keeping the string taut (figure A).

In the mathematical analysis of ellipses, the length of the string is equated to $2a$, where a is the length of the *semimajor axis* (see figure A). The letter b is used to indicate the length of the *semiminor axis*. The two points to which the string is attached are said to be the *foci* of the ellipse and their distances from the

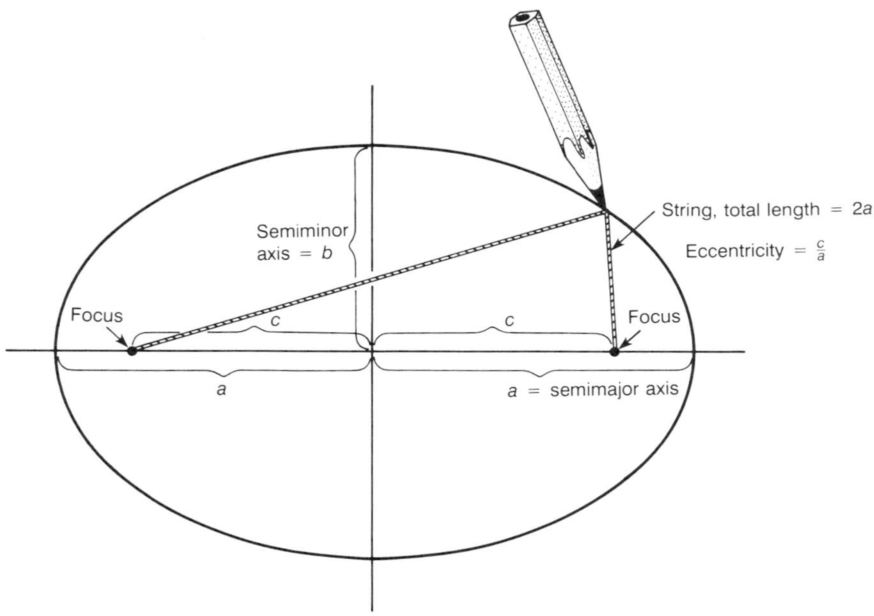

Figure A. Construction of an ellipse.

BOX 4.1 continued

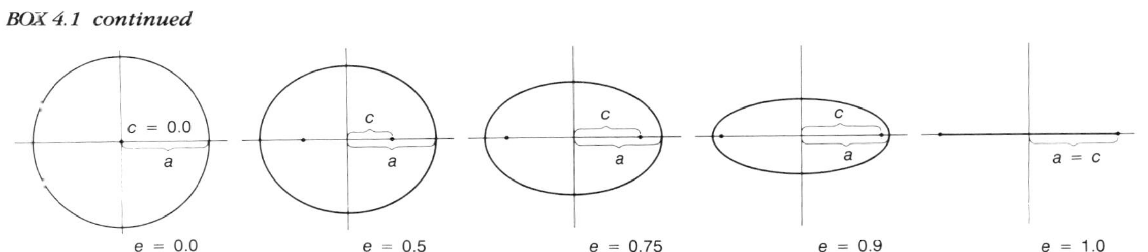

Figure B. Elliptical shapes. On the left is a circle of radius a. The next three ellipses (left to right) have identical semimajor axes, a, but different eccentricities. The straight line segment on the right has length $2a$.

center of the ellipse are labeled c. The ratio c/a is defined as the *eccentricity*, e. For elliptical orbits, e has a value between zero and one.

Differently shaped ellipses result if the two foci are moved closer together or farther apart (that is, by changing c) but the length of the string (that is, $2a$) remains the same. Changing the length of the string (the length of $2a$) changes the size of the ellipse. If the two foci overlap, a circle is obtained. If the two foci are so far apart that the distance between them is equal to the length of the string, a straight line results. Between these two extreme cases lies the full range of elliptical shapes, as illustrated in figure B.

The orbits of most planets have very small eccentricities and, hence, are nearly circular. The orbits of most asteroids have eccentricities in the range from 0.1 to 0.3 and, hence, deviate slightly from circles. Finally, the orbits of most comets that visit the inner regions of the Solar System have eccentricities between 0.5 and nearly 1.0 and, hence, are noticeably elongated. In all cases, the Sun occupies one of the two foci of the body's orbit; the other focus is unoccupied and without any physical significance.

Kepler's *second law* quantifies the fact that an orbiting body travels more slowly far from the Sun and more rapidly close to it. Such changes in orbital speed are yet another consequence of the *conservation of angular momentum*. The speeds of an orbiting planetary body are related to the corresponding distances from the Sun in such a way that the line joining the Sun and the planetary body sweeps out equal areas during equal intervals of time. In other words, if the orbital travel times from position 1 to position 2 ($t_{1,2}$) and from position 3 to position 4 ($t_{3,4}$) are the same, then $Area_{1,2}$ equals $Area_{3,4}$ (figure C).

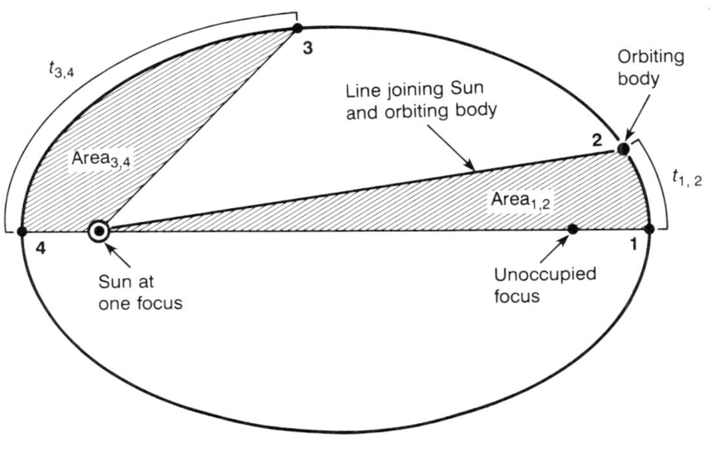

Figure C. Conservation of angular momentum. When the orbiting body is close to the Sun, its orbital speed is fast; when it is far from the Sun, its orbital speed is slow. If the orbital travel time from point 1 to point 2 ($t_{1,2}$) amd from point 3 to point 4 ($t_{3,4}$) are the same, then, according to Kepler's second law, $Area_{1,2} = Area_{3,4}$.

BOX 4.1 continued

Kepler's *third law*, which is also known as the harmonic law, is demonstrated by the entries in the following table. The planets' orbital periods are given in Earth years and their orbits' semimajor axes are in astronomical units. The use of these units makes the proportionality between the square of a planet's orbital period, P, and the cube of the planet's semimajor axis, a, an equality: $P^2 = a^3$ (see columns 4 and 5 of the table):

	P (years)	a (AU)	P^2	a^3
Mercury	0.240 85	0.387 10	0.058 01	0.058 01
Venus	0.615 21	0.723 33	0.378 5	0.378 5
Earth	1.000 0	1.000 0	1.000	1.000
Mars	1.880 9	1.523 7	3.538	3.538
Ceres	4.602	2.766	21.2	21.2
Jupiter	11.862	5.202 8	140.7	140.8
. . .				
Comet Halley	76.1	17.8	5800.	5600.

The small differences between some of the entries in the final two columns are due to inaccuracies in the measurements of a and P and to the fact that the Sun is not the only body in the Solar System exerting a gravitational pull.

the importance of this latter discovery. He had found a new satellite system—a miniature version of the Solar System. Apparently, systems in which one or several objects revolve around a larger central object are common in nature, with the Earth–Moon system being yet another example.

Perhaps the most spectacular sight for Galileo was the surface of the Moon. There he saw mountains, craters, valleys, and large dark areas, which he thought were seas. Obviously, that body is not the perfectly round and smooth sphere envisioned by the ancients. In many ways it resembles the Earth. Nor is the Sun a perfect body. On its surface, Galileo found dark spots, which today we call sunspots. These blemishes had been observed before, but they were dismissed as being either illusions in the Earth's atmosphere or shadows cast by intervening planets. From the motions of the sunspots across the solar disk, Galileo deduced that our star rotates with a period of slightly less than one month. Galileo described these and other telescopic discoveries in his book *Sidereus nuncius* ("The Starry Messenger").

Besides his astronomical discoveries, Galileo made numerous other major contributions to the development of modern science. In particular, through his experiments and mathematical analysis of the acceleration of material bodies, he paved the way for the work of Isaac Newton. Newton was convinced that a fundamental principle governs the motions of all objects and he set out to find it. This eventually led him to realize that whenever a body experiences a *change in motion*—when it is either speeded up or slowed down or when the direction of its motion is changed—a force is acting on it. He expressed his idea in three laws, which he published under the title *Philosophiae naturalis principia mathematica*, known as the *Principia*. With these three laws (more fully described in box 4.2) Newton laid the foundation of modern physics.

In the course of his work on the laws of motion, Newton wondered whether the force responsible for the falling of objects on Earth might not also be the force that keeps the Moon in orbit about the Earth. Legend has it that this idea

BOX 4.2.
Newton's Three Laws of Motion

SINCE the days of the ancient Greeks, scholars have endeavored to understand the motions of material bodies. Until the sixteenth century, however, these efforts were based strictly on geometrical interpretations of movement, without any recourse to experimentation, and therefore they were largely unsuccessful. Then, starting with Galileo, the French philosopher René Descartes (1596–1650), and the two English scientists Sir Christopher Wren (1632–1723) and Robert Hooke (1635–1703), physical reasoning based on observations and experiments was brought to bear on the problem. Each of these men acquired some understanding of motion, but Isaac Newton was the first to understand it thoroughly and to describe it quantitatively. He published his results in the form of three laws in the introductory part of his *Principia* (first edition published in 1687):

> Law I: Every body continues in its state of rest, or uniform motion in a right line, unless it is compelled to change that state by forces impressed upon it.
>
> Law II: The change of motion is proportional to the motive force impressed; and is made in the direction of the right line in which that force is impressed.
>
> Law III: To every action there is always opposed an equal reaction: or, the mutual actions of two bodies upon each other are aways equal, and directed to contrary parts.

All three of these laws are based on the concept of *force*. More specifically, they are concerned with the net force that acts upon a body. *Net force* refers to *the sum of all the forces* that act on the body. It is important to keep this in mind because in general bodies are exposed simultaneously to a number of different forces, often pointing in a number of different directions.

The *first law* describes the response of a body when *there is no net force* exerted upon it: if the body is initially at rest, it will remain at rest; if the body is initially in motion, it will continue with this motion without any change in speed or in the direction of the motion.

We observe the application of the first law while parking or driving a car. When the car is parked and the engine is off, no net force acts on the car and it remains at rest. If we drive the car on a straight, level road and press down on the gas pedal just enough so that the torque of the engine produces exactly the forward push required to overcome the frictional drag, then no net force is acting on the car either (the forward push and frictional drag cancel) and we will drive along at constant speed.

The *second law* describes the case when *there is a net force* exerted upon a body. The effect of the net force is a change in the motion of the body. In particular, this change in motion is proportional to the applied net force and it points in the direction of that force.

The change in motion may be a change in the speed of the body, a change in the direction in which the body is moving, or both. For instance, if the net force is applied in the direction in which the body is moving, the body's speed increases. This would be the case if, in the car example above, we stepped harder on the gas pedal. If the net force is applied in the direction opposite to that in which the body is moving, the body's speed decreases. This would be the consequence if we let up somewhat on the gas pedal or if we let up entirely on the gas and pressed on the brake pedal. Finally, if the net force is applied at an angle to the direction in which the body is moving, the direction of motion will change. This would be the case if we drive the car around a curve. All three of these changes in motion are called *accelerations*, including the slowing down (which is a negative acceleration) and the change in the direction of motion.

As just described, the *direction* of the acceleration of a body depends on the direction in which the net force is applied, but the *magnitude* of the acceleration depends on the strength of the applied force. In fact, the magnitude of the acceleration is proportional to the net applied force. This means that if the net force applied to an object is doubled or tripled, then the acceleration is also doubled or tripled.

In his formulation of the second law, Newton referred only to the proportionality between acceleration and applied force. However, the acceleration also depends on the mass of the body. For example, for a given force, a small mass experiences a much greater acceleration than does a large mass. If we use the car

BOX 4.2 continued

example once more, this statement expresses the common experience that an engine of a given horsepower accelerates a light sports car much more rapidly than a large family car.

With the introduction of the concept of mass, the proportionality between acceleration and applied force can be expressed by the equation

$$F = ma,$$

where F stands for the applied force, a for the acceleration, and m for the mass of the body. Newton's second law is commonly presented in this form today.

Incidentally, Newton's second law also expresses the content of the first law. In the first law, $F = 0$. According to the above equation, this means that $ma = 0$, which, in turn, implies that $a = 0$. But zero acceleration means that the body's speed and direction of motion remain unchanged, as predicted by the first law.

The *third law* states that a force cannot exist by itself, but is always accompanied by a second force. The two forces—called action and reaction—are of equal magnitude and point in opposite directions. Further, the law indicates that the two forces never act on the same body, but aways on two, as the following examples illustrate.

We experience the presence of the two forces when we jump off a small rowboat: our bodies are accelerated in one direction, while the boat is accelerated in the opposite direction. Still other examples in which the presence of the two forces is readily apparent are the firing of a rocket engine (the rocket goes forward, exhaust gases backward), and the shooting of a gun (bullet flies forward, gun recoils). These examples also make it amply clear that the two forces never act on just one body, but always on two: person and boat, rocket and exhaust gases, gun and bullet.

In many situations we are not aware of the presence of both forces as, for example, when we drive a car on an asphalt road. We notice that there is a forward push on the car, but we do not usually notice that there is also a backward push on the road. We don't because the road is in firm contact with the Earth and the resulting backward acceleration is too small to be observed. However, when we accelerate the car on a gravel road, we do notice the backward push because gravel is thrown out in the backward direction.

first occurred to him while he was watching an apple fall from a tree. He thought that, perhaps, the same force is also responsible for the planets' motions around the Sun as well as the Galilean satellites' motions around Jupiter. To test these hypotheses, he substituted the orbital data of the planets and satellites into Kepler's third law. He found that all of the motions could be explained, including the motions of falling objects on Earth, if he assumed that the force (F) varies with the inverse square of the distance (R) separating the bodies: $F \propto 1/R^2$. He found further that the force is also directly proportional to the product of the masses of the bodies (for example, the masses of the Earth and the Moon, the Sun and a planet, M_1 and M_2): $F \propto M_1 M_2$. By combining the two proportionalities, Newton derived a mathematical expression for the *universal force of gravity*:*

$$F \propto \frac{M_1 M_2}{R^2}.$$

(See discussion of gravitational force, Introduction to Part One). *Universal* means that this force applies not only to the effect of the Earth on falling apples and the Moon or of the Sun on the planets, but to the attraction between all matter and energy in the Universe. With this discovery, Newton demonstrated for the first time that there exists but one "design" in nature and that the scientific method allows us to discover it.

*The term *gravity* is derived from the Latin *gravitas*, meaning "heaviness" or "weight."

MODERN PERSPECTIVE OF THE SOLAR SYSTEM

During the three centuries following the pioneering work of Copernicus, Brahe, Kepler, Galileo, and Newton, knowledge of the Solar System grew at a steadily accelerating pace.

Three New Planets

In 1781, the German–English astronomer and musician Sir William Herschel noted, in the course of a systematic survey of the sky, that the image of a faint star was larger than is typical for stars. He observed the object for several weeks and found that it moved relative to the fixed background stars. At first he thought he had discovered a comet. Several months later, with more positional data on hand, the object's orbit was computed. It turned out not to be the elongated orbit of most comets (more about them later), but a nearly circular path around the Sun at a distance of about nineteen times that of the Earth. This is a distance greater than Saturn's and, judging from the object's brightness, meant that it was much bigger than a comet. Herschel had discovered a new planet. He named it *Georgium Sidus* (meaning "Star of George") in honor of George III, England's reigning king, but the name was later changed by J. E. Bode (see below) to *Uranus*, after the Greek god of the heavens (figure 4.3). Incidentally, Herschel was not the first to sight Uranus. The planet had been plotted on charts of the sky on at least seventeen previous occasions since the year 1690 without recognition of its true nature.

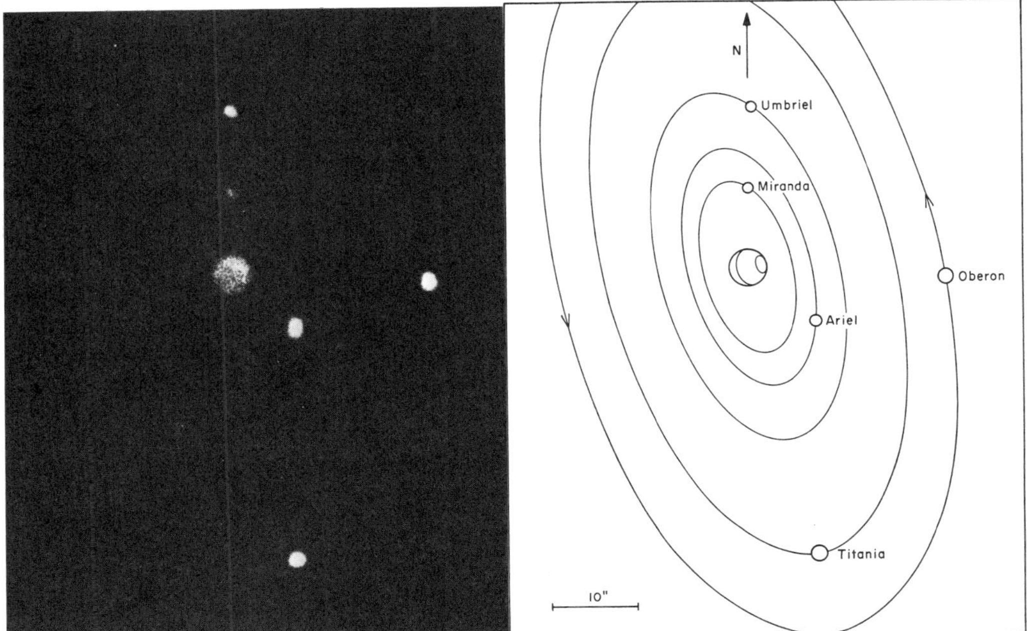

Figure 4.3. Uranus and its five large satellites, photographed at infrared wavelengths from the Mauna Kea Observatory, Hawaii.

Figure 4.4. Neptune and its two known satellites, Triton and Nereid. Triton is visible between the lower two diffraction spikes of Neptune's overexposed image, while Nereid is marked by an arrow. A long exposure was used to capture the image of Nereid, which is a body only 300 km across and extremely faint.

By 1820, it had become clear that the orbit of Uranus is not exactly an ellipse. Over the years, its path in the sky deviated more and more from that predicted by theory, even after the perturbing effects of Jupiter and Saturn were taken into account. It appeared that an unknown and still more distant planet was pulling on Uranus. The question was where in the sky to find the unknown planet. Tentative positions were computed in 1841 by John Couch Adams (1819–1892), an undergraduate at Cambridge University in England, and, four years later, by Urbain Jean Joseph Leverrier (1811–1877) in France. A search was conducted for the unknown planet at the Cambridge and Berlin observatories. It was finally discovered in 1846 by Johann Gottfried Galle (1812–1910), a young German astronomer at the Berlin Observatory, within one degree of the position predicted by Leverrier. It was christened *Neptune*, in honor of the Roman god of the sea (figure 4.4).

As in the case of Uranus, Neptune seems to have been sighted before it was identified as the Sun's eighth planet. Galileo's notebook shows that on several occasions in 1613 he had seen in the vicinity of Jupiter what he thought was a faint star that moved, and recent calculations indicate that it may well have been Neptune.

Early in the twentieth century, it became evident that Neptune alone could not account for the full perturbation that was observed in Uranus's orbit. There had to be yet another planet—*Planet X*, as it became known—that exerted a gravitational tug. Two American astronomers, William Henry Pickering (1858–1938) and Percival Lowell (1855–1916), estimated that Planet X should lie somewhere in the constellation Gemini and that its mass should be between that of the Earth and Neptune. An intense search ensued. Lowell looked for the planet from 1906 until his death, at the observatory he built in Arizona for the purpose of studying the planets. Success finally came in 1930 to Clyde Tombaugh, a 22-year-old Kansas farm boy and assistant at the Lowell Observatory. For many months he photographed sections of the night sky and compared photographic plates taken a few days apart. His aim was to see if any of the stellar images showed displacement and, thereby, revealed itself as the missing planet. The work was tedious and time consuming. Then, on February 18, 1930, Tombaugh noted that the image of what he thought was a faint star had moved the expected amount on photographs taken six nights apart earlier in January. He had discovered the Sun's ninth planet. It lay within 6 degrees of the position predicted by Lowell. The new planet was called *Pluto*, after the Greek god of the underworld. It was exceedingly faint and not sighted visually until 1950.

There is an ironic twist to Pluto's discovery that became evident only 48 years later. In 1978, James Christy, an astronomer at the U.S. Naval Observatory, noted that on high-quality photographs Pluto's image has a bump and that this bump moves systematically around the planet. The bump turned out to be a satellite. Its orbit is so close to Pluto that on photographs the two bodies are not fully resolved, but merge into a single bumpy image (figure 4.5). The satellite has a

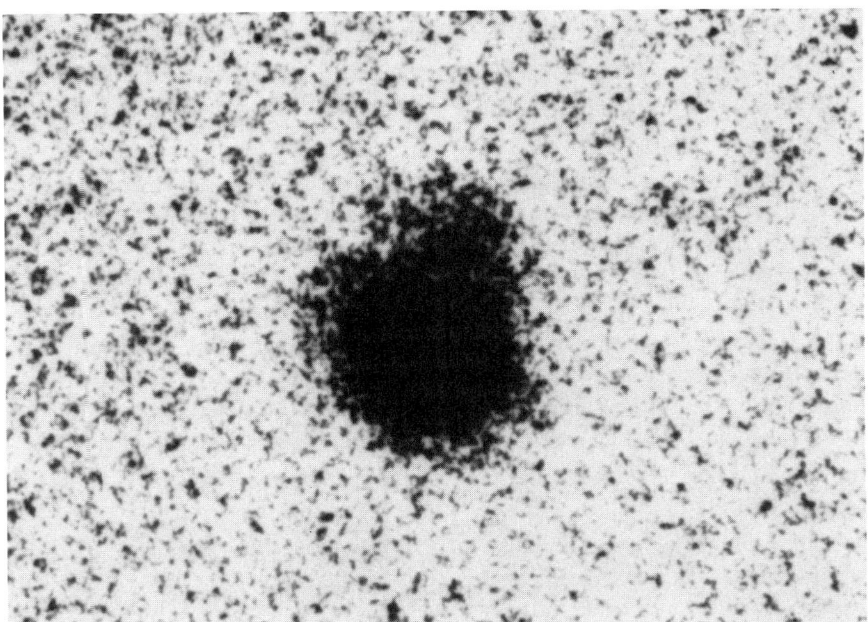

Figure 4.5. The discovery photograph of Pluto's satellite Charon, taken on July 2, 1978, with the 1.54-m reflector of the U.S. Naval Observatory. Charon orbits Pluto at a distance of a mere 19,700 km once every 6.4 days and appears in this photograph as a bump on the upper edge of Pluto's image.

period of 6.4 days, the same as Pluto's rotation period, and it circles Pluto at a distance of about 20,000 km (equal to three earth radii). Christy named the satellite *Charon*, for the boatman of Greek mythology who ferries the souls of the dead to Hades, the underworld and Pluto's domain. Charon's orbital period and distance imply—through Kepler's third law—that Pluto's mass is only 0.002 earth masses, which is considerably less than the mass assumed by Lowell in making his prediction of Pluto's position. It is, in fact, such a small mass that it cannot possibly be responsible for the observed residual perturbation in Uranus's orbit. Therefore, Pluto's discovery on the basis of Lowell's prediction was pure coincidence. It was, however, a lucky coincidence: without it we might still not be aware of Pluto's existence.

If Pluto is not Planet X, perhaps there exists a larger and as yet undiscovered planet at a still greater distance from the Sun that is the true source of the residual perturbation of Uranus's orbit. The case for this hypothesis is strengthened by small and unaccounted for discrepencies in the orbits of the other planets, from Jupiter to Neptune. In fact, the case for a tenth planet is so strong that astronomers in the United States and Europe are now actively searching for it. They do not presently know where exactly in the sky it is located, if, in fact, it does exist; but they estimate that it has a mass between two and five earth masses and an orbit beyond Pluto's, at about 50 to 100 astronomical units* from the Sun.

The Planets' Orbital Characteristics

Before continuing with the description of the Sun's planetary system, a summary of the planets' orbital characteristics is in order (see figures 4.6, 4.7, and table 4.1):

1. All planets revolve around the Sun in nearly the same plane. Relative to the Earth's orbital plane—also known as the *plane of the ecliptic*—the other planets' orbits are inclined by only three degrees or less. The only exceptions are Mercury and Pluto, whose orbits are inclined by 7 and 17 degrees, respectively.
2. The orbits of all but two of the planets are nearly circular. Again, the two exceptions are Mercury and Pluto, whose orbits have eccentricities of 0.21 and 0.25, respectively (see box 4.1 for a definition of eccentricity).
3. All planets orbit the Sun in the same direction, namely from west to east. This direction is called *prograde*, in distinction to *retrograde*, which refers to the opposite direction.
4. With two exceptions, the planets rotate in the same direction in which they orbit. That is why, for example, on Earth the Sun rises in the east and sets in the west. Only Venus and Uranus are different. Both rotate in the retrograde

*The *astronomical unit*, or AU, is a measure of length equal to the semimajor axis (see box 4.1) of the Earth's orbit about the Sun:

$$1 \text{ AU} = 1.50 \times 10^8 \text{ km};$$
$$1 \text{ LY} = 63{,}000 \text{ AU}.$$

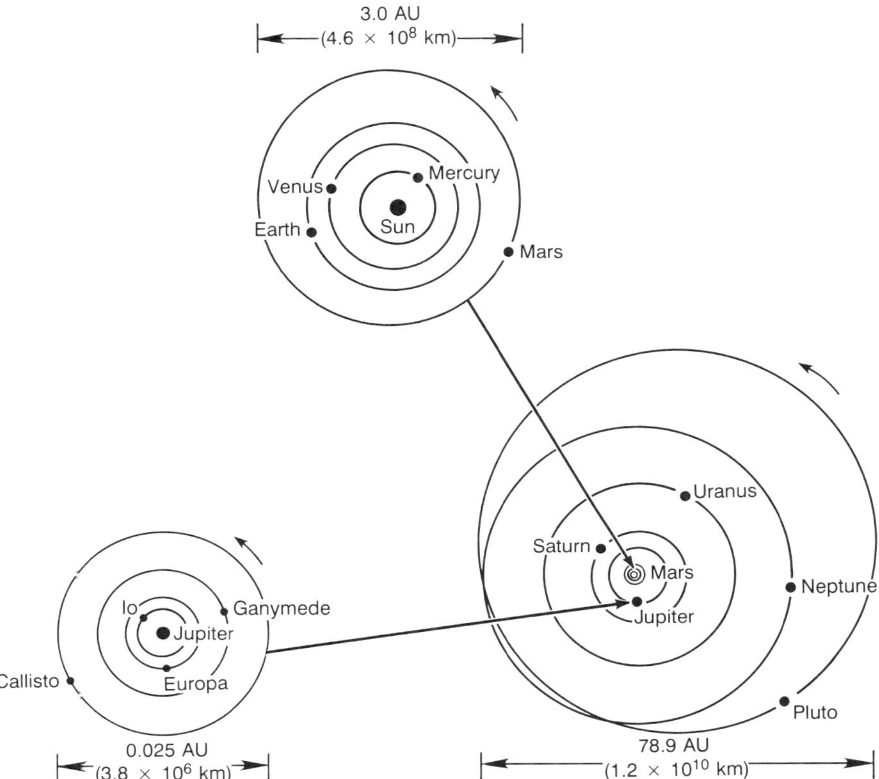

Figure 4.6. The relative sizes of the orbits of the planets, including those of Jupiter's Galilean satellites. The Sun is not drawn to scale (it is 80 times smaller than Mercury's orbit).

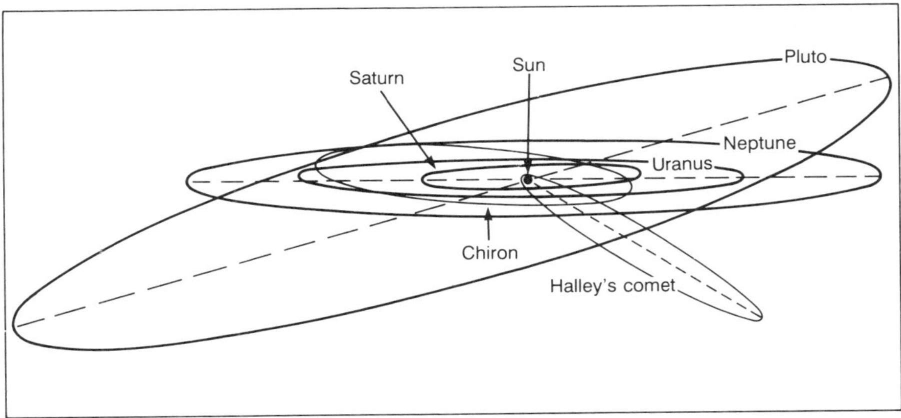

Figure 4.7. Diagram of the orbits of Saturn, Uranus, Neptune, and Pluto; the asteroid 2060 Chiron; and Halley's comet; as viewed from 5° above the ecliptic. Pluto, Chiron, and Halley's comet have noticeably elongated and inclined orbits.

Table 4.1. Orbital and Physical Data of the Sun's Planets.

	Orbital characteristics					Physical characteristics				
	Semi-major axis[a] (AU)[b]	Sidereal period[c] (years)	Average orbital velocity (km/sec)	Eccentricity[d]	Inclination to ecliptic[e] (degrees)	Mass (earth masses)[f]	Equatorial radius (earth radii)[g]	Average density (g/cm^3)	Sidereal rotation period[c]	Inclination of equator to orbit
Mercury	0.39	0.24	48.	0.21	7	0.056	0.38	5.4	58.7 days	0°
Venus	0.72	0.62	35.	0.007	3	0.82	0.95	5.3	243 days (retrograde)	2°
Earth	1.00	1.00004	30.	0.02	—	1.00	1.00	5.5	23.9345 hours	23°.44
Mars	1.52	1.88	24.	0.09	2	0.11	0.53	3.9	24.6 hours	24°
Jupiter	5.20	11.9	13.	0.05	1	318.	11.2	1.3	9.8 hours	3°
Saturn	9.55	29.5	9.6	0.06	2	95.	9.5	0.69	10.2 hours	29°
Uranus	19.2	84.0	6.8	0.05	1	14.5	4.0	1.2	17.3 hours (retrograde)	82°
Neptune	30.1	165.	5.4	0.009	2	17.	3.8	1.7	ca. 16 hours	29°
Pluto	39.4	248.	4.7	0.25	17	ca. 0.002	ca. 0.24	ca. 0.9	6.4 days	≥ 50°

Source: Data from Beatty, et. al., *The New Solar System*.

[a]Semimajor axis: half the major axis of a planet's elliptic orbit (see box 4.1).

[b]Astronomical unit (AU): semimajor axis of Earth's orbit; equal to 1.5×10^8 km.

[c]Sidereal period: period of rotation or revolution relative to the fixed stars.

[d]Eccentricity: measure of deviation of ellipse from a circle; the eccentricity of a circle is zero, that of a straight line segment is one (see box 4.1).

[e]Ecliptic: plane of Earth's orbit.

[f]Earth mass: 6.0×10^{24} kg

[g]Earth radius (equatorial): 6,380 km the west.

direction. However, while Venus' equator is inclined by only 2° to its orbit, Uranus' equator is tilted almost perpendicularly to its orbit.

5. The planets orbital distances from the Sun are so regular that they can be approximated by a very simple numerical "law," discovered in 1781 by Johann Daniel Titius (1729–1796) of the University of Wittenberg in Germany. Titius found that taking the sequence of nine numbers 0, 3, 6, 12, . . . , 384, adding 4 to each, and dividing the sums by 10, yields approximately the distances from the Sun to Mercury, Venus, and so forth in astronomical units (with the single exception of Neptune), as shown in the table on the following page. This law is named after its discoverer and after Johann Elert Bode (1747–1826) who, as director of the Berlin Observatory, published and popularized it.

Symbol	Planet	Distance from the sun according to the Bode-Titius Law (AU)	Actual average distance from the Sun (AU)
☿	Mercury	(0 + 4)/10 = 0.4	0.39
♀	Venus	(3 + 4)/10 = 0.7	0.72
⊕	Earth	(6 + 4)/10 = 1.0	1.00
♂	Mars	(12 + 4)/10 = 1.6	1.52
	Asteroids	(24 + 4)/10 = 2.8	2.70
♃	Jupiter	(48 + 4)/10 = 5.2	5.20
♄	Saturn	(96 + 4)/10 = 10.0	9.55
♅	Uranus	(192 + 4)/10 = 19.6	19.2
♆	Neptune		30.1
♇	Pluto	(384 + 4)/10 = 38.8	39.4

Asteroids

At the time of the discovery of the Bode–Titius law, the space between Mars and Jupiter was marked by a glaring gap. The law predicted a planet at a distance of 2.8 AU, but none was known to exist. The problem was solved 20 years later in 1801, when a small body approximately one-third the size of the Moon was sighted orbiting the Sun at almost the exact distance predicted. It was named *Ceres*, after the Roman goddess of agriculture. Soon other similar bodies were found, most of them in the gap between Mars and Jupiter, and today we know of thousands of them. Most are less than one kilometer across and their total mass is no more than one-tenth the mass of our Moon. These bodies are known as the *asteroids* or *minor planets*.

The word *asteroid* comes from the Greek and means "starlike," although these bodies are not like stars at all. Most of them are rocky objects, and all but the largest ones are irregularly shaped and heavily marked by collisions they have suffered over the eons. Some appear to be double or multiple bodies and are thought to be the broken-up pieces of formerly larger bodies. All asteroids are believed to be leftover debris from the formation of the Solar System. A later discussion will show why they never managed to condense into one single planetary mass.

Most asteroids orbit the Sun in a broad belt, the *Asteroid Belt*, at an average distance from the Sun of 2.7 AU (see table 4.2). However, not all asteroids are so confined. Some have very elliptical orbits that extend inward past the orbits of Mars and Earth. They are the *Amor* (Mars-crossing) asteroids and *Apollo* (Earth-crossing) asteroids. Asteroid Icarus, which is a chunk of rock 700 m across, has an orbit that carries it to within 0.19 AU of the Sun (in comparison,

Table 4.2. Orbital and Physical Data of the Moon, the Galilean Satellites, Titan, Triton, and Some of the Small Bodies in the Solar System.

			Orbital characteristics				Physical characteristics		
	Total number of known satellites or bodies		Average distance from central body	Sidereal period	Eccentricity	Inclination to planet's equator (degrees)	Mass (kg)	Radius (km)	Average density (g/cm^3)
Earth	1	Moon	384,400 km	27.3 days	0.055	23	7.4×10^{22}	1740	3.3
Mars	2	Phobos	9380 km	0.3 days	0.02	1	9.6×10^{15}	14×10	1.9
		Deimos	23,500 km	1.3 days	0.002	2	2.0×10^{15}	8×6	2.1
Asteroids	about 2000	Vesta	2.4 AU	3.6 years	0.09	7	2.0×10^{20}	263	3.0
		Iris	2.4 AU	3.7 years	0.23	6	1.5×10^{19}	105	3.0
		Hebe	2.4 AU	3.8 years	0.20	15	2.0×10^{19}	98	5.0
		Eunomia	2.6 AU	4.3 years	0.19	12	4.0×10^{19}	124	5.0
		Juno	2.7 AU	4.4 years	0.26	13	2.0×10^{19}	134	2.0
		Ceres	2.8 AU	4.6 years	0.08	11	1.0×10^{21}	487	2.0
		Pallas	2.8 AU	4.6 years	0.24	35	2.0×10^{20}	269	2.0
		Psyche	2.9 AU	5.0 years	0.14	3	4.0×10^{19}	119	5.5
		Hygiea	3.2 AU	5.6 years	0.10	4	6.0×10^{19}	211	1.5
		Davida	3.2 AU	5.7 years	0.18	16	3.0×10^{19}	159	1.5
Jupiter	16 (4 in retrograde orbits), Ring System	Io	413,000 km	1.8 days	0.00	0	8.9×10^{22}	1820	3.6
		Europa	671,000 km	3.6 days	0.00	0.5	4.9×10^{22}	1560	3.0
		Ganymede	1,070,000 km	7.2 days	0.00	0.2	1.5×10^{23}	2640	1.9
		Callisto	1,880,000 km	16.7 days	0.01	0.3	1.1×10^{23}	2410	1.8
Saturn	23 (1 in retrograde orbit), Ring System	Mimas	186,000 km	0.9 day	0.02	2	4.5×10^{19}	196	1.4
		Enceladus	238,000 km	1.4 days	0.00	0	8.4×10^{19}	250	1.2
		Tethys	295,000 km	1.9 days	0.00	1	7.6×10^{20}	530	1.2
		Dione	377,000 km	2.7 days	0.00	0	1.1×10^{21}	560	1.4
		Rhea	527,000 km	4.5 days	0.00	0.4	2.5×10^{21}	765	1.3
		Titan	1,220,000 km	15.9 days	0.03	0.3	1.4×10^{23}	2580	1.9
		Iapetus	3,560,000 km	79.3 days	0.03	15	1.9×10^{21}	730	1.2
Uranus	15, Ring System	Miranda	130,000 km	1.4 days	0.00	3	7.5×10^{19}	240	1.3
		Ariel	191,000 km	2.5 days	0.00	0	1.4×10^{21}	580	1.7
		Umbriel	266,000 km	4.1 days	0.00	0	1.3×10^{21}	595	1.4
		Titania	436,000 km	8.7 days	0.00	0	3.5×10^{21}	805	1.6
		Oberon	583,000 km	13.4 days	0.00	0	2.9×10^{21}	775	1.5
Neptune	2, Ring System	Triton	355,000 km	5.9 days (retrograde)	0.00	20	3.4×10^{22}	1750	1.5
		Nereid	5,560,000 km	365 days	0.75	28	2.8×10^{19}	150	2.0
Pluto	1	Charon	19,700 km	6.4 days (retrograde)	0.0	86	1.6×10^{21}	ca. 750	ca. 0.9
Comets	about 100 about 500 about 200 billion	short-period long-period Oort cloud	<35 AU >35 AU 10^3–10^5 AU	<200 years >200 years 10^4–10^7 years	0.1 to 0.97 (?) (?)	mostly <30 all values all values	10^9–10^{17}	0.1–25	1–2

Sources: Data from Beatty, et al., 1982; Allen 1973; Brown, R. H. and D. P. Cruikshank 1985; Stone and Miner 1986.

Note: The table includes data for all satellites and asteroids with masses greater than 10^{-4} times the mass of the Moon. The satellites of Mars and comets are included also, even though their masses are less.

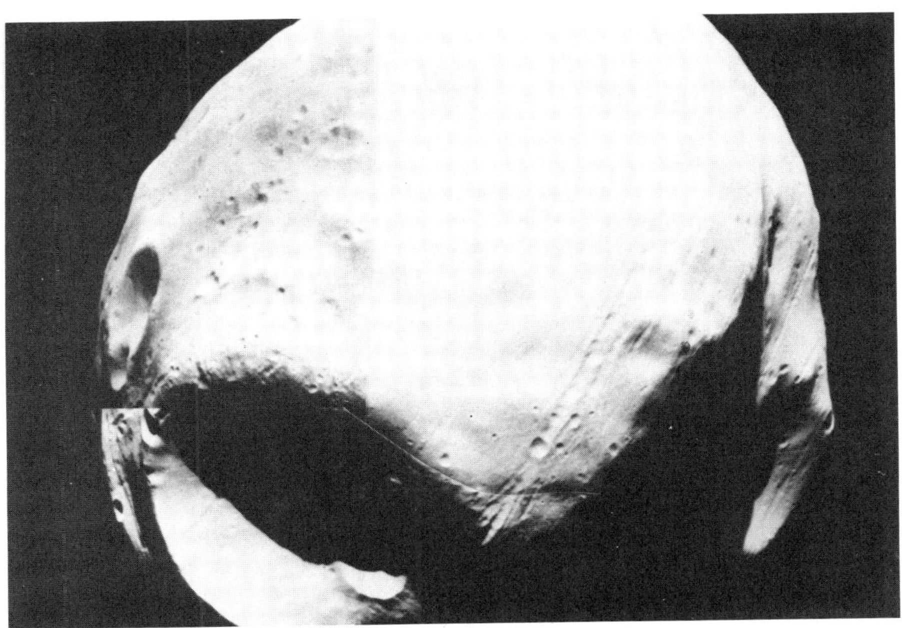

Figure 4.8. Phobos, one of the two asteroid-like moons of Mars. This potato-shaped satellite measures roughly 14 × 10 km and revolves about Mars in a nearly circular orbit at a distance of 9380 km. The crater near the bottom of Phobos (called Stickney) is evidence of the frequent violent collisions among planetary bodies that characterized the early history of the Solar System.

Mercury orbits the Sun at roughly twice that distance) and outward past the orbit of Mars. Other asteroids stray outward to beyond the orbits of Saturn and Uranus as, for example, Chiron (or Asteroid 2060), discovered in 1977. Still others, the *Trojan Asteroids* (most of which are thought to be icy rather than rocky objects), are firmly locked into orbits with the same period as Jupiter's, with one group traveling 60 degrees ahead and the other group 60 degrees behind Jupiter, as seen from the Sun's position.

Many of the smaller satellites of the planets are thought to be asteroids that were captured by the planets (see table 4.2). These small satellites have masses, sizes, and compositions like asteroids. For example, the two moons of Mars, Phobos (figure 4.8) and Deimos, have dimensions of only 14 × 10 km and 8 × 6 km, respectively. They are irregularly shaped and photographs reveal surfaces marred by many large and small impact craters, indicating countless collisions in the past with other rocky debris in interplanetary space. Eleven of Jupiter's 16 known satellites are less than 100 km across and the outer four of them are in retrograde orbits. Twelve of Saturn's 23 known satellites are also quite small (100 km in size or smaller), and so are 10 of Uranus's 15 known satellites. No doubt, as astronomers continue their explorations of the outer Solar System, they will discover many additional small asteroid-like satellites circling the giant planets.

Comets

Far beyond the planets, in the dark and cold outer fringes of the Solar System, many billions of small, primordial bodies of rock and ice are suspected to circle the Sun. They constitute the *Oort Cloud* of dormant cometary bodies, which, like the asteroids, are thought to be leftover debris from the formation of the Solar System. This cloud is believed to extend up to 100,000 AU outward from the Sun, which is more than a third of the way to the Alpha Centauri system (the nearest stars, at a distance of 270,000 AU; see chapter 1).

The existence of this cloud of cometary bodies was first proposed in 1950 by the Dutch astronomer Jan Oort, in order to explain the frequency of appearance of comets in the inner part of the Solar System. According to Oort, the occasional passage of nearby stars perturbs the orbits of some of these bodies and deflects them toward the inner part of the Solar System (see box 4.3). There, bombardment by radiation and charged particles from the Sun brings them to "life." Within about 3 AU of the Sun their ices start vaporizing, and gas and dust particles stream outward to form an envelope—the *coma* and *hydrogen cloud*—around the central body, the *nucleus* (see figure 4.9). Pressure exerted by the solar radiation and solar wind drives the gases and dust many millions of kilometers through interplanetary space to form the cometary *tail*. Due to the way it is formed, the tail always points away from the Sun. The main features of the components of comets are summarized in table 4.3.

BOX 4.3.

Nemesis: Companion Star of the Sun?

IN 1984 a group of scientists proposed that the gravitational perturbations that regularly deflect comets from the Oort Cloud toward the inner part of the Solar System may, in fact, arise from a small and very faint companion star of the Sun, with an orbital period of 26 million years. They made this proposal because of the recent realization that during the past 250 million years life on Earth has experienced episodes of mass extinctions roughly every 26 million years. It is thought that some of these extinctions may have been caused by comets that collided with Earth. Such collisions would have thrown enormous quantities of dust into the upper atmosphere and prevented sunlight from reaching the ground. This, in turn, would have lowered the average ground temperature over much of the globe, slowed or stopped plant growth, and severely interrupted the food chain (see box 9.4 for a discussion of the mass extinction 65 million years ago). Perhaps these periodic comet showers are triggered every time the postulated companion star—which has been named *Nemesis*, after the Greek goddess of doom—is on its closest approach to the Sun in what may be a highly elliptical orbit. (But the point of closest approach would be still far beyond Pluto's orbit.)

Few scientists are convinced that the case for a companion star of the Sun is very strong. If the star's period is 26 million years, the major axis of its orbit would be 2.8 LY, which is a goodly fraction of the average distance between stars in the solar neighborhood. This means that Nemesis's orbit would be exposed to strong tidal forces from other stars and from the Galaxy as a whole, and that the orbit would probably not be stable for much more than about a billion years (which is much less than the age of the Solar System). Still the possible existence of Nemesis cannot be entirely ruled out. For example, the difficulties with orbital stability would be eased if the star had been captured by the Sun during the past few hundred million years (an improbable event) or if it originally had a smaller orbit, for which the tidal disturbances were much smaller. In any case, several astronomers are now actively searching for Nemesis.

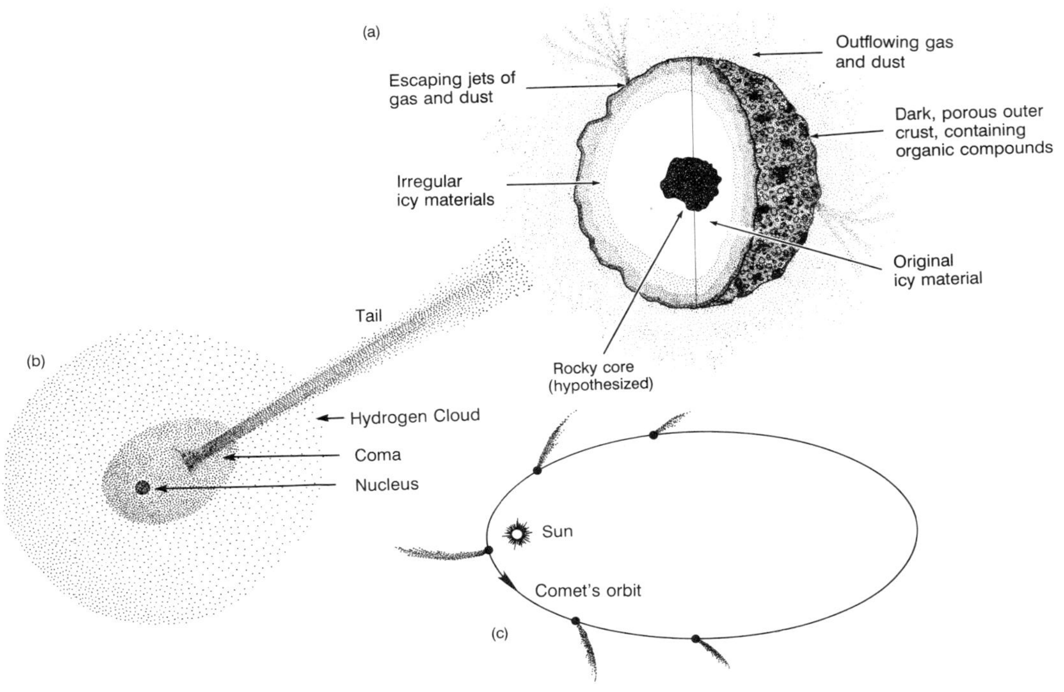

Figure 4.9. The structure of comets. (*a*) Model of comet nucleus; (*b*) Diagram of major components of comets; (*c*) Shape and growth of tail of typical comet as it passes near the Sun.

Table 4.3. Features of Components of Comets.

Component	Composition	Size
Nucleus: central solid body of comet	rocks and ices of water (H_2O), methane (CH_4), and ammonia (NH_3), source of all the dust and gas of the comet's coma, hydrogen cloud, and tail	100 m–25 km irregularly shaped; *mass*: 10^9–10^{17} kg (10^6–10^{14} tons)
Coma: envelope of gas and dust surrounding nucleus	*gas*: neutral atoms and molecules rich in hydrogen, carbon, oxygen, nitrogen, sulfur, and various metals (H, C, O, S, OH, CH, CN, CO, CS, NH, NH_2, HCN, CH_3CN, Na, Fe, K, Ca, V, Cr, Mn, Co, Ni, Cu) *dust*: microscopic silicate minerals (similar to terrestrial sandy dust but less compact)	10^5–10^6 km (approx. the size of the Sun)
Hydrogen Cloud: enormous nearly spherical envelope	atomic hydrogen, derived from the disintegration (by sunlight) of OH and other hydrogen-rich radicals of the coma	up to many million km (much larger than the Sun)
Tail: elongated structure of dust and gas	*dust*: primarily silicate minerals *gas*: charged particles (e^-, CO^+, CO_2^+, H_2O^+, OH^+, CH^+, CN^+, N_2^+, C^+, Ca^+)	*length*: up to 1 AU

Sunlight reflected from the gas and dust particles produces the image that we commonly have of comets—the long, sweeping tail, extending out from the bright, globular coma. We obtain this image of comets mostly from photographs. With the unaided eye comets can usually be seen only as inconspicuous, faintly glowing objects, right before sunrise or after sunset. However, occasionally comets do become spectacular. Then they arouse people's interest and imagination, and even strike awe and fear in some of them. That was the effect Comet Halley had during its apparition in 1910.

About four comets visit the inner part of the Solar System every year. Most of them come from far beyond the orbit of Pluto, approach the Sun in highly elongated orbits, whip around it, and then recede again. Hundreds of thousands of years may go by before they return once again. There are, however, a number of comets that by near encounters with a planet were placed into orbits closer to the Sun. If their orbital periods are less than 200 years (corresponding to about 35 AU for the length of the semimajor axis of their orbits), they are called *short-period comets*. If their periods are greater than 200 years, they are called *long-period comets*.

Comet Halley (named after the British astronomer and mathematician Edmond Halley [1656–1742]) is a short-period comet. It completes an orbit every 75 to 80 years, approaching the Sun to within 0.59 AU and receding to 35 AU. This comet has been sighted regularly since 240 B.C., when Chinese records first make mention of it. Its most recent approach to the inner part of the Solar System occurred in 1985–86, when it was extensively investigated and photographed by five unmanned spacecraft (see box 4.4).

Every time a comet passes through the inner part of the Solar System it loses some of its constituent gases, dust, and rocks. After hundreds to thousands of such passages, nothing is left but a fragile rocky body that will eventually break up due to the Sun's tidal forces. The rocky rubble that results continues to travel along the comet's original orbit. In the course of time, the combined gravitational effects of the Sun and the planets disperse the cometary debris further and further until it isspread over the entire orbit. This debris is the source of shooting stars or meteors (see chapter 5).

BOX 4.4.
Comet Halley, 1985–86

FOR most nonscientists, the 1985–86 apparition of Comet Halley turned out to be disappointing. Even in the Southern Hemisphere, where viewing the comet was optimal, most people who looked for it were not sure whether they had actually seen it. This disappointment was in stark contrast to the excitement Comet Halley had created in 1910, when its coma became as bright as Polaris, the North Star, and its tail stretched more than halfway across the sky. On May 6 of that year, the Earth encountered rocky debris from the comet, which caused a meteor shower, and on May 19 and 20 the Earth passed either through or very close to the comet's tail.

The differences in the visual displays of Halley's comet between 1910 and 1985–86 (figure A) were mainly due to differences in the relative positions of the comet, Earth, and Sun during these two apparitions. For example, in 1910 the closest approach between Comet Halley and Earth was 0.14 AU and it occurred one month after the comet's perihelion passage, when the coma and tail were still substantial and very bright. In 1986 the closest approach was 0.42 AU and it occurred two months after perihelion passage, when the coma and tail had already begun to fade.

In contrast, professional astronomers were excited by the unprecedented wealth of data, including the first photographs of a cometary nucleus ever taken, that were sent back to Earth from five spacecraft: *Vega 1* and *2* from the Soviet Union, *Suisei* and *Sakigake* from Japan, and *Giotto* from the European Space Agency. Besides television cameras, these spacecraft carried instruments for measuring the energy and chemical composition of ions and neutral gas particles, the rate of impacts by dust particles, the mass and chemical composition of individual dust particles, the strength of magnetic and electric fields, the brightness of the coma at different wavelengths,

Figure A. Comet Halley photographed during its 1985–86 apparition.

BOX 4.4 continued

and the turbulent interaction between the outer coma and the solar wind.

Sending spacecraft to Halley's comet was a scientific adventure of the highest order and a feat that probably even the most optimistic observers of 1910 would not have thought possible. For example, *Giotto*, which weighed 520 kg and was equipped for the most ambitious scientific program of the five probes, was launched on July 2, 1985, atop an Ariane rocket, and sent into an orbit that for the first two months carried it just slightly beyond that of Earth. It then began falling sunward to roughly halfway between the orbits of Earth and Venus, while its forward momentum carried it more than three-quarters of the way around the Sun toward its distant rendezvous with the comet (see figure B).

When *Giotto* was launched, Comet Halley was still beyond the asteroid belt, traveling at approximately 20 km/sec in a retrograde orbit toward the Sun (see figure 4.7). As the comet approached the inner part of the Solar System and the Sun's gravitational pull became stronger, it speeded up rapidly. Late in November, the comet crossed the orbit of Mars. On December 31 it crossed the orbit of Earth, traveling now at slightly more than 40 km/sec. On February 9 it passed perihelion at 0.59 AU from the Sun and a speed of 55 km/sec. It then began its long climb out of the Sun's gravitational well. Finally, on March 13, the comet and *Giotto* passed each other at a distance of 605 km, traveling in opposite directions at a relative speed of nearly 70 km/sec (about 250,000 km/hr, or 50 times the speed of a rifle bullet). The rendezvous point was 0.88 AU from the Sun and 0.99 AU from Earth.

Giotto's designers thought that the craft's journey so close to Halley's comet was going to be a suicide mission, in that cometary dust particles and chunks of debris were likely either to destroy the craft or at least incapacitate some of its instruments. Therefore, they shielded the spacecraft with an inner blanket of Kevlar (material used in bulletproof vests) surrounded by a sheet of aluminum. To be on the safe side, all data

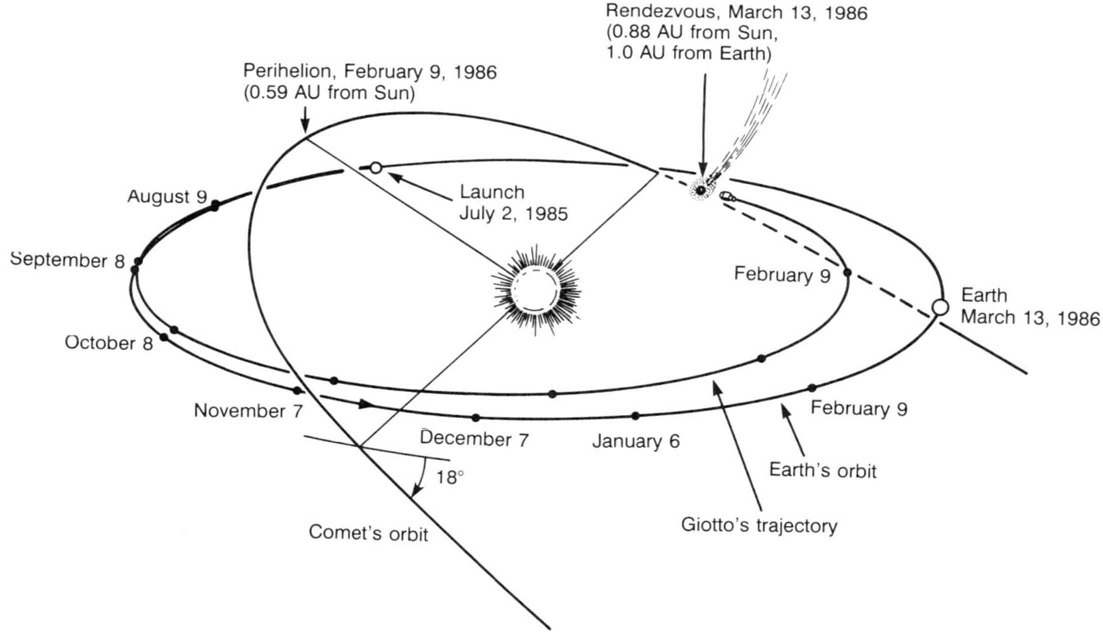

Figure B. The *Giotto* trajectory from launch on July 2, 1985, to its rendezvous with the nucleus of Halley's comet on March 13, 1986.

BOX 4.4 continued

were sent back to Earth in real time; the craft carried no tape recorders for storing information at a fast rate to be transmitted more slowly to Earth after the encounter, as is common with interplanetary probes.

The first indication that *Giotto* was nearing Comet Halley came roughly a day and a half before the rendezvous when, at a distance of 8 million km from the nucleus, the craft began encountering ionized gas particles from the comet, mostly H^+. At 4 million km away from the nucleus, the spacecraft passed through a region of magnetic turbulence, probably caused by interaction between the solar wind and cometary ions. At 3.5 million km, the ionized gas had thickened sufficiently to slow the solar wind by measurable amounts. At 1.1 million km and just 4 hours from closest approach, *Giotto* crossed the comet's bow wave, a 250,000 km wide region of ionized gases created by deflection of the solar wind around the inner part of the comet's coma (somewhat analogous to the bow wave in water in front of and around a moving boat). A variety of atoms, molecules, and radicals were detected, among them singly ionized oxygen (O^+), hydroxyl radicals (OH), carbon dioxide (CO_2), and water (H_2O).

At 280,000 km from the nucleus, *Giotto* encountered its first dust particles. They were microscopic in size, with masses of less than a microgram. Roughly 90 percent of them contained carbon compounds, while some of the rest contained iron. All of the early dust particles were held off by the craft's outer shield. However, at 8000 km distance, the dust particles were bigger and more concentrated, and they started to penetrate the craft's outer shield. In fact, the steady impact of cometary gas and dust at 50 times the speed of bullets slowed *Giotto* noticeably.

At 4300 km from the nucleus, instruments on *Giotto* detected a big decrease in the number of ions and a corresponding increase in neutral particles. The strength of the magnetic field, which tends to follow the motions of charged particles, began to diminish sharply as well. This was the point where purely cometary particles held off the last of the solar wind particles.

About 3 hours before closest approach, and still some 750,000 km distant, *Giotto* turned on its television camera and began taking pictures. They showed a peanut or potato-shaped nucleus, 15 km by 8 km in size, silhouetted against the bright background of the coma. They also showed two very bright jets of dust spurting outward from the nucleus.

The nucleus appears to be pitch-black. Its surface reflects only about 1 to 2 percent of the incident sunlight, making Halley the darkest known object orbiting the Sun. (In comparison, the Moon reflects 7 percent of the incident sunlight.) The dark color took most experts by surprise, for they had thought of cometary nuclei as being dirty snow or ice balls, and the presence of ices should raise the reflectivity. The concept of a dirty snow ball is probably still correct for the interior of Halley's nucleus. But the surface layers are depleted in ices. The ices that originally were present there have been vaporized and were driven off into interplanetary space during the comet's many close passages around the Sun. This repeated loss of volatiles probably produced a porous, rocky surface crust or loosely bound "rubble pile,"* as one scientist suggested, containing tarry material rich in organic compounds that give rise to the observed blackness.

Giotto took its last picture 1350 km from the nucleus, when all data reception ceased. Apparently, dust jolted the spacecraft and caused it to wobble violently. This knocked its narrow radio beam out of alignment with Earth and interrupted transmission. A few seconds later, intermittent signals were received again, as *Giotto*'s transmitter swung in and out of alignment. A little over half an hour later, the onboard damping device had steadied the craft sufficiently for continuous transmission of data to resume. Surprisingly, all of the craft's 10 instruments survived the incident, though six suffered some damage. The European Space Agency is now making plans to boost *Giotto* into a slightly altered orbit that would bring it by Earth in 1990 and then to use our planet's gravitational field for redirecting the craft toward another comet or asteroid.

The other space probes sent to Halley's Comet passed the nucleus a few days earlier than *Giotto* and at greater distances. Qualitatively they confirmed the findings of the *Giotto* mission, though there were differences. For example, the two *Vega* spacecraft encountered more dust than *Giotto*. Apparently, much of the gas and dust is emitted from the nucleus in jet-like bursts that vary in time and direction. Because the spacecraft followed courses that differed

**Science News*, March 22, 1986, p. 181.

BOX 4.4 continued

in space and time, they encountered different dust conditions.

In addition to emitting gas and dust in bursts and unevenly from its surface, Halley's nucleus is rotating. This was deduced from pictures sent back by *Suisei*, one of the Japanese craft, which indicate that the comet's extensive hydrogen cloud was growing and waning regularly. This is best explained by assuming that new gas is injected into the hydrogen cloud from a nucleus that releases the gas preferentially from one side and that also rotates. The rotation period was calculated to be 53 hours.

All of the data sent back to Earth from the five spacecraft were freely shared with the international scientific community. In fact, only through this sharing was *Giotto*'s close fly-by of the nucleus possible. Before any of the spacecraft had come close to Comet Halley, the position of the nucleus was known only to within several thousand kilometers within the haze of the coma's scattered light. However, by using information from several sources—precise measurements of the trajectories of each of the *Vega* spacecraft, obtained by Soviet and United States tracking stations, and position estimates of the comet's nucleus relative to the *Vegas*, deduced from the photographs sent back by these two spacecraft—European scientists were able to reduce *Giotto*'s targeting error to within about 100 km. Perhaps this spirit of international cooperation will remain Comet Halley's major legacy.

Exercises

1 (*a*) Name the seven planets or "wanderers" known to the ancients.
(*b*) Using a planet chart locate as many of these objects on the sky as conditions allow. (Monthly planet charts are included in each issue of such popular astronomy magazines as *Astronomy* and *Sky and Telescope*.)

2 Draw a diagram showing the Moon's orbit around Earth and the Moon's dark and sunlit hemispheres at several positions. Use this diagram to explain the Moon's phases. In particular,
(*a*) Why is the first-quarter Moon overhead at sunset?
(*b*) Why does the full Moon rise at sunset and set at sunrise?
(*c*) Why is the last-quarter Moon overhead at sunrise?
(*d*) Why do solar eclipses occur only when the Moon is new?
(Note that the directions of the Moon's orbital motion and the Earth's rotation are the same.)

3 Mercury and Venus can only be seen in the evening and morning. Why?

4 Convert the Bode–Titius law into one that gives the planets' distances from the Sun in units of 5.2 AU, the Sun–Jupiter distance.

5 (a) From the values of the astronomical unit and the light-year given in the text, calculate how many astronomical units equal one light-year.
(*b*) Calculate in AU the distance from Earth to α Centauri (the distance equals 4.38 LY).

6 According to one theory, Planet X orbits the Sun beyond Pluto, at about 50 to 100 AU from the Sun. If the Bode–Titius law applies to this hypothetical planet, what would its orbital distance be?

7 (*a*) Using a planet chart, check which of the following planetary objects are presently visible in the night sky: Jupiter, Saturn, Ceres, Vesta, and Pallas. Locate as many of these objects, including Jupiter's Galilean satellites, as conditions allow. (Use binoculars in the case of the asteroids and Jupiter's satellites.)
(*b*) If Jupiter is visible, record the positions of its Galilean satellites over the course of several weeks and determine their orbital periods.

8 (*a*) Where in their orbital journeys, and how, do comets acquire their comae, hydrogen clouds, and tails?
(*b*) Describe the composition and structure of each of these cometary components.

Suggestions for Further Reading

Beatty, J. K. 1986. "A Place Called Uranus." *Sky and Telescope*, April, 333–37. An article about *Voyager 2*'s close-up view of Uranus in January 1986.

Beatty, J. K., B. O'Leary, and **A. Chaikin,** eds. 1982. *The New Solar System.* 2d ed. Cambridge: Cambridge University Press. An in-depth, yet very readable, treatment of the Solar System, consisting of 20 chapters written by research scientists and covering planetary observations through the *Voyager* missions to Jupiter and Saturn.

Boss, A. P. 1986. "The Origin of the Moon." *Science*, 231: 341–45. The author suggests that the Moon originated as a result of a collision of the proto-Earth with a roughly Mars-sized body; the ejecta from this collision created a circumterrestrial disk of matter from which the Moon later accreted. The author also discusses the older fission, capture, and binary accretion theories for the Moon's origin and presents arguments why they are physically either impossible or highly improbable. For the advanced student.

Brandt, J. C., and **M. B. Niedner, Jr.** 1986. "The Structure of Comet Tails." *Scientific American*, January, 48–56.

Briggs, G., and **F. Taylor.** 1982. *The Cambridge Photographic Atlas of the Planets.* Cambridge and New York: Cambridge University Press.

Brown, R. H., and **D. P. Cruikshank.** 1985. "The Moons of Uranus, Neptune, and Pluto." *Scientific American*, July, 38–47. A preview of what *Voyager 2* might find when it flies by Uranus in January 1986 and Neptune in 1989.

Chaikin, A. 1986. "Voyager Among the Ice Worlds." *Sky and Telescope*, April, 338–43. The rings and satellites of Uranus as they were observed by *Voyager 2* during its January 1986 flyby.

Chapman, R. D., and **J. C. Brandt.** 1984. *The Comet Book: A Guide for the Return of Halley's Comet.* Boston: Jones and Bartlett Publishers. An excitingly written, easy-to-read book about all aspects of comets—their observations past and present, structure, composition, origin, eventual death, and legends about them.

Cohen, I. B. 1981. "Newton's Discovery of Gravity." *Scientific American*, March, 166–79.

———. 1955. "Isaac Newton." *Scientific American*, December, 73–80.

Cuzzi, J. N., and **L. W. Esposito.** 1987. "The Rings of Uranus." *Scientific American*, July, 52–66. *Voyager 2*'s observations of Uranus' rings suggest that they may be only "a fleeting stage in a continuing saga of creation and destruction."

Dietz, R. S., and **J. C. Holden.** 1970. "The Breakup of Pangaea." *Scientific American*, October, 30–41.

"The Dynamic Earth." 1983. *Scientific American*, September. Eight articles on the structure, dynamic changes, ocean, atmosphere, and biosphere of our planet.

Gehrels, T. 1985. "Asteroids and Comets." *Physics Today*, February, 32–41.

Gingerich, O. 1982. "The Galileo Affair." *Scientific American*, August, 132–43. The article describes the prevailing pattern of thinking when Galileo defended the Copernican system, the mode of reasoning introduced by Galileo, and the new method of hypothesis testing that developed from it.

Gore, R. 1981. "*Voyager 1* at Saturn, Riddles of the Rings." *National Geographic*, July, 3–31. A well-illustrated article about the drama of *Voyager 1*'s flyby and observations of Saturn, its ring system, and its moons.

———. 1986. "Uranus: *Voyager* Visits a Dark Planet." *National Geographic*, August, 178–94.

Haberle, R. M. 1986. "The Climate of Mars." *Scientific American*, May, 54–62. The climate of Mars started out much like Earth's, with temperatures sufficiently high for water to flow on the planet's surface. But Mars evolved differently, and today the Martian surface is so cold that carbon dioxide freezes at its poles.

Hess, W., et al. 1969. "The Exploration of the Moon." *Scientific American*, October, 54–72.

Horowitz, N. H. 1977. "The Search for Life on Mars." *Scientific American*, November, 52–61. Experiments designed to detect life on Mars, carried out by the two *Viking* landers in 1976.

Ingersoll, A. P. 1981. "Jupiter and Saturn." *Scientific American*, December, 90–108. Description of two competing models of the Sun's two largest planets.

Ingersoll, A. P. 1987. "Uranus." *Scientific American*, January, 38–45. Results from *Voyager 2*'s flyby of the planet in January 1986.

Johnson, T. V., and **L. A. Soderblom**. 1983. "Io." *Scientific American*, December, 56–67.

Kasting, J. F., O. B. Toon, and **J. B. Pollack**. 1988. "How Climate Evolved on the Terrestrial Planets." *Scientific American*, February, 90–97. The authors address the question, "Why is Mars too cold for life, Venus too hot, and the Earth just right?"

Lawless, J. G., C. E. Folsome, and **K. A. Kvenvolden**. 1972. "Organic Matter in Meteorites." *Scientific American*, June, 38–46.

Lewis, R. S., and **E. Anders**. 1983. "Interstellar Matter in Meteorites." *Scientific American*, August, 66–77. Carbonaceous chondrites incorporate material that originated outside the Solar System, including matter expelled by supernovae and other stars.

Maxwell, J. C. 1985. "What is the Lithosphere?" *Physics Today*, September, 32–40. An authoritative review of our current understanding and of some of the unresolved problems of the theory of plate tectonics.

McCloskey, M. 1983. "Intuitive Physics." *Scientific American*, April, 122–30. Newton's laws are well known, but tests show many people think moving objects act otherwise.

Moore, P., and **J. Mason**. 1984. *The Return of Halley's Comet.* New York: W. W. Norton and Co. A well-written book, with excellent discussions of Comet Halley's 1910 and earlier apparitions, its orbital characteristics, and the spacecraft sent to meet it in 1986.

Nance, R. D., T. R. Worsley, and **J. B. Moody**. 1988. "Supercontinent Cycle." July, 72–79. Plate tectonics keeps rearranging the Earth's surface, forming a supercontinent roughly every 500 million years and then breaking it up again. The process affects climate and biological evolution.

Pollack, J. B., and **J. N. Cuzzi**. 1981. "Rings in the Solar System." *Scientific American*, November, 104–29. The structure and composition of the rings of Jupiter, Saturn, and Uranus, the processes that shape them, and their possible origins.

Press, F., and **R. Siever**. 1982. *Earth*. 3d ed. San Francisco: W. H. Freeman and Co. A college-level geology text.

Prinn, R. G. 1985. "The Volcanoes and Clouds of Venus." *Scientific American*, March, 46–53.

Runcorn, S. K. 1987. "The Moon's Ancient Magnetism." *Scientific American*, December, 60–68. Lunar fossil magnetism points to an eventful past, including the existence once of a liquid iron core and collisions with orbiting satellites.

Russell, D. A. 1982. "The Mass Extinctions of the Late Mesozoic." *Scientific American*, January, 58–65. The mass extinction of 65 million years ago.

Schramm, D. N., and **R. N. Clayton**. 1978. "Did a Supernova Trigger the Formation of the Solar System?" *Scientific American*, October, 124–39. Unusual isotope abundances in carbonaceous chondrites offer strong observational evidence that the formation of the Solar System was triggered by a nearby supernova.

Schubert, G., and **C. Covey**. 1981. "The Atmosphere of Venus." *Scientific American*, July, 66–74.

Soderblom, L. A., and **T. V. Johnson**. 1982. "The Moons of Saturn." *Scientific American*, January, 100–116.

Soderblom, A. 1987. "The Moons of Uranus." *Scientific American*, April, 48–60. *Voyager 2*'s results suggest that Uranus' five major satellites experienced vigorous geological activity early in their histories.

Weaver, K. F. 1986. "Meteorites: Invaders From Space." *National Geographic*, September, 390–418.

Wetherill, G. W. 1981. "The Formation of the Earth from Planetesimals." *Scientific American*, June, 162–74. Computer simulations show how the terrestrial planets might have formed. This article is one of the key sources for the last section of chapter 4.

Additional References

Allen, C. W. 1973. *Astrophysical Quantities*, 3rd ed. London: The Athlone Press.

Hughes, D. W. 1982. "Was Galileo 2000 Years Too Late?" *Nature* 296 (18 March): 199.

Kepler, J. 1981. *Mysterium Cosmographicum*, trans. A. M. Duncan. New York: Abaris Books.

National Geographic Atlas of the World, 5th ed. 1981. Washington, D.C., 30–31.

Press, F., and **R. Siever**. 1978. *Earth*, 2d ed. San Francisco: W. H. Freeman and Co.

Stone, E.C., and **E. D. Miner**. 1986. "The *Voyager 2* Encounter with the Uranian System." *Science* 233 (4 July): 39.

Erupting volcano on Io, photographed from a distance of 500,000 km by *Voyager I* during its historic encounter with Jupiter, March 4–5, 1979. ➤

5

The Solar System: Structure of Planetary Bodies and Origin

In chapter 4 we have primarily concerned ourselves with the orbital layout of the Solar System. In order to progress toward our goal of understanding the origin of the Sun and the planets, we must also learn about the physical makeup of the system's member bodies. The structure of the Sun was discussed in chapter 3. Now let us look at other bodies in the Solar System and, in the last section of this chapter, at some of the current theories of its origin and evolution.

THREE FAMILIES OF BODIES IN THE SUN'S PLANETARY SYSTEM

The bodies circling the Sun and constituting the planetary system range from dust and pebble-sized debris to planets with radii of thousands to tens of thousands of kilometers. The lower part of this range includes the asteroids, comets, many (but not all) of the satellites of the planets, the particles composing the rings of Jupiter, Saturn (figure 4.1), Uranus and Neptune, and the interplanetary dust and rock fragments that have resulted from collisions among the bodies. I am including among these *small bodies* of the Solar System those objects that have radii and masses up to approximately 1000 km and 1/1000 earth masses (a few times 10^{21} kg; figures 5.1, 5.2, 5.3). Many of them have orbits that are highly elliptical and inclined to the ecliptic. Together they form a family of bodies with a common history and with compositions that, on the average, vary systematically with distance from the Sun. They are believed to be the leftover debris from the original material out of which the planets and their large satellites were assembled. In the theory of Solar System formation, such small bodies are referred to as *planetesimals*.

A second family of Solar System bodies includes the terrestrial planets—Mercury, Venus, Earth, and Mars. In this book, the Moon, the Galilean satellites, Saturn's satellite Titan, and Neptune's satellite Triton are also included in this family. I refer to them as the *intermediate-sized planetary bodies*. They have radii between roughly 1500 km and 6500 km (figure 5.4); their masses range from about 1/200 to 1 earth mass; they have similar structures; and related physical processes shaped them. The orbits of most of them are nearly circular, with inclinations either close to the ecliptic or to the equatorial plane of the central body about which they revolve. The major exceptions are Mercury, as already noted, and Triton. Triton is in a retrograde orbit that is inclined by 20° to Neptune's equator. As in the case of the small bodies of the Solar System, the compositions of the intermediate-sized planetary bodies also vary systematically with distance from the central body about which they revolve.

In the past, Pluto has been included in the family of intermediate-sized planetary bodies; but, as this family is defined here, Pluto is a marginal case. Pluto's size is comparable to Europa, the smallest of the Galilean satellites; yet its mass is only 1/460 earth masses, which places it somewhere between a large asteroid and a small intermediate-sized planetary body. Furthermore, Pluto's

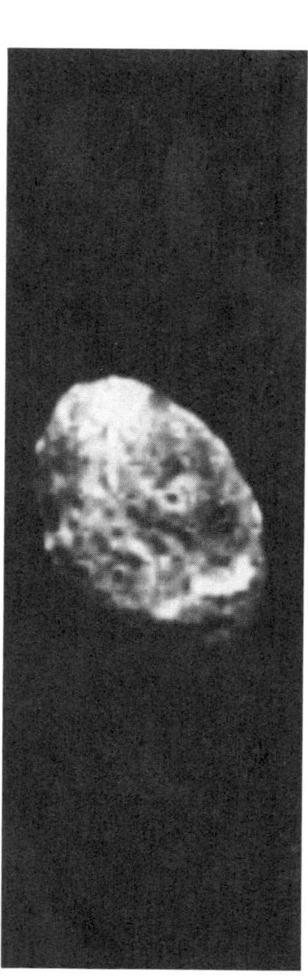

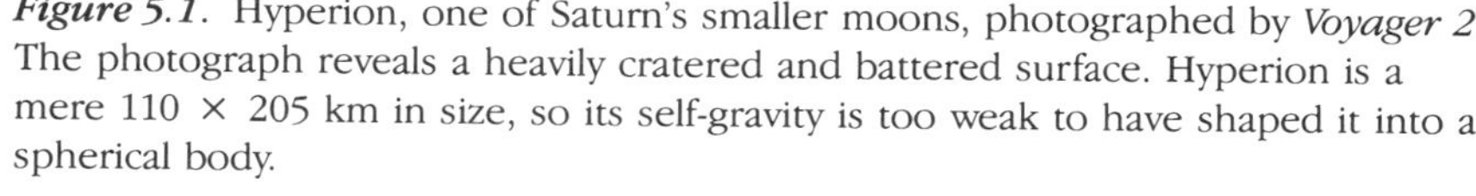

Figure 5.1. Hyperion, one of Saturn's smaller moons, photographed by *Voyager 2*. The photograph reveals a heavily cratered and battered surface. Hyperion is a mere 110 × 205 km in size, so its self-gravity is too weak to have shaped it into a spherical body.

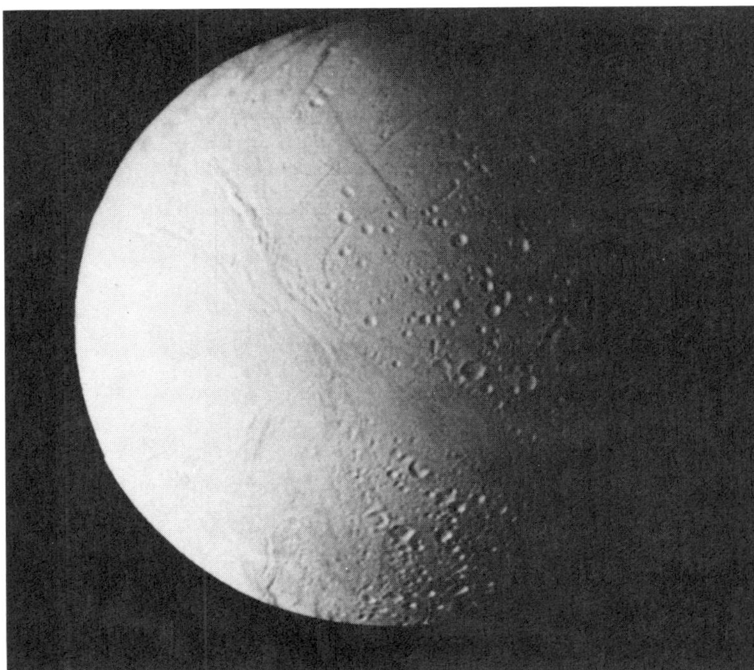

Figure 5.2. Dione, Saturn's fifth largest satellite, with a radius of roughly 560 km. Its density is about 1.4 g/cm^3, indicating that it consists largely of ices intermingled with some rocks. Thousands of craters cover the satellite's surface. A fracture can be seen near the terminator in this photograph, suggesting that Dione has experienced some geologic evolution since its formation. (*Terminator* refers to the dividing line between the illuminated and dark parts of a satellite's or planet's disk.)

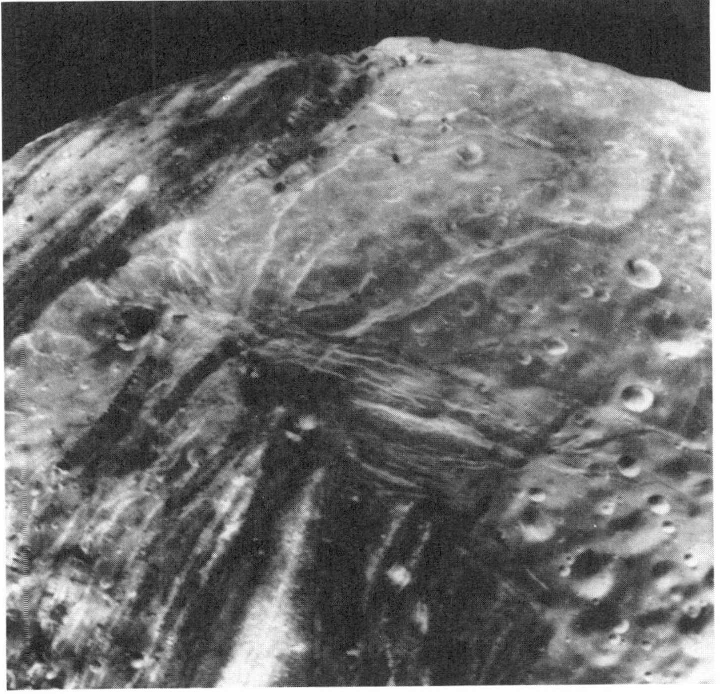

Figure 5.3. Miranda, one of Uranus's five large satellites (about 480 km in diameter), photographed by *Voyager 2* from a distance of 31,000 km. This satellite is one of the most unusual bodies in the Solar System. Its surface is characterized by a bizarre patchwork of ridges and grooves (a chevron near the center and oval "racetracks" near the upper-right and lower-left limbs), jagged cliffs, and crater-dotted regions resembling the lunar highlands. The grooves are up to several kilometers deep and the cliff visible at the lower right is 20 km high. No one knows what forces and events gave rise to this strange moon.

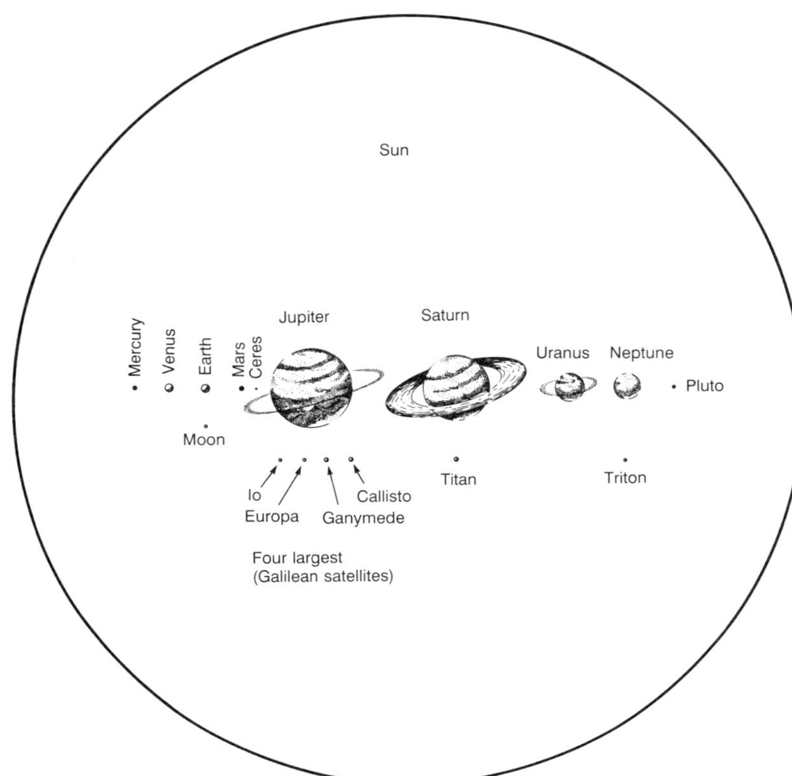

Figure 5.4. Relative sizes of the Sun, the planets, the largest of their satellites, and Ceres.

orbit is quite elliptical and inclined, similar to the orbits of asteroids and comets. Hence in this book Pluto (as well as its companion, Charon) is regarded as a member of the family of small bodies of the Solar System, even though in the tables it will be listed among the planets.

Of course, despite the division made here, there is no really natural or absolute borderline between the families of small and intermediate-sized planetary bodies. Together these bodies constitute a continuum from microscopic dust through small irregularly shaped boulders, to large, roundish bodies of rocks, metals, and volatiles. The choice of the borderline between the two families is quite arbitrary and mainly one of convenience.

The third family of Solar System bodies are the *giant* or *Jovian planets*—Jupiter (figure 5.33), Saturn (figure 4.1), Uranus, and Neptune. Their radii extend from roughly 25,000 to 70,000 km and their masses range from approximately 15 to more than 300 earth masses. Their structures are largely gaseous and liquid, and they have been formed by physical processes that were somewhat different from those of the smaller bodies. Their sizes, structures, and origins place these giant planets somewhere between the terrestrial planets and a small star, though none of them is even close to generating energy by nuclear fusion. Unlike the intermediate-sized planetary and small bodies in the Solar System, they do, however, radiate more energy than they receive from the Sun

Table 5.1. Viewing Data and Orbital Characteristics of the Four Meteor Showers Whose Orbits Are Represented in Figure 5.5.

Meteor shower	Date of maximum display	Normal period of visibility	Constellation in which meteor tracks appear	Period of original comet (years)	Velocity relative to Earth (km/sec)	Eccentricity of orbit	Inclination of orbit to ecliptic (degrees)
Quadrantids	Jan 3	Jan 2–4	Boötes	7	43	0.71	75
Perseids	Aug 12	July 29–Aug 18	Perseus	110 (retrograde)	61	0.96	65
Leonids	Nov 16	Nov 14–19	Leo	33 (retrograde)	72	0.92	17
Geminids	Dec 13	Dec 8–15	Gemini	1.6	37	0.90	26

Source: Adapted from Allen 1973, 158.

by processes that are at present not fully understood. (Summaries of the physical characteristics of members of the three families of Solar System bodies are found in tables 4.1–4.3 and 5.1–5.3.)

SMALL BODIES OF THE SOLAR SYSTEM

The best way to study the small bodies of the Solar System would be to send space probes to the Asteroid Belt and comets and to collect samples. That is what the United States did with its *Apollo* missions to the Moon. Similar missions to the asteroids and comets are technically feasible, but would be so expensive that to date they have not been funded. Until there is the opportunity to collect samples of asteroids and comets directly, scientists have to make do with the information they can obtain through remote sensing, flyby missions, and by collecting and analyzing whatever debris from these smaller bodies happens to collide with our planet and survives the fall to the ground.

Meteorites

As it turns out, many asteroids or fragments of asteroids, as well as the rocky remnants of past comets, are in elongated Earth-crossing orbits (figure 5.5), and every year Earth collides with and sweeps up roughly ten thousand tons of this interplanetary rubble. Most of these encounters are harmless, for the rubble is usually microscopic in size and burns up by friction as it passes through the Earth's upper atmosphere, briefly creating fiery trails across the sky at night. Such trails are popularly known as *shooting stars* and technically as *meteors*. Some meteors arrive in groups of thousands regularly every year and can give rather spectacular displays. Viewing data and orbital characteristics of a few *meteor showers* are given in table 5.1 and figure 5.5 on the next page.

Occasionally, large bodies weighing several kilograms or even tons enter the atmosphere. Typically they travel at speeds of a few times 10 km/sec (about 10^5 km/hr) and friction with the air sometimes produces so much heat that they may become brighter than the full Moon. Often they are sheared apart into two or more fragments. The largest of them, weighing thousands of tons or more, create sonic booms that can be heard for many miles. These large bodies do not

Table 5.2. Physical Characteristics of Meteorites

		Dominant characteristics	Density (g/cm^3)	Elemental composition	Comments
Undifferentiated Meteorites[a]	Ordinary chondrites	Fine-grained earthy matrix containing two distinct components: (1) hard irregularly shaped *mineral inclusions* with high concentrations of Ca, Al, Ti, and (2) round or oval *mineral chondrules*[b] with high concentrations of Fe, Mg, Si.	about 3.7	Solar composition, except volatiles are underrepresented.	*Ca–Al–Ti inclusions* in chondrites are the most refractory constituents found in meteorites, meaning they were at one time exposed to high temperatures (1500 K or higher) that vaporized and drove off the volatile compounds. *Fe–Mg–Si chondrules* were once molten droplets, probably created during high-speed collisions of planetesimals.
	Carbonaceous chondrites	Similar to ordinary chondrites including the presence of Ca-Al-Ti inclusions, but chondrules are fewer in number or absent, and H_2O (2 to 20% by mass) and *organic compounds* (up to 5%) are present.	2.2–3.6		Organic compounds of carbonaceous chondrites include amino acids and tarlike chains of hydrocarbons. Among the amino acids, glycine, alanine, valine, proline, and glutamic acid are abundant, as well as amino acids not present in biologically synthesized protein (see the Introduction to Part Two and chapter 6). Both L- and D-amino acids are present, indicating an abiotic origin.
Differentiated Meteorites[a]	Achondrites	Similar to terrestrial and lunar igneous rocks.	3.0–3.5	Composition very different from solar due to process of differentiation.	Differentiation is believed to have occurred in parent bodies (that is, the planetesimals) of solar or ordinary chondritic composition and with diameters exceeding several ten kilometers.
	Stony irons	Roughly equal proportions of silicate minerals and Fe-Ni metals.	5.5–6.0		
	Irons	About 90% Fe, 5–10% Ni.	7.5–8.0		

[a]*Differentiation* in this context refers to the same process that separated the Earth early in its history into an iron–nickel core, a mantle of high-density silicate rocks, and an outer crust of low-density rocks.

[b]*Chondrite* and *chondrule* are derived from the Greek *chondros*, meaning "grain."

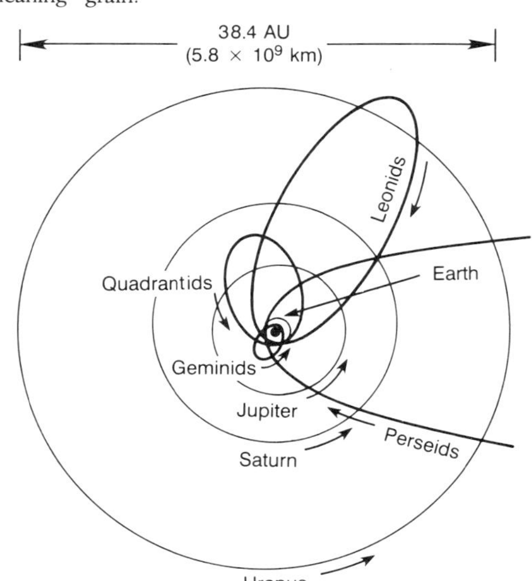

Figure 5.5. Diagram of the orbits of four meteor showers. The shapes (that is, eccentricities) and sizes of the orbits are drawn to scale. The inclinations of the orbits to the ecliptic are given in table 5.1.

Table 5.3. Characteristics of the Atmospheres of Five Intermediate-Sized Planetary Bodies.

	Composition of gases (fractions by number)	Surface pressure P_{Planet}/P_{Earth}	Average surface temperature (K)	(C)	Surface gravity (m/sec^2)
Venus	CO_2 (0.96) N_2 (0.035) H_2O (0.0001) SO_2 (0.0001)	90.	730	457	8.9
Earth	N_2 (0.77) O_2 (0.21) H_2O (0.01) Ar (0.0093) CO_2 (0.00033)	1.	288	15	9.8
Mars	CO_2 (0.95) N_2 (0.027) Ar (0.016) O_2 (0.0013) CO (0.0007) H_2O (0.0003)	0.007	218	−55	3.7
Titan	N_2 (0.80–0.95) CH_4 Organic Molecules Smog Particles	1.6	93	−180	1.4
Triton	CH_4	0.0001	(?)		0.9

Note: O_2 is molecular oxygen, SO_2 is sulfur dioxide.

burn up, but pass right through the atmosphere with very little deceleration. Most of these *meteorites** fall into the ocean, but about one-third of them do come down on land where they carve out craters that may be several meters to many tens of kilometers across.

Fortunately, impacts by meteorites are relatively rare today, and the probability of being struck by one is extremely low. Still, on every continent there exist a number of craters, such as the Barringer Crater near Winslow, Arizona (see figure 5.6), that give evidence of violent collisions between Earth and interplanetary bodies. Another example of a collision of Earth with an interplanetary body is the violent explosion that shook the Tunguska region of Siberia on the morning of June 30, 1908 (see figure 5.7). Apparently, a cometary fragment entered the Earth's atmosphere and exploded just above the ground. People as far as 750 kilometers away saw the event, and the energy released by the explosion scorched a 30-kilometer wide area and leveled trees like matchsticks.

Planetary scientists get greatly excited about new finds of meteorites because they are presently the only samples of asteroid and comet material. Because

*Interplanetary fragments that have actually fallen to the ground are known as *meteorites*, those that burn up in the Earth's atmosphere are called *meteors*, and the bodies still in space are termed *meteoroids*.

Figure 5.6. The Barringer Crater near Winslow, Arizona. The volcanic area of the San Francisco Peaks, near Flagstaff, Arizona, is visible on the horizon, at a distance of 70 km. The crater measures 1.2 km across, has a depth of 180 m, and its rim rises 60 m above the surrounding plain. It was formed about 25,000 years ago by the impact of an iron meteorite estimated to have weighed several thousand tons.

Figure 5.7. The Tunguska region of Siberia, showing the destruction wrought by the violent explosion of a meteorite on June 30, 1908. The photograph was taken a number of years after the explosion.

asteroids and comets are thought to be part of the original material from which the terrestrial planets and the cores of the Jovian planets were assembled, analysis of their chemical and structural makeup is expected to reveal much about the early history of the Solar System.

There exist two fundamentally distinct types of meteorites: meteorites that have changed very little since their formation and meteorites that have experienced major alterations. Meteorites that have changed very little are said to be *undifferentiated* or *primitive*, and they have compositions resembling that of the Sun, except for a paucity in the volatiles (such as the noble gases and compounds containing H, C, N, and O). Among these meteorites astronomers recognize two subspecies—the *ordinary chondrites* and the *carbonaceous chondrites*. Meteorites that have experienced major alterations are called *differentiated* meteorites. Among them astronomers recognize three subspecies—the *achondrites*,* the *stony-iron*, and the *iron* meteorites. The main features of these five kinds of meteorites are summarized in table 5.2.

Planetesimals

The composition of the meteorites indicates that they—or their parent bodies, the *planetesimals*—formed by the coalescence of fine- and coarse-grained mineral matter, metals, and gases of various kinds, including organic compounds (see figure 5.8). This coalescence probably took place in the *protoplanetary disk*, which theory suggests surrounded the nascent Sun and initially was of the same chemical composition as the material that went into the making of the Sun.

Two different mechanisms are believed to have been at work in the formation of planetesimals. One mechanism produced the undifferentiated planetesimals, while the other mechanism produced the differentiated planetesimals:

1. *Origin of the undifferentiated planetesimals (ordinary and carbonaceous chondrites)*. The undifferentiated planetesimals were bodies that never grew much larger than a few ten kilometers across, which was too small for internal melting and differentiation by gravitational forces. However, these planetesimals acquired rather different compositions, depending on where in the protoplanetary disk they formed—compositions resembling those of ordinary chondrites in the inner part of the protoplanetary disk to compositions resembling those of carbonaceous chondrites in the middle and outer parts. These differences in composition resulted from the differential heating the protoplanetary disk experienced, with temperatures ranging from above 1500 K near the young or still forming Sun to below 100 K hear the outer edge of the disk. The matter near the inner part of the disk, which was exposed to the highest temperatures, lost all but the most refractory minerals.† Only the calcium–aluminum–titanium (Ca–Al–Ti) inclusions found

*The chondrites and achondrites together are often referred to as *stony meteorites*.

†The term *refractory* refers to elements or compounds that vaporize only at high temperatures (generally above 500 K), such as the silicate minerals and metals. The term *volatile* means the opposite of refractory. It refers to elements or compounds that vaporize at low or moderate temperatures, such as the noble gases (He, Ne, Ar, Kr, Xe, Rn), H_2O, molecular nitrogen (N_2), carbon monoxide (CO), carbon dioxide (CO_2), and organic compounds.

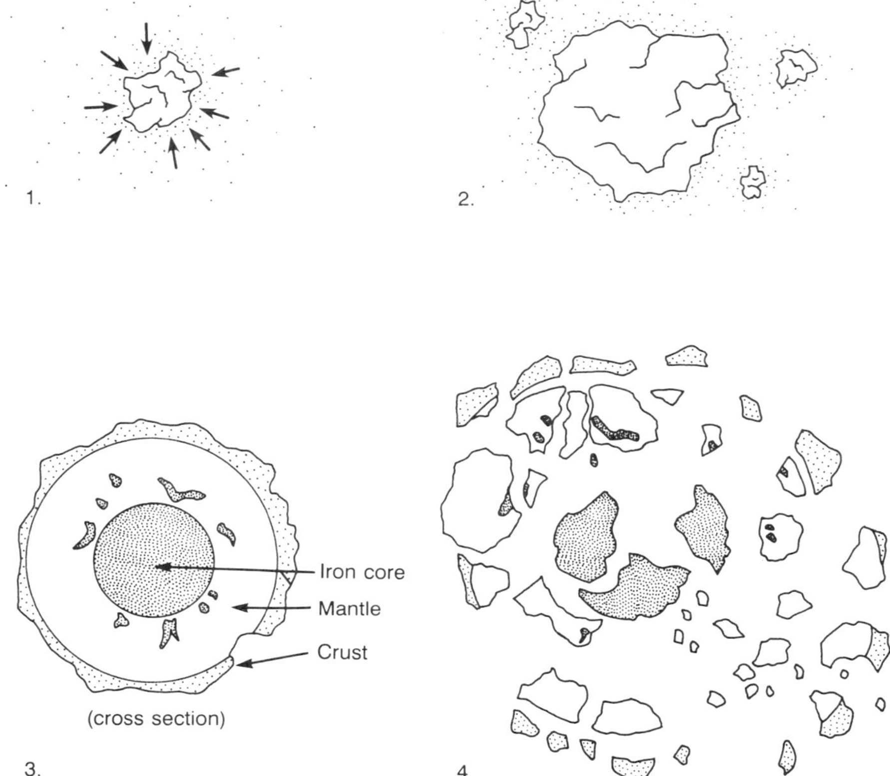

Figure 5.8. Four major stages in the formation history of stony, stony-iron, and iron meteorites.

1. Dust grains in the protoplanetary disk collided and coalesced first into pebble-sized and then into boulder-sized and larger objects.
2. Additional collisions with dust grains and with other larger objects led to the growth of planetesimals. The force holding planetesimals together was mainly gravity.
3. After a planetesimal had grown to several ten kilometers in size, heating by the decay of radioactive isotopes (such as ^{26}Al) melted the interior and led to differentiation. Iron, nickel, sulfur, and other heavy compounds sank toward the center to form a core. Lighter compounds (mostly silicate rocks and volatiles) rose to form a mantle and crust.
4. Violent collisions with other large planetesimals caused shattering and produced fragments containing stone, stone mixed with iron, and iron–nickel. The larger of these fragments are called asteroids when they are in orbit. When they fall to Earth, we call them stony, stony-iron, and iron meteorites.

in chondrites (see figure 5.9) and a few other silicates survived there. Matter more distant from the Sun, which was exposed to somewhat lower temperatures, was able to retain the less refractory minerals as well as some of the volatiles, including carbonaceous compounds. Finally, matter in the outer regions of the disk, which never became heated much at all, retained all elements (with the possible exception of the noble gases), including ices of water, ammonia, and methane.

2. *Origin of the differentiated planetesimals (achondrites, stony-iron, and iron meteorites).* The bodies that gave rise to the differentiated meteorites grew initially by the same process that formed the undifferentiated planetesimals. However, in time they grew to much larger sizes and they became heated to the point of partial melting. The densest of the elements—mostly iron and nickel—sank toward the centers of the bodies, while the lighter materials—the silicates and volatiles—rose to the outer layers. Thus, the bodies became differentiated into iron-nickel cores surrounded by mantles

Figure 5.9. A large fragment from the Allende meteorite shower, which fell over a 10- × 50-km area near the small village of Pueblito de Allende in north-central Mexico, on February 8, 1969. The Allende meteorite fragments are classified as carbonaceous chondrites. The near end of this sample was broken off to expose unaltered meteoritic material, revealing mineral inclusions and chondrules (one to a few millimeters across) embedded in a matrix of fine-grained, dark gray, and relatively dense (3.67 g/cm^3) carbonaceous matter.

whose compositions varied from mixtures of iron and stone to pure stone. Subsequent collisions or rotational instabilities broke the differentiated bodies into fragments and created the achondrites and stony-iron and iron meteorites. (The same mechanism of differentiation operated in all of the terrestrial planets and larger satellites.)

The variation in composition acquired by the planetesimals due to the temperature gradient in the protoplanetary disk is still in evidence today among the asteroids and comets. For example, asteroids near the inner edge of the Asteroid Belt (at distances of about 2 AU from the Sun) generally consist of rocks (comparable to ordinary chondrites), rocks intermixed with iron (stony-iron meteorites), and iron–nickel (iron meteorites). The densities of the iron asteroids are 7.5 to 8.0 g/cm^3, while those of the rocky and stony-iron asteroids range from 3.0 to 6.0 g/cm^3, depending on the amount of iron they contain. Many of the asteroids in the middle of the belt (at about 3 AU from the Sun) are carbonaceous chondrites, with densities of 3 g/cm^3 or somewhat less, though there are still some stony and stony-iron as well as an occasional iron asteroid among them. In the outer belt and beyond (that is, beyond 4 AU), virtually all of the asteroids are of the carbonaceous type. Many are reddish in color, indicating increased admixtures of organic matter, including amino acids. Finally, the space beyond the planets is occupied by comets, which consist of rocks intermixed with ices of H_2O, NH_3, and CH_4 as well as many other volatiles. They have densities of about 1 to 2 g/cm^3.*

Deviations from this general pattern occur. Objects of cometary origin can be found close to the Sun, and stony or stony-iron asteroids exist beyond the Asteroid Belt. These exceptional cases are thought to be the result of orbital perturbations that the bodies experienced by past collisions among themselves, by gravitational interactions with Jupiter and the other planets, or, in the case of the comets, by gravitational interactions with passing stars.

What could have been the sources of heat that were responsible for the melting of the larger planetesimals? There is at present no sure answer to this question, though it is known that the gravitational energy released during the formation of the planetesimals was far too insignificant to have caused melting of bodies of only a few or even several hundred kilometers in size.

However, one source of heat that may have been very important in the melting of at least some of the planetesimals was the decay of the radioactive isotope aluminum-26 (^{26}Al) into magnesium-26 (^{26}Mg). The telltale sign of this decay is a slight excess of ^{26}Mg, relative to its abundance in solar material, in the Ca–Al–Ti inclusions of ordinary chondrites. Presumably, the excess ^{26}Mg was not incorporated into the inclusions as magnesium, but as ^{26}Al, which subsequently decayed into ^{26}Mg and released energy (see box 5.1 for details concerning the decay of unstable isotopes).

*Information about the composition of interplanetary bodies is obtained by analyzing the sunlight reflected from them. For example, asteroids that resemble ordinary chondrites and stony-iron meteorites are quite light in color; carbonaceous chondrites are very dark, mainly because of the tarry compounds they contain; and carbonaceous chondrites that are particularly rich in organic matter tend to be reddish, in addition to being dark, probably due to the presence of amino acids. In the case of comets, the analysis is made of the sunlight reflected from their comas and tails as they pass through the inner part of the Solar System.

The ^{26}Al must have come to the protoplanetary material within a few million years of being synthesized, as we can infer from this isotope's relatively short halflife of 720,000 years. Had ^{26}Al been synthesized much earlier, most of it would have decayed into ^{26}Mg and given off its energy before it ever became incorporated into planetesimals.

The only known sources of ^{26}Al are red giants, novae, and supernovae. We do not know which of these potential sources actually supplied the ^{26}Al to the protosolar material. However, Alastair G. W. Cameron of Harvard and James W. Truran of the University of Illinois made the interesting suggestion that if the ^{26}Al came from a supernova—perhaps one that exploded within a few dozen light-years from the interstellar gas cloud out of which the Solar System formed—the shock wave from the explosion may have been the trigger of the cloud's collapse.

Age of the Solar System

The decay products of long-lived unstable isotopes in meteorites allow us to deduce yet another important characteristic about the Solar System, namely its age. For example, potassium-40 (^{40}K) decays into argon-40 (^{40}Ar) with a halflife of 1.3×10^9 years. The amount of ^{40}Ar present in chondrites indicates that ^{40}K has been decaying in these meteorites for the past 4.5 to 4.6 billion years. Because the chondrites have experienced no changes in their composition other than by radioactive decay of unstable isotopes, this must be the age of these meteorites. It is also regarded to be the age of the Sun and the planets, for it is believed that all of these bodies formed together out of the same interstellar cloud. In addition, similar ages are obtained from the decay products present in some lunar rocks (figure 5.10) as well as from theoretical model studies of the Sun. (In comparison, minerals and rocks from geologically active planets like the Earth, in which the composition is continually altered physically and chemically, give widely varying ages, from more than 4 billion years to very recent ages.)

Figure 5.10. Lunar rock, collected by the *Apollo 17* astronauts. The age of this rock, as determined by comparing traces of the long-lived radioactive element rubidium-87 and its decay product strontium-87, is estimated to lie between 4.5 and 4.6 billion years. It is among the oldest known rocks of the Solar System and is believed to date back to the time of its formation.

BOX 5.1.
Decay of Radioactive Isotopes

IN 1896, the French physicist A. Henri Becquerel (1852–1908) performed a series of simple, yet significant experiments. He was intrigued by the fact that X rays, which had been discovered in the previous year by the German physicist Wilhelm C. Röntgen (1845–1923), are capable of inducing fluorescence in nearby materials.* He thought that there might exist some fundamental relation between fluorescence and the emission of X rays, and he set out to investigate this possibility.

Because some compounds of uranium fluoresce when exposed to sunlight, Becquerel began by concentrating on them. He placed samples of the compound on photographic plates wrapped in black paper and let sunlight shine on them. Upon developing the plates, he noticed that they were blackened. Then, after a succession of cloudy days, during which there was insufficient sunlight to induce fluorescence, he developed a photographic plate that had been left lying near some of the uranium. Much to his surprise, the plate turned out to be blackened, too, with the image of the uranium crystal clearly standing out. When he placed other, nonfluorescent compounds of uranium, including the metal itself, upon photographic plates, they were blackened as well. Clearly, the blackening had nothing to do with the fluorescence, but was a property of uranium itself.

During the subsequent years, Becquerel and others—among them Marie Curie (1867–1934), Pierre Curie (1859–1906), Ernest Rutherford (1871–1937), and Paul Villard (1860–1934)—discovered that many other substances have properties similar to that of uranium and emit "radiation." Three kinds of radiation were recognized and labeled *alpha-, beta-, and gamma-radiation* after the first three letters of the Greek alphabet. Within about a dozen years of Becquerel's first experiments with uranium, the alpha-radiation was identified as streams of ^{4}He nuclei (alpha-particles), beta-radiation as streams of electrons or positrons (beta-particles), and gamma-radiation as short-wavelength electromagnetic radiation. Furthermore, it was discovered that these particles and electromagnetic radiation originate when certain elementary particles or nuclei change spontaneously into other kinds of particles by processes now called *radioactive decay* or *nuclear fission*. Thus, the modern nuclear age was born.

The simplest example of radioactive decay is the decay of free neutrons. These particles are not stable, but decay spontaneously into protons, with the simultaneous emission of electrons and neutrinos:

$$n \rightarrow p^+ + e^- + \nu.$$

Because this decay emits electrons, it is called *beta-decay*. Another example of radioactive decay is the spontaneous transformation of uranium-238 (^{238}U) into thorium-234 (^{234}Th), with the simultaneous emission of alpha-particles:

$$^{238}_{92}\mathrm{U} \rightarrow {}^{234}_{90}\mathrm{Th} + {}^{4}_{2}\mathrm{He}.$$

This kind of decay is known as *alpha-decay*. Incidentally, the product of this decay, ^{234}Th, is itself radioactive and decays in a sequence of steps to lead-206 (^{206}Pb), which is stable.

The decays of neutrons and of ^{238}U take place at very different rates. For example, if we could contain a number of neutrons, we would find that after approximately 13 minutes half of them would have undergone beta-decay, leaving only half of the original number of neutrons. After another 13 minutes, half of those would have decayed, leaving only one-quarter of the original. And during every succeeding 13-minute interval, the number of neutrons remaining would again be reduced by a factor of one-half. The process stops only after all neutrons are gone and have been converted into protons, electrons, and neutrinos. If we started with a number of ^{238}U atoms, however, we would have to wait 4.5 billion years before half of them decayed into ^{234}Th. Yet another 4.5 billion years would go by before half of the remaining ones would decay, and so forth. The two time intervals—13 minutes and 4.5×10^9 years—are called the *halflives* of neutrons and ^{238}U, respectively. Every unstable elementary particle and unstable atomic nucleus has a characteristic mode of decay and halflife, and in table 5.4 these characteristics are summarized

**Fluorescence* refers to the emission of light by certain substances when they are exposed to some energy source. The most common example of this phenomenon is the fluorescent lamp, which emits light when an electric current (the energy source) is sent through the gas inside the glass tube. Other examples are the dials of many alarm clocks, which keep glowing for hours after they have been exposed to light.

BOX 5.1 continued

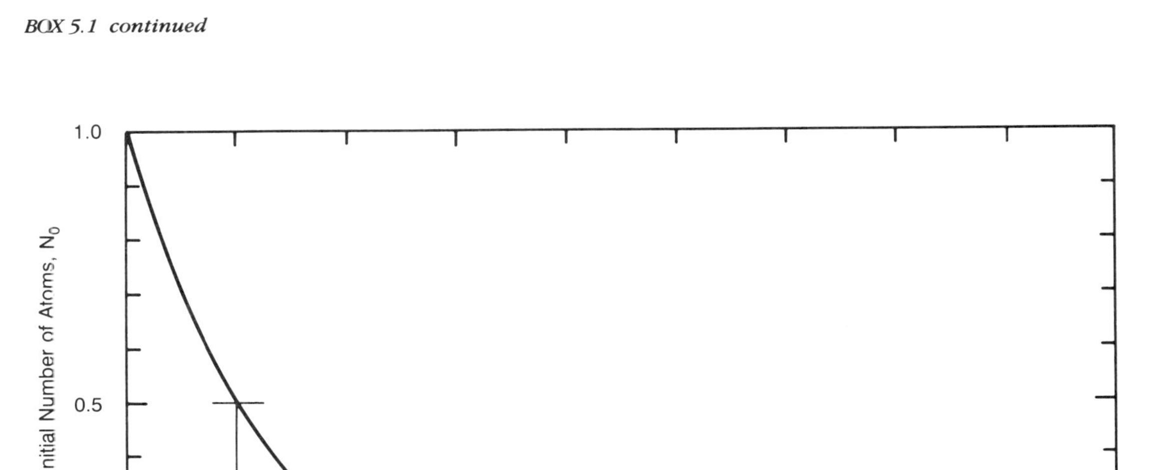

Figure A. Decay of radioactive atoms. At time zero, there is a given number of radioactive atoms, N_0. The atoms decay into their offspring products at rates such that after one halflife, half the N_0 atoms remain; after two halflives one-quarter of the N_0 atoms remain; and so forth.

for those radioactive isotopes that are of greatest interest to us.

Radioactive decay is illustrated in figure A. In particular, the figure shows the fraction of the initial number of radioactive atoms, N_0, remaining after one, two, three, and so forth halflives. The decay is a probabilistic phenomenon. It is not possible to predict exactly *which* atoms will decay during a given time interval. One can predict only *how many* will decay during a given time interval (provided that the number of atoms is large, so that the laws of probability apply). Initially, when there are still many radioactive atoms present, the curve drops steeply (that is, many atoms decay per unit time). Later when the number of radioactive atoms has diminished, the curve flattens (that is, fewer atoms decay per unit time). From the fraction of radioactive atoms remaining and a knowledge of the halflife one can calculate how long the decay has been going on.

The obvious question now is, "Why are some particles and nuclei—such as electrons, protons, alpha-particles, and the nuclei of ^{12}C and ^{16}O—stable, while others are not?" The answer is, in principle, quite simple. Stable particles and nuclei are in their lowest possible energy states and, therefore, last forever unless perturbed by outside forces (for example, by forces arising from collisions with other particles traveling at high speeds). Unstable particles, in contrast, are in higher energy states and it is only a matter of time—as indicated by their halflives—before they decay into lower, more stable energy states. The emission of electrons, positrons, alpha-particles, and gamma rays during such decays constitutes the energy difference between the initial, higher energy states (those of the unstable particles) and the final, lower energy states (those of the stable particles).

For example, the energy liberated by the decay of ^{26}Al into ^{26}Mg is thought to have caused the internal

BOX 5.1 continued

melting of the larger planetesimals during the formative stages of the Solar System. The decays of the long-lived radioactive isotopes ^{40}K, ^{232}Th, ^{235}U, and ^{238}U were and still are responsible for much of the heat content of the intermediate-sized planetary bodies. In the case of the Earth today, these sources of energy make up roughly 80% of the heat that flows from the crust to the surface (the geothermal energy); only 20% comes from stored-up, primordial heat.

Radioactive decay may also affect us in more direct ways than by contributing to the Earth's heat budget. The particles and gamma-rays emitted by radioactive materials may do damage to human tissues, cause cancer, and induce genetic mutations. But, when properly applied by trained medical personnel, they may also have beneficial uses, as in the treatment of cancer and other tumors.

The decay of radioactive isotopes has yet another important application. It allows geologists and archaeologists to deduce the ages of rocks and organic matter. For example, the decay of potassium-40 (^{40}K) into argon-40 (^{40}Ar) may be used to determine the ages of chondrites, which are meteorites that have not changed since their origin during the birth of the Solar System. This is done by assuming that when these meteorites formed the rock crystals contained only ^{40}K, but no ^{40}Ar. This is a reasonable assumption because ^{40}Ar is a noble gas, which does not readily combine with other elements and, hence, would not have become incorporated into meteorites in significant amounts except by the decay of ^{40}K. Next, the relative abundance of ^{40}K and ^{40}Ar is measured in chondrites today and, from the known rate of decay of the former into the latter, it is a simple matter to compute the chondrites' ages. The decay of the other long-lived radioactive isotopes listed in table 5.4 may be used likewise in estimating the ages of chondrites and is useful for obtaining independent checks. The same isotopes may also be used to date terrestrial rocks. The ages so obtained refer only to the epochs when the rocks last experienced melting or other compositional changes, however; they do not tell us the age of the Earth. (The same is true if the method is applied to meteorites that have undergone compositional changes since their formation.)

The dating methods just described, which are based on the decay of long-lived radioactive isotopes, are useful only for estimating ages of hundreds of thousands of years or longer because over shorter time spans too few of the isotopes decay to yield dependable measurements. To estimate ages of shorter duration, radioactive isotopes that decay more rapidly must be used. One such short-lived radioactive isotope is carbon-14 (^{14}C). This isotope is continuously created in the Earth's upper atmosphere as a result of bombardment of nitrogen by neutrons. The newly synthesized ^{14}C reacts with atmospheric oxygen to form CO_2 and is subsequently incorporated into organic matter through photosynthesis. By this sequence of processes, approximately one out of every 10^{12} carbon atoms in living matter turns out to be ^{14}C rather than ^{12}C. Because ^{14}C is unstable, its presence in dead organic matter (matter that no longer takes up fresh ^{14}C from the food chain) diminishes with time. By comparing the relative abundance of ^{14}C in a piece of fossil bone, cloth, leather, wood, or the like with that in living matter, it is possible to obtain age estimates of such objects. This *radiocarbon dating method* works well for time intervals of up to 50,000 years. In objects older than that, too little ^{14}C remains for accurate measurements, and other dating techniques need to be used.

INTERMEDIATE-SIZED PLANETARY BODIES: INTERNAL STRUCTURE

The family of intermediate-sized planetary bodies includes Mercury, Venus, Earth, Mars, the Moon, the Galilean satellites, Titan, and Triton. These bodies have similar sizes, masses, and other characteristics, as noted earlier and shown in tables 4.1 and 4.2. They also have similar, though not identical, structures. The differences that do exist between them can be explained in terms of their masses, their different distances from the Sun or, in the case of the Galilean satellites, from Jupiter, and the kinds of materials that were available for their assembly at those distances. Let us examine the compositions and physical structures of these bodies, starting with the terrestrial planets.

The single most meaningful parameter concerning the composition of a planetary body is its density. A body made entirely of iron has a density of about 7.5 g/cm^3; one made mainly of silicate rocks has a density of about 3.0 to 3.5 g/cm^3; one made of rocks plus some admixture of volatiles has a density of between 2.0 and 3.0 g/cm^3; and one containing mostly ices has a density between 1.0 and 2.0 g/cm^3. (In comparison, basalt and granite, the major constituents of the Earth's rocky crust, have density ranges from 2.7 to 3.4 and 2.5 to 2.8 g/cm^3, respectively.)

Care must be taken in using density as an indicator of a celestial body's internal constitution. If a body's radius exceeds a few thousand kilometers, its self-gravity becomes so strong that even rocks and metals become appreciably compressed. In such cases it is not the actual density that is meaningful, but the body's *uncompressed* density, namely the density it would have if it were not compressed by self-gravity.

Terrestrial Planets

The distances from the Sun, radii, and actual and uncompressed densities of the terrestrial planets are shown in the following table.

	Distance From Sun (10^6 km)	Equatorial Radius (km)	Actual Density (g/cm^3)	Uncompressed Density (g/cm^3, approximate)
Mercury	58	2440	5.4	5.4
Venus	108	6050	5.3	4.3
Earth	150	6380	5.5	4.2
Mars	228	3400	3.9	3.8

The relatively high values of their uncompressed densities suggest that the terrestrial planets are not made of rocks alone, but must contain substantial amounts of iron and other high-density elements. Furthermore, there is a general trend of decreasing values with increasing distance from the Sun. Mercury, the planet closest to the Sun, has the highest uncompressed density and, hence, contains the greatest proportion of iron, in addition to rocks. Venus, Earth, and Mars have progressively lower densities and, hence, contain less iron and more rocks and volatiles. This is the same pattern observed in the case of the asteroids and comets, and it is yet another indication of the existence of a protoplanetary disk that became differentially heated. Near the Sun the temperatures rose to such high values that only the metals and most refractory minerals survived and became incorporated into the planets. With increasing distance from the Sun the temperatures stayed progressively lower and greater

Table 5.4. Decay Modes and Halflives of Some Important Radioactive Isotopes.

Radioisotope	Decay mode	Decay reaction	Halflife (years)	Applications or effects
Carbon-14	β-decay	${}^{14}_{6}C \rightarrow {}^{14}_{7}N + e^- + \nu$	5730	Dating of organic matter.
Aluminum-26	β-decay	${}^{26}_{13}Al \rightarrow {}^{26}_{12}Mg + e^+ + \nu$	7.2×10^5	Heating and melting of interiors of planetesimals.
Potassium-40	β-decay e-capture[a]	${}^{40}_{19}K \rightarrow {}^{40}_{20}Ca + e^- + \nu$ ${}^{40}_{19}K + e^- \rightarrow {}^{40}_{18}Ar + \nu$	1.3×10^9	Heating and melting of interiors of the intermediate-sized planetary bodies, including Earth.
Rubidium-87	β-decay	${}^{87}_{37}Rb \rightarrow {}^{87}_{38}Sr + e^- + \nu$	5.0×10^{10}	
Thorium-232	α-decay	${}^{232}_{90}Th \rightarrow {}^{228}_{88}Ra + {}^{4}_{2}He$ ($\rightarrow {}^{208}_{82}Pb$)[b]	1.4×10^{10}	Dating of rocks and meteorites.
Uranium-235	α-decay	${}^{235}_{92}U \rightarrow {}^{231}_{90}Th + {}^{4}_{2}He$ ($\rightarrow {}^{207}_{82}Pb$)[b]	7.1×10^8	
Uranium-238	α-decay	${}^{238}_{92}U \rightarrow {}^{234}_{90}Th + {}^{4}_{2}He$ ($\rightarrow {}^{206}_{82}Pb$)[b]	4.5×10^9	

[a]Electron-capture is a decay mode (also called inverse β-decay), in which one of the nuclear protons captures an orbital electron and changes into a neutron.
[b]The lead isotopes are the final products, reached by decay reactions beyond those shown explicitly.

amounts of volatiles survived, giving the more distant planets their lower densities.

The terrestrial planets are not homogeneous bodies of metals, rocks and volatiles. They are differentiated into high-density cores, mainly of iron and nickel, surrounded by rocky mantles of intermediate densities and surface crusts of low-density rocks and volatiles. These differentiations took place early in the planets' formation by mechanisms similar to those that differentiated the larger planetesimals, though the release of energy from gravitational coalescence and from the decay of long-lived radioactive isotopes was much more severe. Figure 5.11 depicts the internal structure of the Earth, which is typical (but not identical in details) of all terrestrial planets.

Moon

The Moon is an oddity among the intermediate-sized planetary bodies. Because it is close to Earth and, presumably, was formed nearby, its density might be expected to be similar to the uncompressed density of Earth, about 4.2 g/cm^3. Instead it is 3.3 g/cm^3, lower even than the density of Mars. This indicates that the Moon has either no iron core at all or only a very small one.

There are other anomalies about the Moon's makeup. Its rocks are depleted in the volatiles (H, C, N, O, and the noble gases) relative to Earth. For example, none of the lunar rocks brought back by the Apollo astronauts shows any trace of water, whereas terrestrial igneous rocks (like basalt and granite) generally contain between 0.2 and 0.5% water. The lunar rocks are also depleted in iron and all elements that tend to combine with iron—the "siderophiles" or "iron-loving" elements—such as nickel, cobalt, rubidium, iridium, and platinum.

Clearly, an explanation is required why the Moon has no substantial iron core and why its rocks are depleted in the iron and siderophiles that once must have been part of the material from which it formed. John A. O'Keefe, a NASA

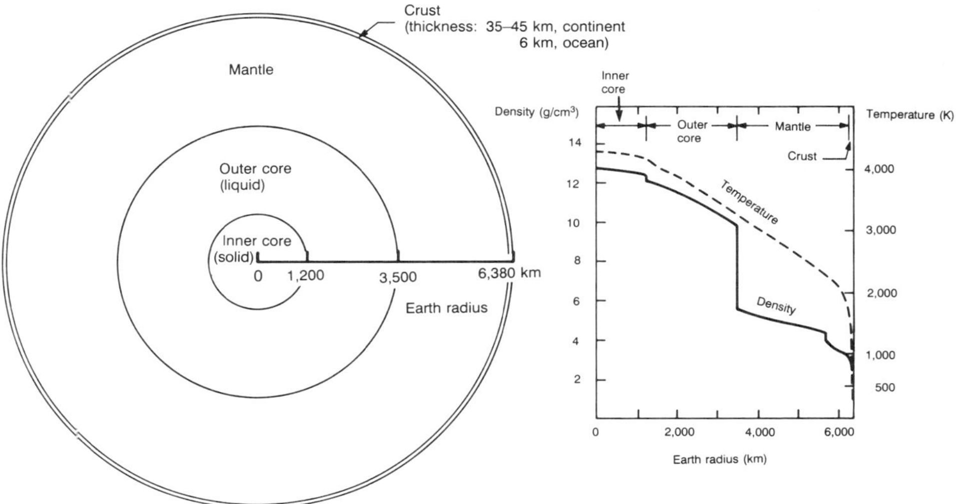

Figure 5.11. Model of the structure of the Earth. On the left are the major components of the Earth: the solid inner core and molten outer core, the mantle, and the crust (the latter is not drawn to scale). Both cores consist largely of iron and some nickel; the mantle mainly of iron and magnesium silicates; and the crust mainly of basaltic and granitic rocks. The graph on the right shows the approximate distribution of density (from 12.6 to 2.6 g/cm^3) and temperature (from 4300 to 280 K). The density profile displays marked discontinuities at the edges between the inner and outer cores, between the outer core and the mantle, and within the mantle at a depth of about 670 km.

astronomer and geophysicist, gives a simple and rather compelling answer (personal communication 1979): "The Moon's iron and siderophiles are in the core of the Earth." He suggests that the material that today comprises the Moon was at one time part of a rapidly spinning Earth. After the Earth's iron and siderophiles had sunk toward its center to form the terrestrial iron core, the outer layers were torn off by the strong centrifugal forces resulting from the rapid rotation. Others have suggested that the Earth's outer layers were thrown into space by the impact of a very large (perhaps Mars-sized) planetesimal. However the layers were ejected, part of them—depleted in iron, siderophiles, and volatiles (presumably, many of the volatiles were outgassed and lost due to tidal friction and heating)—formed the Moon, and the rest escaped into interplanetary space.

These theories of the origin of the Moon by fission from Earth, originally proposed by Sir George H. Darwin (1845–1912, son of Charles Darwin), or by the impact of a large planetesimal, are not without some difficulties. In particular, they require that either the young Earth rotated unusually rapidly or that it was hit by an uncommonly large planetesimal. Hence, some planetary scientists prefer the *accretion* theory. According to it, the Moon accreted from material that was in Earth orbit from the outset, rather than part of the Earth. However, this theory cannot explain the Moon's many composition anomalies.

Therefore, in my opinion, further research will eventually prove either the fission or the impact theory to be the correct one. (The theory that the Moon formed somewhere else in the Solar System, perhaps beyond the orbit of Mars, and was captured by the Earth is now generally discounted.)

Galilean Satellites

The densities shown in table 4.2 for the Galilean satellites of Jupiter indicate that both Io and Europa have largely rocky structures and, possibly, small iron cores. Photographs show that Io's surface is rocky and covered by sulfur-containing compounds, while Europa is enveloped by a crust of water ice. The low densities of Ganymede and Callisto suggest that they contain substantial amounts of water (perhaps as much as 50% by mass), the rest being rock. They probably have rocky central cores that are overlain by thick mantles of water, either in liquid or solid (ice) forms. On top of the mantles float hard and rigid crusts of water ice with a thin covering of rock debris from meteorite impacts.

The decrease in density in the Galilean satellites with increasing distance from Jupiter follows the same pattern noted for the smaller bodies and the terrestrial planets. It suggests that Jupiter was at one time surrounded by its own protoplanetary—or protosatellitic—disk, that became heated as a result of the planet's formation. For reasons similar to the ones discussed above, the ensuing temperature gradient was probably responsible for the composition and density differences observed in the Galilean satellites today. Thus, Galileo's suggestion that Jupiter and its four giant satellites are a miniature Solar System was correct beyond the conditions he could envision.

Figure 5.12. Oblique view of intercrater plains on Mercury. Numerous large and small impact craters, many with central peaks, alternate with terrain of a gently rolling nature. This terrain is believed to be the result of planetwide surface melting during Mercury's final accretion phase.

Titan and Triton

The last two planetary bodies to be discussed are Saturn's Titan and Neptune's Triton. The data in table 4.2 show that Titan resembles Ganymede in size, distance from its central planet, and density. Hence, astronomers expect that its structure, too, is similar to Ganymede's, though they are not sure. But they do know that Titan has a substantial atmosphere, consisting of 80 to 95% molecular nitrogen, with the rest being mainly CH_4 and traces of H_2 and organic molecules. It is believed that these gases are the result of evaporation of ices of H_2O, CH_4, and NH_3 from the satellite's surface, followed by chemical reactions.

Our knowledge of Triton is even more limited than that of Titan, and it will remain so until *Voyager 2* makes its rendezvous with Neptune in 1989. We know that Triton is roughly of the same size as Europa, but its density is smaller. Consequently, ices of H_2O, CH_4, and NH_3 are probably major constituents of this satellite, too. The presence of CH_4 in Triton's thin atmosphere is further evidence of this supposition.

INTERMEDIATE-SIZED PLANETARY BODIES: SURFACE FEATURES

Descriptions of the surface features of the intermediate-sized planetary bodies require selectivity because two decades of space exploration have amassed far more information than can possibly be covered here. Hence, the surfaces of each of the bodies will not be described separately. Instead, they will be compared with each other, paying attention to four topics: (1) impact craters, (2) geological activities, (3) presence of water, and (4) atmospheres. This comparative coverage is consistent with grouping the intermediate-sized planetary bodies into a single family and with the modern trend of regarding the forces that shaped them as part of a continuum.

Impact Craters

The most heavily cratered intermediate-sized planetary bodies are the Moon, Mercury, Callisto, and Ganymede, as well as the southern hemisphere of Mars. Earth, Venus, the northern hemisphere of Mars, and Europa show relatively few impact craters. Io has none at all. Titan and Triton may be cratered similarly to Ganymede and Callisto, but at present they are not well enough explored for us to be sure.

On Callisto, Ganymede, and the highlands of the Moon and Mercury, the crater density is close to the saturation point, with craters edge-to-edge or overlapping and larger craters having smaller ones within them. Some of these impact craters are truly gigantic as figures 5.12–5.15 and figure 5.28 demonstrate. Others are much smaller, and lunar rocks show that they exist even on the microscopic level (see figure 5.16). Very little geological or weathering activity has disturbed these heavily cratered regions, and therefore, the cratering record, much of which dates back nearly 4 billion years, has been preserved.

Figure 5.13. View of the Caloris Basin of Mercury, situated near the planet's terminator in this Mariner 10 mosaic. This enormous impact feature is 1300 km in diameter and was formed about 3.7 billion years ago. It is surrounded by a pattern of concentric rings of mountains, interspersed with lava plains. The name *Caloris* is derived from the Latin *calor*, meaning "heat," and refers to the fact that the basin lies near one of Mercury's "hot poles," which alternately face the Sun when the planet is at its closest distance from the Sun.

The Moon and the other heavily cratered bodies are, however, not without some evidence of past geological activity. For example, from about 3.3 to 2.5 billion years ago, lava flows filled in most of the large impact basins on the Moon and formed the dark colored and relatively smooth "seas" or *maria*. Similar, though somewhat more ancient, lava flows have created the intercrater plains on Mercury.

At the other extreme with regard to crater density lies Io, geologically the most active body in the Solar System. Its volcanic ejectae and lava flows erase craters as rapidly as they form. On Earth, geological activity is somewhat less severe but is still sufficiently dynamic that only a few impact features older than about 500 million years remain. The same is true of Europa, though on this body it is not the movement of rock that obscures impact craters, but the flow of ice. Mars is interesting. Its heavily cratered southern hemisphere also contains intercrater plains like those on the Moon and Mercury. However, its northern hemisphere is dominated by volcanoes and other features of geological activity, which have erased most of the older impact craters.

The collective record of impact craters on the intermediate-sized planetary bodies indicates that up to approximately 3.8 billion years ago all of them were

Figure 5.14. Oblique view of lunar craters Eratosthenes (right of center) and Copernicus (on horizon, at left), and of Sinus Aestuum (foreground) and Mare Imbrium (upper right). Eratosthenes has a diameter of about 60 km. It was formed approximately 3.4 to 3.2 billion years ago, probably by the impact of a large planetesimal. However, the absence of obvious ejection features, such as rayed ejection patterns, has prompted the suggestion that it may be a caldera, the collapsed center of a former volcano.

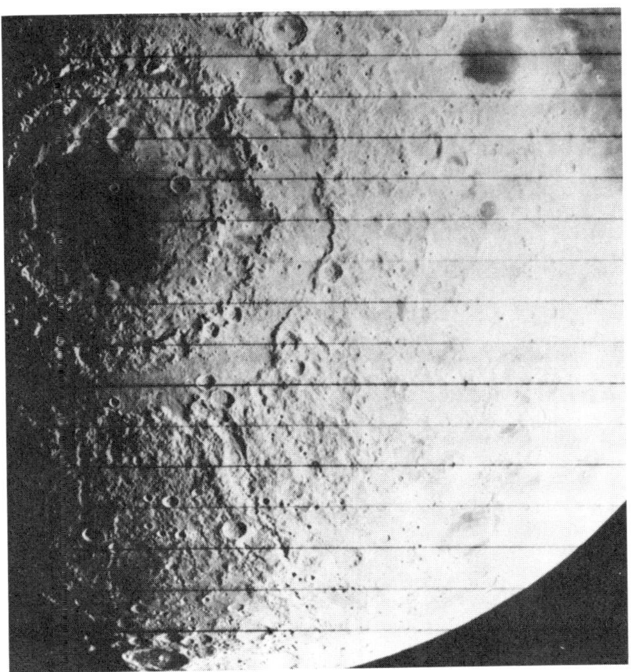

Figure 5.15. Mare Orientale, as viewed by *Lunar Orbiter 4* in 1967. This gigantic impact feature, with a central crater and surrounding multi-ring structures, was formed when an object perhaps 25 km across struck the Moon roughly 4 billion years ago. The outermost ring of mountains, named *Cordillera Mountains*, is nearly 900 km in diameter and rises about 6 km above the surrounding plains.

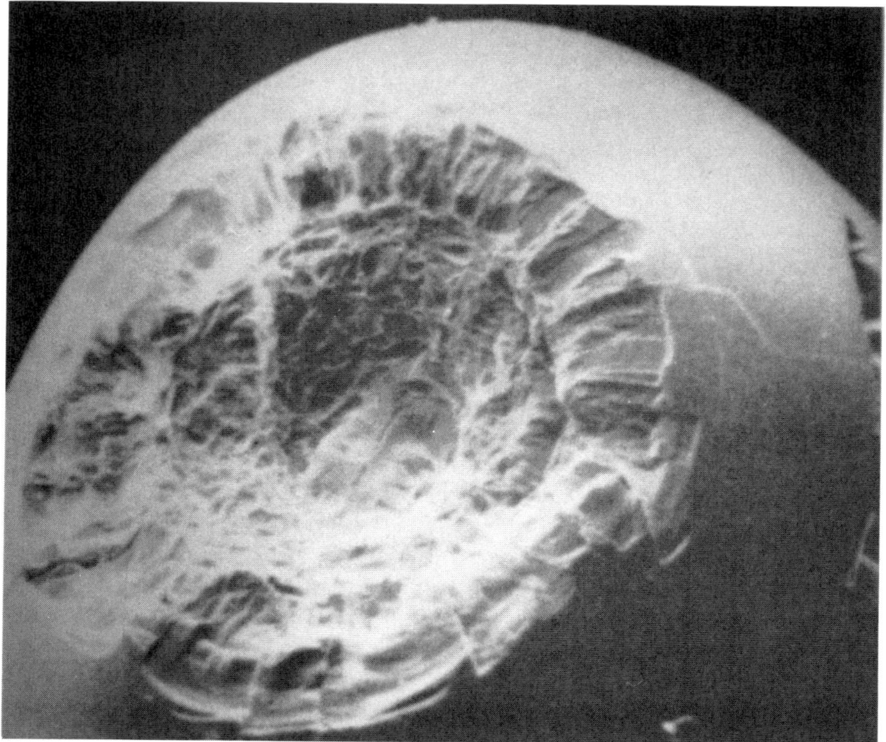

Figure 5.16. Microscopic crater on a spherule of glass, from a lunar rock sample. The crater is less than 1 mm in size and was carved out by the impact of a tiny particle of cosmic dust. The crater exhibits a multi-ring pattern of fractures similar to that found in lunar impact craters many kilometers in size.

heavily bombarded by large and small planetesimals (see figure 5.17). Most likely, this bombardment constituted the terminal phase of the accretion that created these bodies in the first place. From about 3.8 billion years forward, the bombardment declined rapidly; though even today it has not yet reached the zero point, as noted in the discussion of meteorite falls on Earth.

Geological Activity

Geological* activity on planetary bodies is caused by surface forces that are sufficiently strong to bring about movements in the crust. The forces may be in the horizontal or vertical directions or both, and they may be pushing, pulling, twisting, contracting, or lifting kinds of forces. They usually are the result of gravity, convective motions inside the body, weathering activities on the surface,

*The term *geological* is derived from the Greek root *geo*-, meaning "earth" or "land," and, strictly speaking, refers only to the Earth. In particular, it refers to changes in the Earth's surface features due to forces from within and from outside, as revealed by the rock record. However, with the development of the science of planetology in recent years, the term is now also used to refer to changes experienced by the other planetary bodies with solid surfaces.

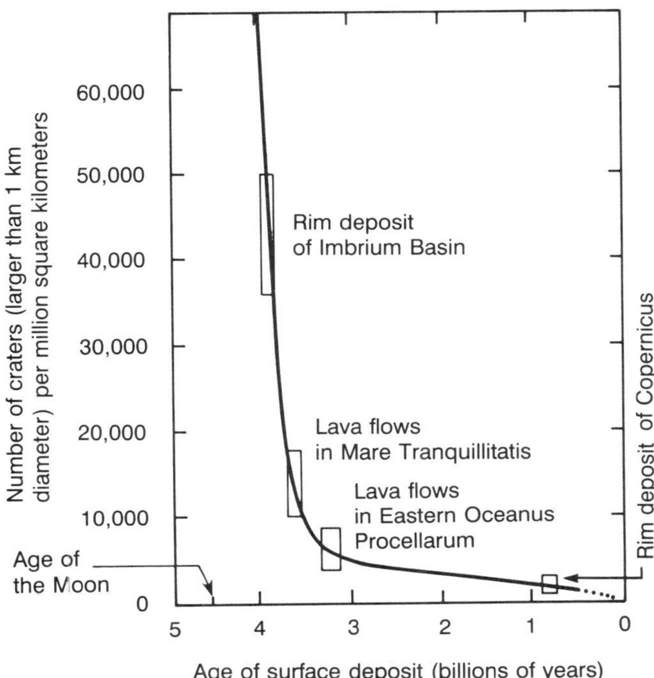

Figure 5.17. Density of impact craters at four different regions of the lunar surface plotted against the regions' ages. The slope of the curve drawn through the data rectangles shows that the rate of massive impact cratering diminished rapidly between 3.8 and 3.3 billion years ago. It was followed by a lower and much more slowly diminishing rate of impact cratering that continues to this date. (Adapted from Beatty et al. 1982, 39.)

and tidal interactions with another nearby body. The forces are most likely to produce movement if the surface crust is thin and weak, and if the interior is hot and in a molten or plastic state. The interior then acts as a lubricant that enhances movement. In contrast, if the surface crust is thick and strong and if there is little or no melting inside, the surface forces are unlikely to bring about movement. The body is then geologically inert.

The major sources of heat for melting the interior of the intermediate-sized planetary bodies are the gravitational energy that was released during their formation and the energy liberated since then by the decay of long-lived unstable isotopes, such as uranium-235 and -238, thorium-232, and potassium-40 (see box 5.1). Both of these heat sources are most effective in the largest bodies, for their surface-to-volume ratios are relatively small (surface $\propto$ radius^2, volume $\propto$ radius^3; hence, surface-to-volume ratio $\propto$ 1/radius), so that they lose their internal heat slowly. Also, much more heat was released early in the bodies' histories than, for example, today because the stored-up heat from their formation and the amount of unstable isotopes were greatest then.

Io. Io is one of the few bodies in the Solar System whose geological activity is mainly due to tidal heating. This heating is brought about by gravitational interactions with nearby Jupiter, which creates an appreciable tidal bulge on Io. As Io revolves about Jupiter and alternately approaches and recedes from it due to a slight eccentricity in its orbit (the eccentricity is maintained by gravitational interaction with Europa), the tides raise and lower different parts of Io. This causes friction by rock rubbing against rock and produces enough heat to melt part of Io's interior. Unbalanced forces, probably also due to the tides, open up cracks in Io's crust and allow the molten materials of the interior, consisting largely of sulfur and sulfur-containing compounds, to push to the surface. There they gush in geyserlike fashion into the open, creating enormous lava flows and huge volcanoes, giving the satellite its mottled appearance (figures 5.18, 5.19). Unquestionably, Io deserves its reputation as the Solar System's strangest and geologically most active planetary body.

Earth. After Io, the geologically most active intermediate-sized planetary body is Earth. This is not surprising. It is the most massive body among them. Thus it has the greatest amount of stored-up internal heat, is molten to a greater extent than the others (except for Io, because of its tidal heating, and, possibly, Venus), and its outer solid layer, the lithosphere, is relatively thin. In

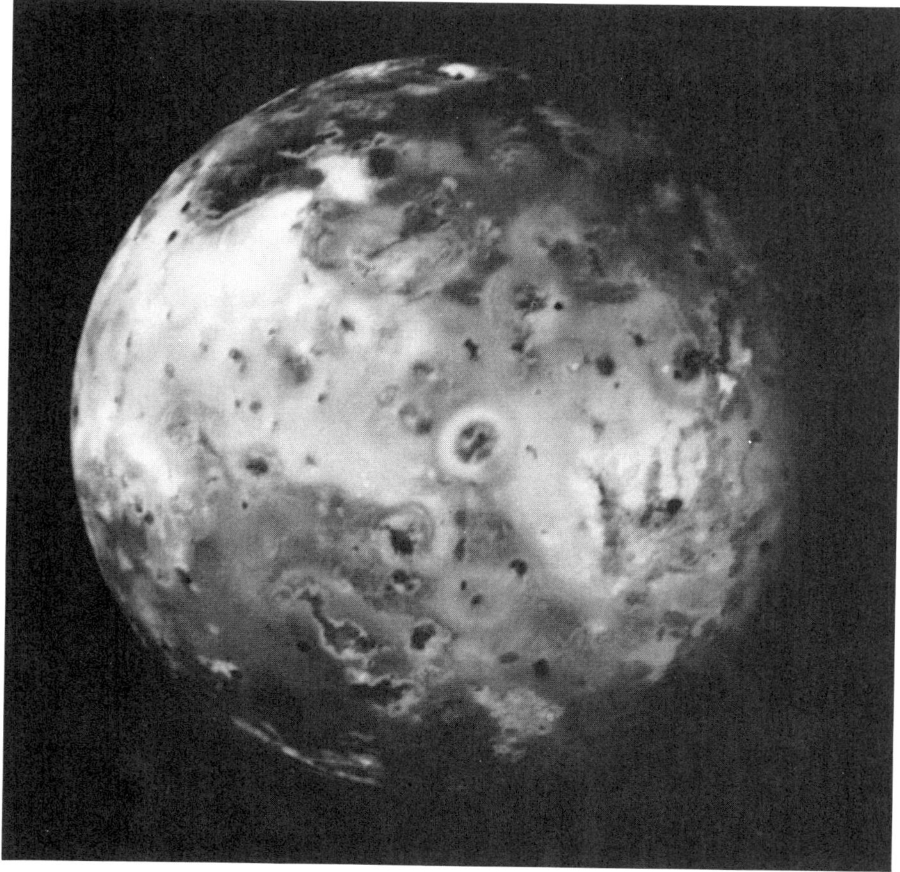

Figure 5.18. Io, the innermost of Jupiter's four Galilean satellites. Nearly all features seen on Io appear to be linked to volcanic activity, including lava flows and volcanic caldera up to 200 km across. The doughnut-shaped structure near the center of this photograph is the plume of an erupting volcano.

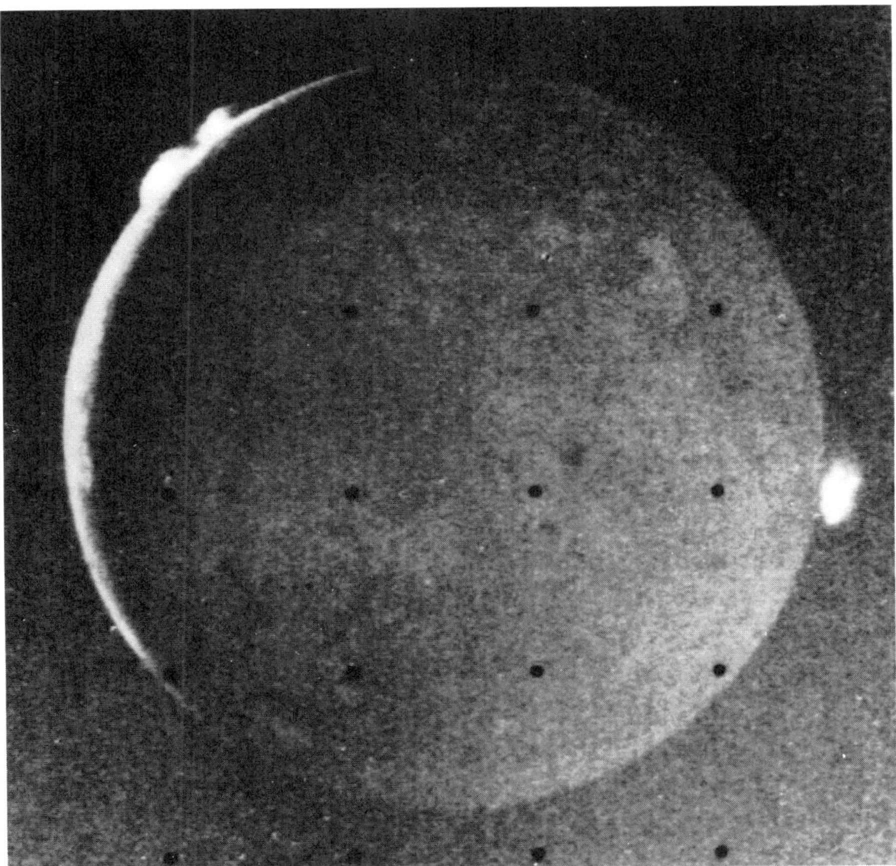

Figure 5.19. Three of the eight active volcanoes of Io, detected by the Voyager spacecrafts.

fact, the terrestrial lithosphere is a mere 100 km thick on the average, compared to a total radius of 6380 km.

The terrestrial lithosphere is composed of a number of distinct and rigid tectonic plates that float like rafts on the denser underlying asthenosphere. Forces within the Earth push and pull on the plates, slowly moving them about relative to each other. This pushing and pulling is called *plate tectonics* and is responsible for most volcanism (figure 5.20) and earthquakes, as well as for the building of mountains and the shaping of continents and ocean basins.

In addition to plate tectonics, the surface of the Earth is continuously altered by weathering. Flowing water, glaciers, and wind erode the land, and rivers carry the resulting debris toward the oceans, where they deposit it as sediments. Over time, the layers of sediments become compressed and solidified as more sediments are piled on top of them. When subsequent uplifting and erosion expose these ancient deposits, their banded structures reveal their mode of

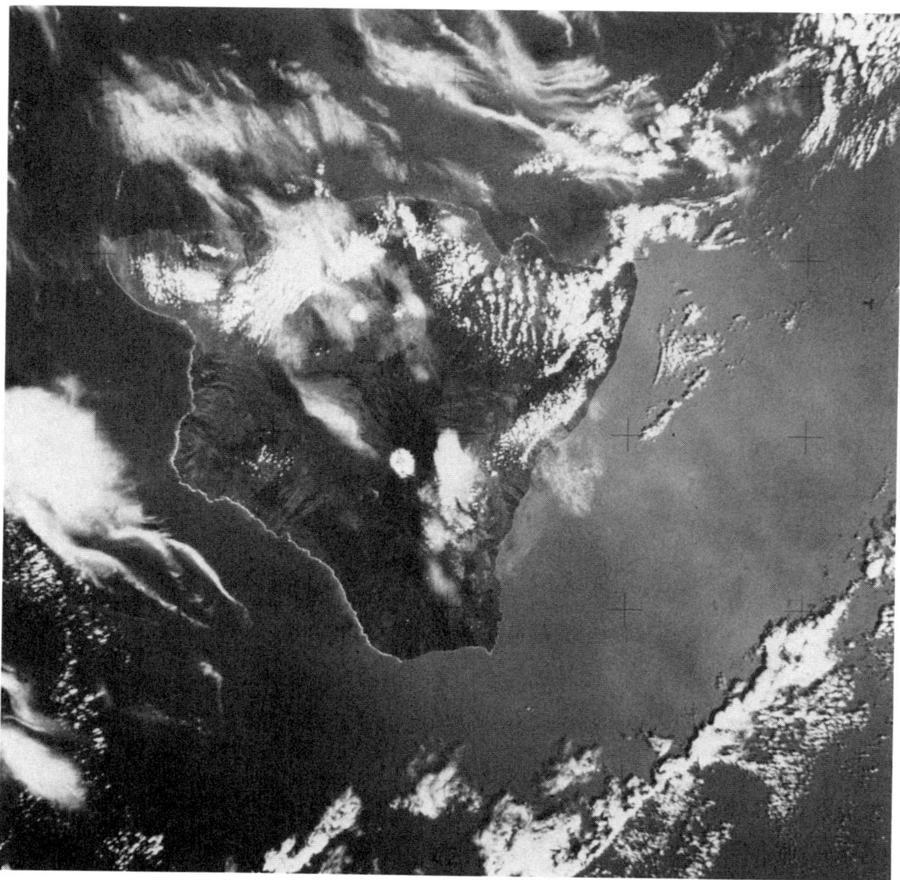

Figure 5.20. LANDSAT photograph of Mauna Kea (just above center in this photograph, 4205 m) and Mauna Loa (center, 4169 m), the two giant shield volcanoes of the Island of Hawaii. These two volcanoes rise roughly 9000 meters above the ocean floor and are 120 km wide at their base on the ocean floor. They were formed, along with the rest of the island, by the rising from the ocean floor of basaltic lavas and are probably the closest terrestrial analogues to the large volcanic mountains on Mars.

origin. Examples are the banded structures of the Canadian Rocky Mountains and the Grand Canyon.

Plate tectonics and weathering have been going on on Earth for at least 4 billion years. The oldest surviving rock formations of these processes are the *continental shields*, which are present on every continent and comprise about 10% of the Earth's surface (more about them in figure 7.3). These shields are the closest terrestrial analogues to the cratered highlands of the Moon, Mercury, and the southern hemisphere of Mars, though few craters remain.

Venus. Venus has nearly the same mass and density as Earth. Hence, we might expect that plate tectonics and volcanism were, and, perhaps, still are, also important in sculpturing its surface. Unfortunately, we do not yet know for certain if this is the case because the planet's surface remains hidden under a thick, opaque atmosphere. However, radar soundings bounced off Venus from Earth stations and orbiting spacecraft reveal a topography not unlike that of Earth and hence point to geological activity (figure 5.21). There are two massive highlands, Ishtar Terra and Aphrodite Terra, that resemble terrestrial continents. There also are several smaller islandlike regions and broad depressions comparable to the ocean basins on Earth. Most of these landforms are shallower than the contours on Earth, but not all. For example, Ishtar Terra contains a number of substantial mountain ranges. The highest of them, Maxwell Mons, rises 12,000 m above the planet's average radius and is comparable to the Himalayan Mountains.

Mars. Like Earth and Venus, Mars has had a complex and interesting geological history. Its southern hemisphere is heavily cratered and shows evidence of interspersed flooding by lava (see figures 5.22, 5.23). The northern hemisphere also contains craters and intercrater lava plains, but to a much smaller extent. Its most conspicuous features are enormous mountain ranges and canyon systems that dwarf their counterparts on Earth. The largest of the mountain ranges is a 6000-km-long uplift called the Tharsis Ridge that is punctuated by four gigantic volcanoes. The volcanoes bear a striking resemblance to the shield volcanoes on Earth, such as Mauna Kea and Mauna Loa of Hawaii. The largest of them, Olympus Mons (see figure 5.24), measures 600 km across at its base and rises 26,000 m above the surrounding terrain. To the east of the Tharsis Ridge lies a 4000-km-long system of interconnected canyons called Valles Marineris (figure 5.25). Some individual canyons are more than 200 km wide and 7000 m deep. They show modifications by landslides, wind erosion, and flowing water, mud, or, possibly, glaciers (though no liquid water exists on the Martian surface today; see below). In comparison, Earth's Grand Canyon is 450 km long, 30 km across at its widest point, and 2000 m deep.

The lava flows, mountains, and canyons of Mars indicate that, with regard to geological activity, this planet lies somewhere between the Earth and the Moon. After having been heavily cratered early in its history, its geology began to evolve like that of the Earth. However, before the entire planet was affected, its geological activity came to a halt. Most likely, the reasons were the relatively small mass and size of Mars, which, respectively, are one-tenth and one-half

Figure 5.21. The first television image of Venus's surface, made by the Soviet soft-landing spacecraft *Venera 9*, on October 22, 1975. The rocky landscape is strewn with boulders, most of them tens of centimeters in size.

those of Earth. Consequently, Mars was heated less initially; its crust became thicker than that of Earth; and its volcanism and tectonic activities stopped early in its history.

Europa. The only intermediate-sized planetary body besides Earth, Venus, and Mars that is suspected to have had recent and, maybe, current tectonic activity is Europa. The basis for this speculation is its lack of any large (over 100 km) and, hence, old craters. Furthermore, Europa's surface is marked by a vast tangle of light and dark streaks that are probably filled-in cracks and indicate the presence of tensional forces pushing and pulling on the crust (see figure 5.26). If tectonism does exist on Europa, the tectonic plates would not be

Figure 5.22. Mars, photographed by the *Viking 2* spacecraft from a distance of 418,000 km in 1976. Near the bottom of the photograph and just above the southern pole lies the Argyre Basin, covered here by thin frost. Near the middle of the planet's crescent and close to the terminator is the enormous canyon system of Valles Marineris. And near the top of the photograph lies the extinct volcano Ascraeus Mons (27,000 m), which is part of the Tharsis Ridge and responsible for the cloud streamers in its vicinity.

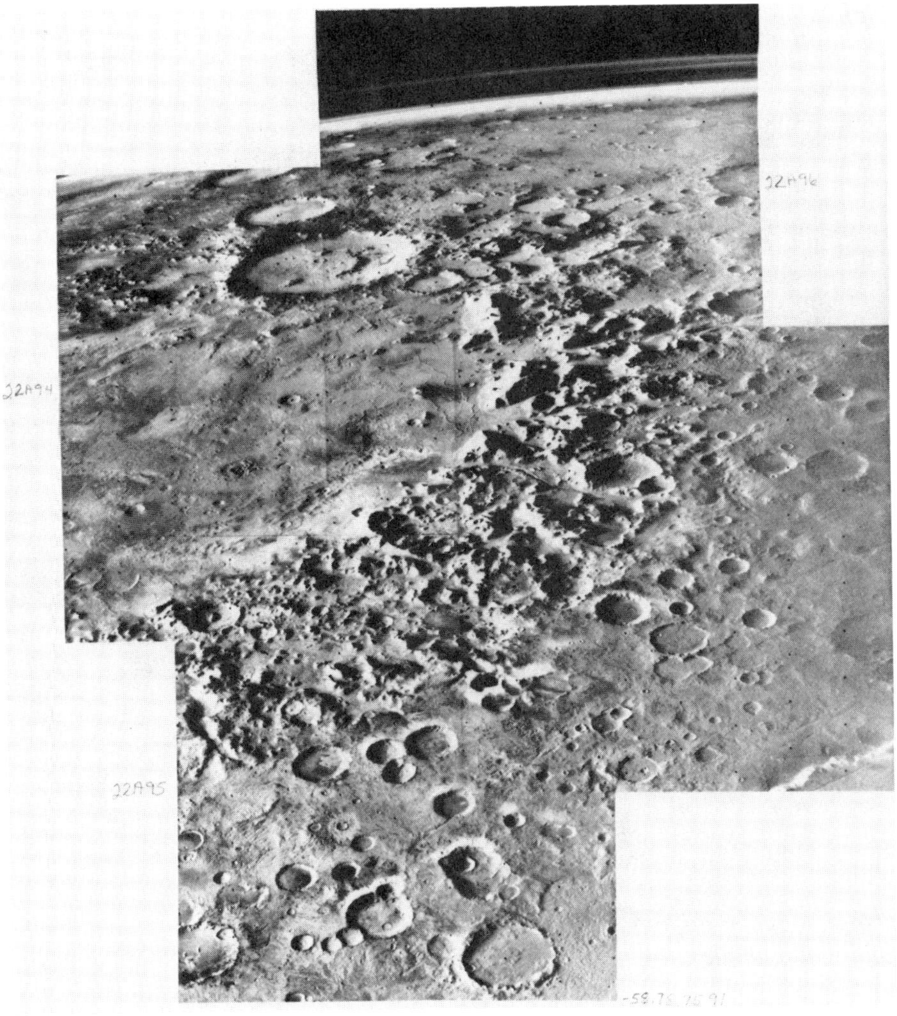

Figure 5.23. Oblique view across Argyre Planitia, one of the large lava-flooded basins of Mars. The basin, which was carved out by an asteroid impact, is bordered on the right by the old, cratered mountains of Chartium. Beyond it lies the crater Galle (210 km in diameter). Notice the thin haze of CO_2 above the Martian horizon.

rock, like those on Earth, but layers of water ice. The heat necessary for keeping the underlying rocky layers molten so that tectonic forces can produce crustal movement would largely come from tidal interactions with Jupiter, though to a much smaller degree than on Io.

Mercury, Moon, Ganymede, Callisto. The remaining intermediate-sized planetary bodies—Mercury, Earth's Moon, Ganymede (figure 5.27), and Callisto (figure 5.28)—are *one-plate planets* (too little is known of Titan and Triton to say anything about their geological activities). No large-scale tectonism deforms their surfaces today, nor is there much evidence of volcanism. However, the

Figure 5.24. Olympus Mons on Mars, the largest known volcano in the Solar System. Differences in crater densities around its flanks indicate that Olympus Mons grew over a considerable period of time.

presence of ancient intercrater lava plains on Mercury and the Moon does indicate that during the first half of their existence molten magma managed to penetrate from their interiors to the surface. Likewise, the rather complex system of grooves and ridges that crisscross Ganymede, with estimated ages of about 3.5 billion years, suggests that this body, too, has experienced some form of geological activity. It appears that by the time the Solar System was 2.5 billion years old, the crusts of all of these bodies began to harden into single, unyielding plates and their geological histories drew to a close.

Presence of Water

Water is present today on all intermediate-sized planetary bodies with the exception of Mercury, the Moon, and Io.

Mercury, Moon, Io. Mercury and Io have no water because of their proximities to the Sun and Jupiter, respectively, and the heating they experienced during their formation. The Moon is without water probably because of the way it formed.

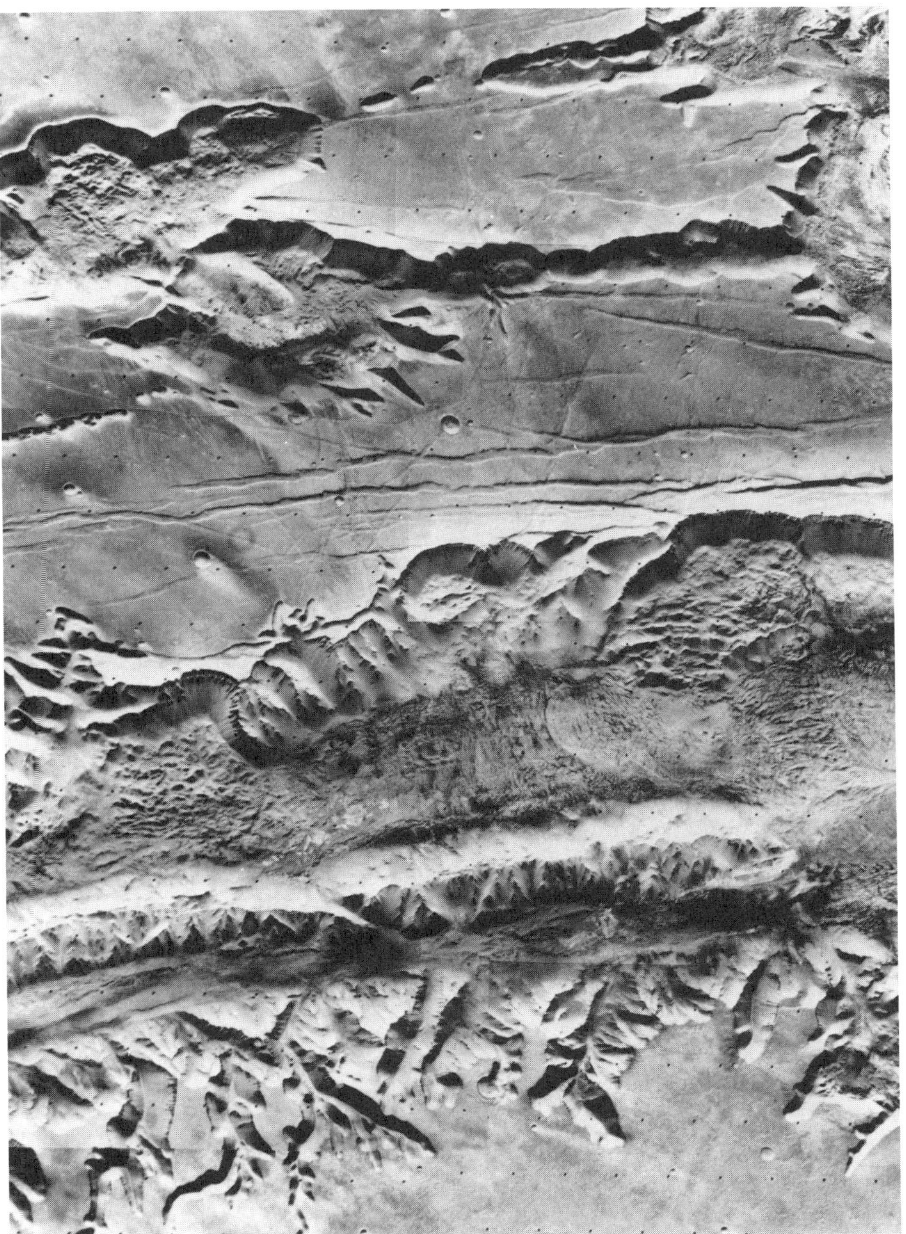

Figure 5.25. View of the central section of Valles Marineris, the enormous canyon system just south of the Martian equator. Numerous huge landslides are visible on the north walls of the canyons, where they have created alcoves up to 100 km across (such as the landslide near the right edge and half way between top and bottom of this photo). Some of the larger landslides extend all the way across the canyon and ride part way up on the opposite wall.

Venus. Venus is also very dry. The little water it does possess exists as vapor in its atmosphere. However, at one time it may have had as much water as Earth, but lost most of it as a result of the evolution of its atmosphere. The evidence comes from the abnormally high ratio of deuterium to hydrogen (^2H:^1H) in the water that is still present. This discovery was made in 1978 with an instrumented capsule that descended through the Venusian atmosphere, after it was delivered to the planet by the United States spacecraft *Pioneer Venus Multiprobe*. The high ^2H:^1H ratio suggests that early in Venus's history UV radiation from the Sun broke the planet's water molecules into their atomic constituents, ^1H, ^{16}O, and an occasional ^2H. The oxygen reacted with other compounds and is now locked up in the crust of Venus. The heavy isotope of hydrogen, ^2H, also was retained, but most of the ^1H escaped into space because of its small mass. This raised the ^2H:^1H ratio in the remaining water to roughly 100 times its value on Earth.

Europa, Ganymede, Callisto, Titan, Triton. Of the intermediate-sized planetary bodies that have retained appreciable amounts of water, the greatest amount is found in the outermost ones—Europa, Ganymede, Callisto, Titan,

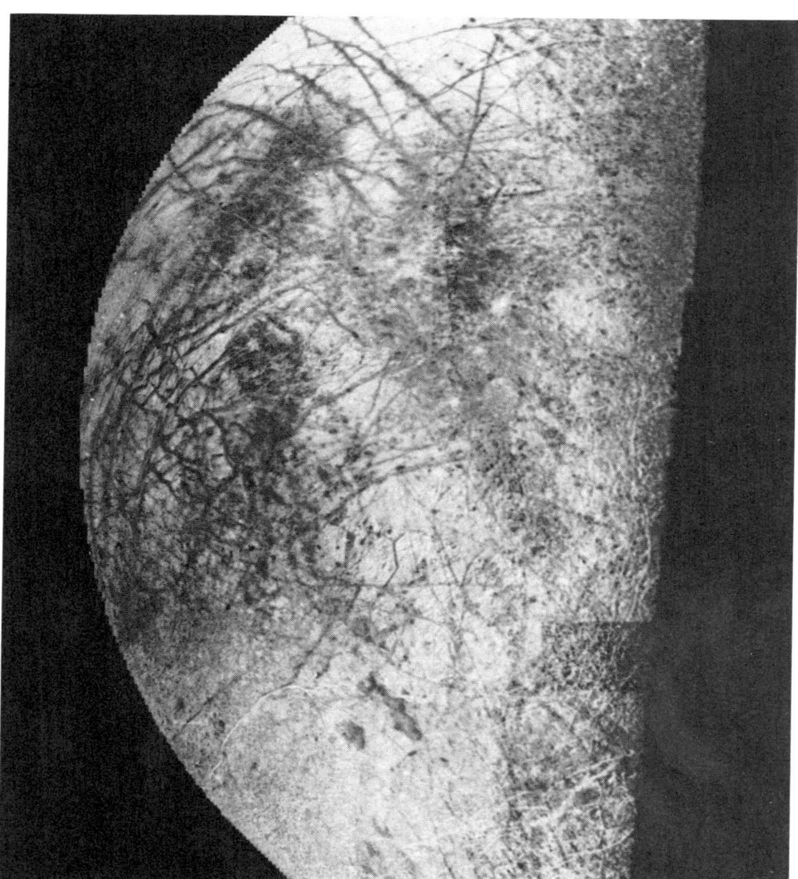

Figure 5.26. Europa, observed by *Voyager 2* from a distance of 241,000 km. The vast tangle of light and dark markings, some up to hundreds of kilometers across, suggest filled-in cracks. They have very little relief, and are primarily albedo markings (that is, they reflect sunlight with differing degrees of intensity).

Figure 5.27. Ganymede, the Galilean satellite that superficially resembles the Moon, with irregular dark regions on a brighter background. The prominent dark area in the lower left part of this photograph, called Galileo Regio, is about 3200 km across and heavily cratered. Like the other dark areas, it is older than the light regions, which exhibit a variety of features (such as complex systems of ridges and troughs) suggestive of past tectonic activity.

and Triton—because of the low temperatures that prevailed during and after their formation. In fact, water may constitute up to 50% of their masses, excepting only Europa. Differentiation early in the histories of these bodies forced the water out of their cores to form massive mantles surrounded by thin crusts. The surfaces are now at temperatures of 150 K or less and are frozen into hard, rigid ice that is capable of retaining the shapes of impact craters and other surface features for considerable lengths of time. For example, Ganymede and Callisto are heavily cratered, much like the rocky crusts of Mercury and the Moon, though the ice craters are considerably shallower (because ice slowly flows and deforms, as glaciers on Earth demonstrate).

Earth. A little more than two-thirds of the surface of Earth is covered by oceans. From this we might conclude that, concerning the abundance of water, Earth falls somewhere in the middle between those bodies that consist half of water and those that have none. That is a serious misconception. The water on the surface of the Earth adds up to approximately 2.3×10^{-2}% of the total mass of our planet. Several times this amount of water may be locked up in the rocks of the crust and mantle (basalt and granite contain between 0.2 and 0.5% water). But the total can hardly be more than a fraction of a percent of the

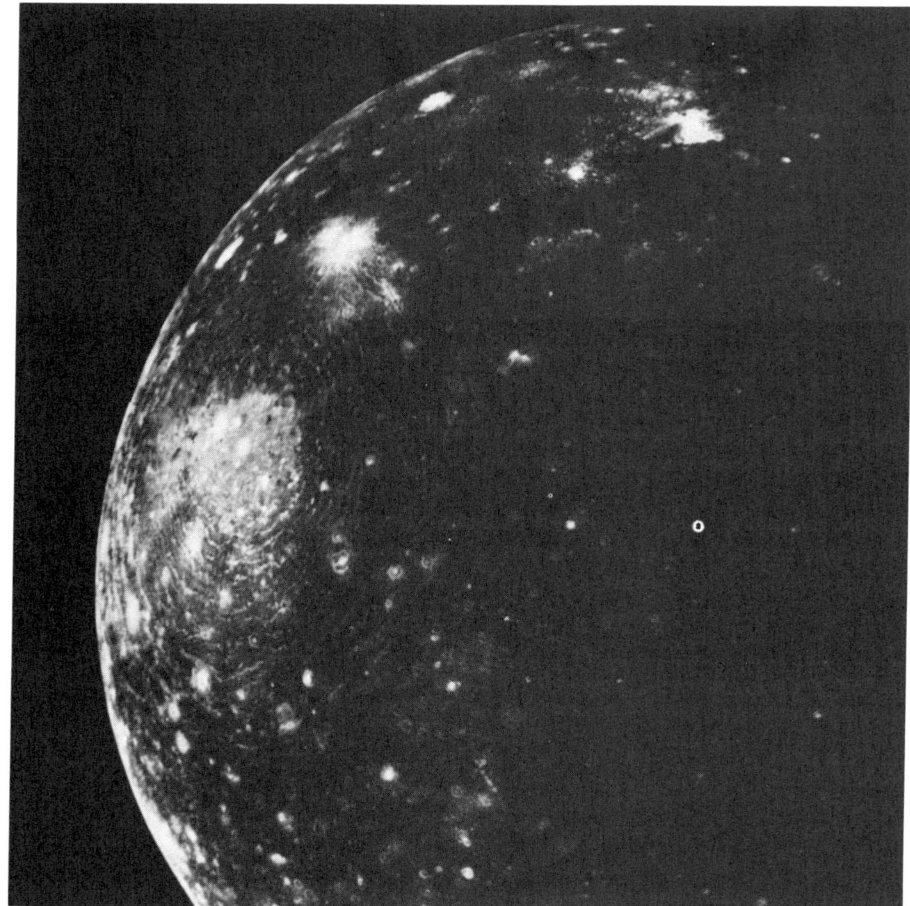

Figure 5.28. Callisto, the most heavily cratered of Jupiter's Galilean satellites. The prominent impact structure seen in this photograph, with an extensive ring system, is known as Valhalla and is similar in many respects to the large impact basins found on Mercury and the Moon. Valhalla's bright central area is about 300 km across, while the system of discontinuous, concentric rings extends radially outward some 1500 km.

total mass of the Earth.* Hence, with regard to water, Earth resembles barren Mercury much more than water-laden Ganymede or Callisto. Clearly, the planetesimals from which Earth was assembled had lost most of their water or never

*The mass of all of the water on the surface of the Earth adds up to approximately 1.4×10^{21} kg, and the total mass of the Earth is 6×10^{24} kg. The Earth's water is distributed approximately as follows (data from Press and Siever 1978, 150):

oceans	97.3%
glaciers and polar ice	2.1
groundwater	0.6
lakes and rivers	0.01
atmosphere	0.001
biosphere	4×10^{-5}

possessed much in the first place. Otherwise Earth should have ended up with roughly the same proportion of water as the outer planets.

Because the water loss was so great, we may assume that any number of minor changes in the protoplanetary disk—such as slightly more or less heating by the nascent Sun—could have significantly altered the amount of water that did end up on Earth, with important consequences for the environment and the origin and evolution of life.

Mars. Though Mars is more distant from the Sun than Earth, and hence should be much wetter, the only water apparent on its surface today is in thin layers of ice that cover its poles. These icecaps grow and wane with the seasons, subliming (that is, changing from the solid to the gaseous state without passing through the liquid state) during the summer and recondensing during the winter. If liquid water were present on the Martian surface, it would not be stable, given Mars's present temperature range of 150–290 K and a barometric pressure of a mere 7/1000 that of the sea level pressure on Earth. It would either boil or freeze, depending on the temperature.

Still, there is ample evidence that at one time liquid water flowed on Mars. For example, the cratered terrain is cut by countless *runoff channels* that bear some resemblance to terrestrial valleys eroded by flowing water (see figure 5.29). These channels start out small, have numerous tributaries, increase in size downstream, and are, on the average, a few tens of kilometers long. Most were probably cut between 3.0 and 3.5 billion years ago, shortly after the heavy initial cratering ended. Other kinds of channels, the *outflow channels*, have the markings of sudden flooding by huge quantities of water or other kinds of fluvial matter (see figure 5.30). These channels are much bigger than the runoff channels, frequently measuring tens of kilometers across, and they meander for hundreds of kilometers across the Martian terrain. They usually are flat-bottomed, often contain islands, and generally start abruptly and full-grown. It is as if huge reservoirs of water had suddenly burst their dams and emptied. In fact, there is some evidence for the previous existence of large standing bodies of water in some regions, in the form of sedimentary layering, but their relation to the outflow channels is ambiguous. Underground reservoirs of water, soil saturated with water that suddenly became unstable, and glaciers have also been suggested as likely sources of the flooding.

The outflow channels are younger than the runoff channels and probably were formed when the geological activity on Mars reached its peak, roughly 2.5 billion years ago. The geological activity was accompanied by outgassing and the release of large quantities of water and other volatiles, as occurred on Earth (see chapter 6). Much of the water was probably lost to space because of the low surface gravity of Mars. A small fraction is still present in the planet's polar icecaps. The rest is suspected to exist as permafrost or possibly in liquid form under the Martian surface.

Atmospheres

Of the intermediate-sized planetary bodies, only Venus, Titan, and Earth have substantial atmospheres, followed by Mars and Triton. The major gases in the

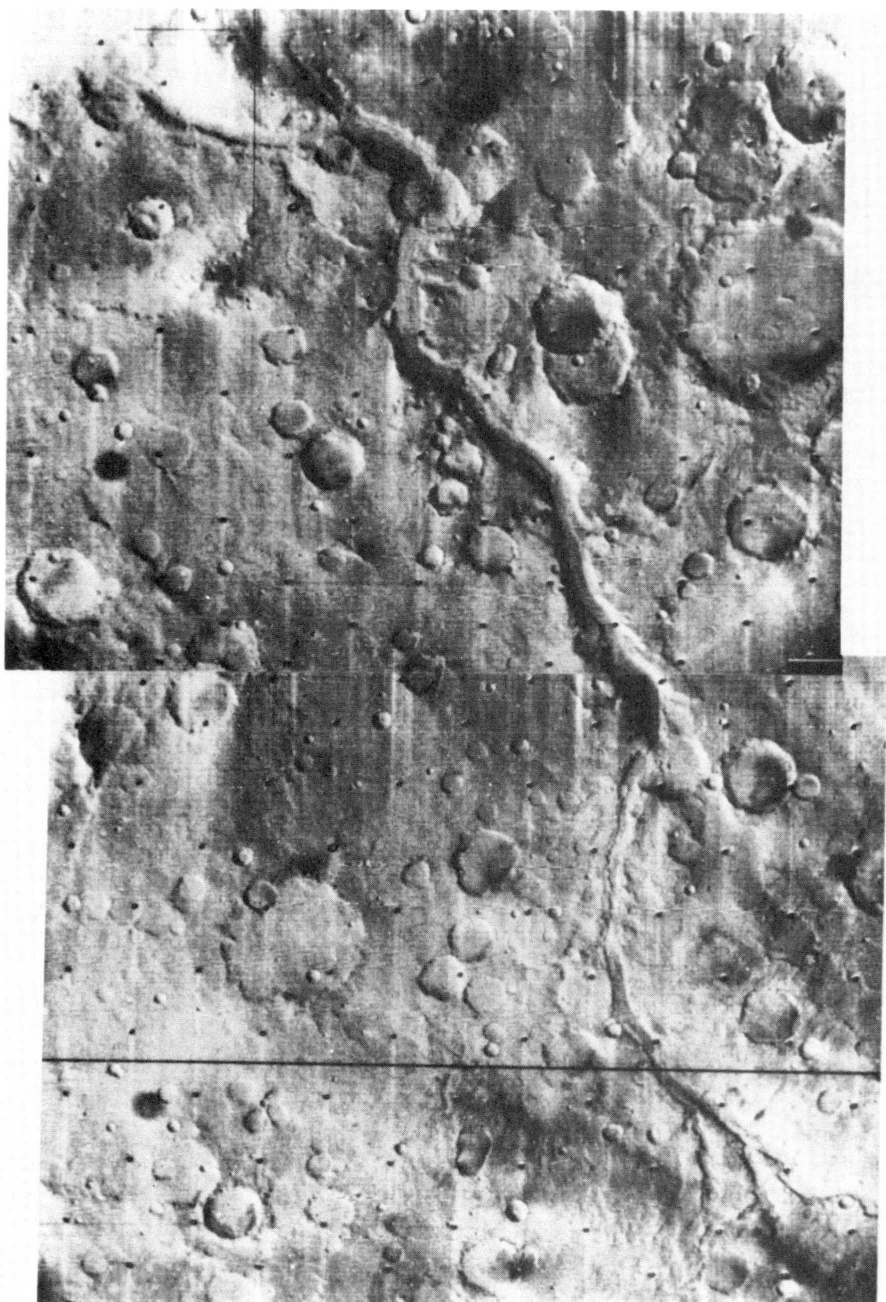

Figure 5.29. Ma'adim Vallis, a 600-km long runoff channel that runs through old cratered terrain on Mars. It contains long, branched tributaries upstream (lower right) and shorter tributaries resembling terrestrial box canyons downstream. Its gentle, meandering form leaves little doubt that it was carved out by a flowing medium, probably water.

atmospheres, the surface pressures, average surface temperatures (in degrees Kelvin and Celsius), and surface gravities of these bodies are listed in table 5.3 (page 201). In comparison, the other intermediate-sized planetary bodies have barren rocky or icy surfaces.

The sizes of planetary atmospheres are determined by two factors: (1) the sources that bring gases to the surface, and (2) the loss mechanisms that tend to remove them. The sources of atmospheric gases are tectonism and volcanism. They deliver severely heated, molten rock to the surface and expose them to low pressures, which allows volatiles trapped in the rocks to escape or outgas into the open. For example, H_2O, CO_2, and N_2 are the most common volatiles trapped in the surface and upper mantle rocks of the Earth, as the compositions of the clouds billowing forth from terrestrial volcanoes demonstrate. Hence, H_2O, CO_2, and N_2 were the major original constituents of our planet's atmosphere. A second source of atmospheric gases is evaporation of liquids or sublimation of ices from the planetary surfaces. The mechanisms by which planets lose atmospheric gases are (1) diffusion into interplanetary space if their surface gravities are small or the atmospheres are very hot, and (2) loss to the ground. The loss to the ground may occur either by condensation, such as the falling of rain, or by chemical reactions that bind the gases to the surface rocks.

The various gain and loss mechanisms of planetary atmospheres are interdependent and far too complex to be discussed here in detail. Briefly, Earth and Venus, for example, have substantial atmospheres because of the extensive geological activities they experienced throughout most of their pasts (figure 5.31). Furthermore, both planets are massive and have relatively strong surface

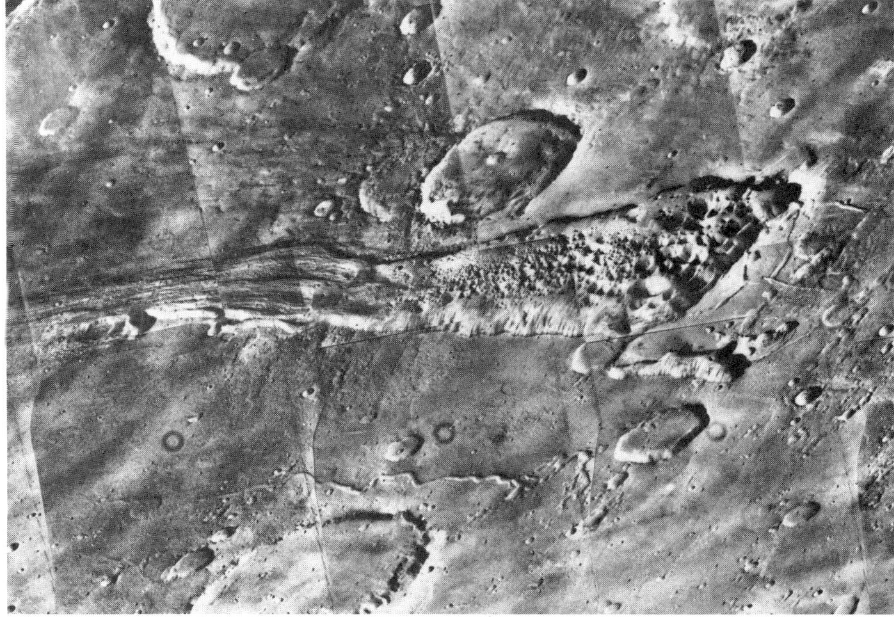

Figure 5.30. Martian outflow channel. The channel emerges full-born from a 40-km wide depression enclosing chaotic terrain.

gravities with which they retain their atmospheres. On Mars, by contrast, the early geographical activity was interrupted and, thereby, so was the growth of its atmosphere. Besides, Mars is less massive than Earth or Venus, with only about one-third their surface gravity. Hence, atmospheric gases escape to space more easily. Mercury is too hot and the other terrestrial planetary bodies have surface gravities that are too weak to have accumulated appreciable atmospheres. The one exception is Titan. Its surface gravity and temperature are comparable to those of Ganymede and Callisto, but unlike these two bodies it has a very thick atmosphere. In fact, above a given area of its surface, Titan's atmosphere exceeds that of Earth by a factor of ten. At present we do not know why this is so.

THE JOVIAN PLANETS

The Jovian planets—Jupiter, Saturn, Uranus, and Neptune—are the "gas giants" of the Sun's planetary system. They possess no solid surfaces and no ancient impact craters or deformations by geological activities mar their surfaces. Instead, thick atmospheres envelope them, whose top layers are characterized by gigantic and often colorful wind and storm patterns. The bottom layers of the atmospheres pass smoothly over into liquid mantles, and near their centers they possess cores of mainly rocks and iron.

Thus, the Jovian planets are quite unlike the intermediate-sized planetary bodies and deserve to be grouped into a family of their own. Their unique structures stem from their large masses and sizes in comparison to the other planetary bodies and from the volatiles they possess in great abundance. For example, Jupiter has roughly 300 times the mass and 10 times the size of the Earth, and Saturn is not much smaller, as table 4.1 indicates. Furthermore, the compositions of Jupiter and Saturn are dominated by hydrogen and helium, followed by carbon, nitrogen, and oxygen. Uranus and Neptune are less massive and smaller than Jupiter and Saturn, but they, too, consist largely of volatiles, though they lost much of their original hydrogen and helium. The large abundances of volatiles in these four planets follow from both their observed surface compositions and from their unusually low average densities (see table 4.1).

The combined mass of the Jovian planets adds up to 0.13% of the mass of the Sun and it exceeds the mass of the other planets, satellites, and smaller bodies of the Solar System by about a factor of 200. Thus, next to the Sun, the four Jovian planets are the Solar System's major components.

Core and Mantle Structure

The three-layered structures of the Jovian planets—gaseous atmospheres, liquid mantles, and solid or liquid cores—differ considerably from one planet to the next because of differences in their masses, compositions, and distances from the Sun.

The smallest difference in structure among the Jovian planets is found in the makeup of their cores. Model calculations indicate that the cores of all four planets consist mainly of silicates and iron, with small admixtures of other heavy

Figure 5.31. The cloud cover of Venus photographed in ultraviolet light.

elements. Clearly, some form of composition differentiation, similar to the one in the intermediate-sized planetary bodies, took place in these giant planets. Uranus and Neptune probably have the largest cores in proportion to their total sizes because they have lost the most volatiles. In contrast, Jupiter and Saturn have the highest core pressures and temperatures because they are the most massive. The estimated conditions at their centers, compared to the Sun's, are as follows:

	Jupiter	Saturn	Sun
Temperature (K)	25,000	20,000	16×10^6
Pressure (earth atmospheres)	8×10^7	5×10^7	3×10^{11}
Density (g/cm^3)	20	15	160

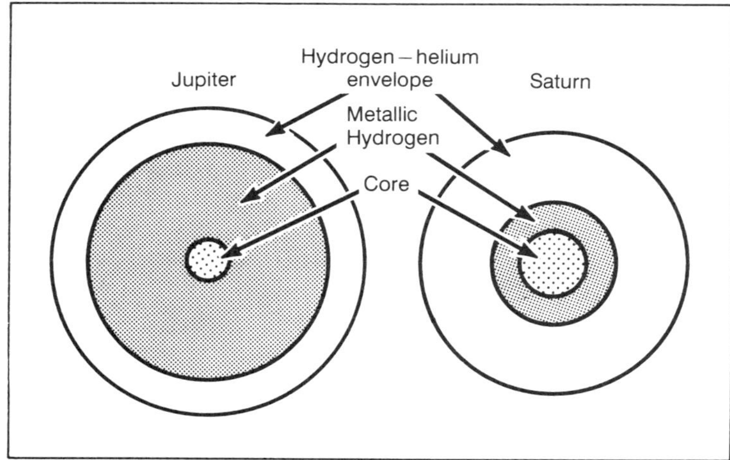

Figure 5.32. Approximate models of the interior structures of Jupiter and Saturn.

The mantles of the four planets are liquid, but beyond that they show a great deal of variation. The mantles of Jupiter and Saturn consist mainly of hydrogen and helium and have two-layered structures (see figure 5.32). In the inner layers, high pressures keep the hydrogen molecules stripped of some of their electrons, which produces metallic characteristics, such as high conductivity of heat and electricity. Hence, these layers are commonly referred to as the *metallic layers*. In the outer mantle layers, which are at lower pressures, the molecules are not stripped of electrons and, consequently, they are not metallic—they are just ordinary liquids. The mantles of Uranus and Neptune are believed to be single-layered and to consist chiefly of liquid water, methane, and ammonia, without any metallic characteristics.

Atmospheres

The greatest differences in structure among the four planets arise in their atmospheres. Jupiter's and Saturn's atmospheres are characterized by well-defined colored zones or belts that encircle the planets and rotate with them (see figures 4.1 and 5.33). These zones are remarkably stable and have not changed their latitudinal positions during the last 80 years. Close-up photographs sent back by the *Voyager 1* and *Voyager 2* spacecrafts show that the zones of both planets are in a general state of turmoil, with large and small eddies, fluctuating atmospheric motions, and elongated fast-moving gas streams abounding. The most prominent of these dynamic features have the appearance of dark or bright spots and are useful markers for tracking the planets' global circulation patterns. The atmospheric compositions are H_2 and He, roughly in solar proportions, followed by small admixtures of NH_3, CH_4, and, in the case of Jupiter, H_2O and various kinds of simple organic compounds such as C_2H_6 (ethane), C_2H_2 (acetylene), and HCN (hydrogen cyanide).

Figure 5.33. Jupiter, photographed by *Voyager 1* from a distance of nearly 33 million km. The planet's atmospheric band structure of alternating bright and dark clouds is clearly visible, as are the Great Red Spot, several large white ovals, and many smaller eddies and vortices.

The atmospheric zones of both Jupiter and Saturn are closely associated with alternating westerly and easterly winds. For example, if in one zone the atmospheric winds blow toward the east, those in the two adjacent zones blow toward the west. These winds are analogues of the jet streams and trade winds on Earth, which also flow in adjacent zones and in opposite directions.* Earth

*The Earth's trade winds are tropical and subtropical surface winds that on the average blow from east to west, with average speeds of about 20 km/hr. The Earth's jet streams are located further north and south of the equator than the trade winds and blow from west to east at heights between 10 and 50 km, reaching maximum speeds of 500 km/hr, though speeds of 100–300 km/hr are more typical.

has only two such wind zones (jet streams and trade winds) in each hemisphere, but Jupiter and Saturn have several and they are much wider. For example, the winds of their equatorial zones are approximately 30,000 km and 80,000 km wide and they blow eastward with speeds of up to 500 km/hr and 1800 km/hr, respectively. In the case of Saturn, this is two-thirds the local speed of sound.

The bright and dark spots on the surfaces of Jupiter and Saturn are eddies of circulating atmospheric gases, comparable to the rotating high- and low-pressure systems that determine the weather on Earth. The smaller eddies do not last long. Within days of their origin they usually are caught between oppositely flowing zonal wind currents, and the resulting shear forces tear them apart. Sometimes they are absorbed by clouds and may be disgorged hours or days later. Complex patterns of oscillating vortex motions are often set up in the wakes of these atmospheric disturbances. Eventually, the eddies and their fragments disappear altogether, only to be followed by new ones that suddenly appear.

The larger eddies last much longer. The most famous of them, the Great Red Spot in the southern hemisphere of Jupiter, has been observed since the invention of the telescope more than 300 years ago and, probably, has existed considerably longer than that (see figure 5.33). The three white ovals just to the south of the Great Red Spot were first observed in 1938; other large eddies have lasted for months or years. They are so stable because instead of being sheared apart, they roll along between the wind currents of adjacent zones.

In contrast to Jupiter and Saturn, very little is known about the atmospheric conditions of Uranus and Neptune. These two planets are too distant and faint to reveal much of their constitution to earth-bound observations. However, in January, 1986, *Voyager 2* passed Uranus and it is now on its way to Neptune, which it will reach in August of 1989.

Uranus is odd in that its equatorial plane as well as its ring system and the orbits of its 15 satellites are nearly perpendicular to the plane of the ecliptic. In contrast, the equatorial planes of all other planets are aligned much more closely with the ecliptic. In fact, at present Uranus's south pole points almost directly toward the Sun. Data sent back by *Voyager 2* indicate that Uranus rotates once every 17.3 hours. Its atmosphere contains hydrogen and helium in proportions similar to those of Jupiter and Saturn. It also contains small admixtures of methane and various other hydrocarbons. There is evidence of cloud structure and atmospheric winds, but *Voyager 2* detected no bands or other detailed surface characteristics. The planet is shrouded in a permanent smoglike haze, which hides most of the atmospheric features that lie below. The Uranian temperature is 50 K at a level where the pressure equals 0.1 earth atmospheres, and it increases with depth. Surprisingly, the temperature varies very little between the planet's sunlit and dark sides. This is further evidence of the existence of atmospheric circulations, which distribute the heat received from the Sun.

Neptune, the most distant and faintest of the Jovian planets, shows up as a greenish disk when viewed through earth-bound telescopes. Its known atmospheric constituents are hydrogen, helium, and methane. Virtually no surface markings are discernible from Earth. Few other details are presently known about this planet.

ORIGIN OF THE SOLAR SYSTEM

Now that we have discussed the orbital distribution and physical structures of the major components of the Solar System, let us turn to their origin. This is a topic of age-long interest and was one of the first to be examined during the early development of western science, more than 300 years ago. In 1644 René Descartes, the great French philosopher, suggested that at one time the Sun was surrounded by a large rotating disk of gas from which the planets and their satellites formed. One hundred years later, this *nebular theory* was challenged by the French naturalist Georges Louis Leclerc de Buffon (1707–1788). He proposed a *collision theory* according to which a massive comet or, as proposed later, a star passed close to the Sun and ripped out the material from which the planets and their satellites then assembled.

Neither of these two theories was based on sufficient facts or theoretical reasoning to be fully convincing, and during the succeeding centuries first one and then the other gained favor. The German philosopher Immanuel Kant (1724–1804) and the French mathematician Pierre Simon Laplace (1749–1827) supported the nebular theory. Laplace pointed out that a rotating disk of gas would slowly contract and develop rings of materials from which planetary bodies could form. Throughout much of the nineteenth century, this remained the accepted theory among most scientists. However, in the second half of that century, the British physicist James Clerk Maxwell (1831–1879) calculated that shear forces resulting from the differential rotation of the disk would prevent the matter from condensing into individual bodies. Thus, Buffon's collision theory was resurrected, and during the first few decades of the present century the English physicist Sir James H. Jeans (1877–1946) and others developed it further. Because a collision between two stars is an extremely rare event, they argued that our planetary system must be unique or at most one among a few.

Maxwell's critique of the nebular theory was based on the mistaken notion that stars and the interstellar material have the same composition as the Earth. Only in the late 1930s was it discovered that in fact hydrogen and helium are the major constituents of the Cosmos and that material like that of rocks adds up to no more than about 2%. From this new knowledge, the German physicist Carl Friedrich von Weizsäcker concluded that the primitive circumsolar disk must have been much more massive than are the planets today, and he demonstrated that Jeans's objections to the nebular theory were inapplicable. It was also realized that material ripped out of the Sun by a passing star would disperse or fall back onto the Sun, rather than condense into planetary bodies. These new theoretical insights led to the abandonment of the collision theory and the general acceptance of the nebular theory.

Since the 1940s, an increasing number of researchers have worked on the nebular theory of the origin of the Solar System, using the growing knowledge about interstellar clouds and star formation, building numerical models with the aid of high-speed computers, and, more recently, applying the wealth of new observational data gathered from space. The theory is still far from complete, with many rival ideas about some of the details. However, there is a general consensus today among the experts how, in broad terms, the Sun and its planetary system formed from an interstellar cloud of gas and dust (see figure 5.34).

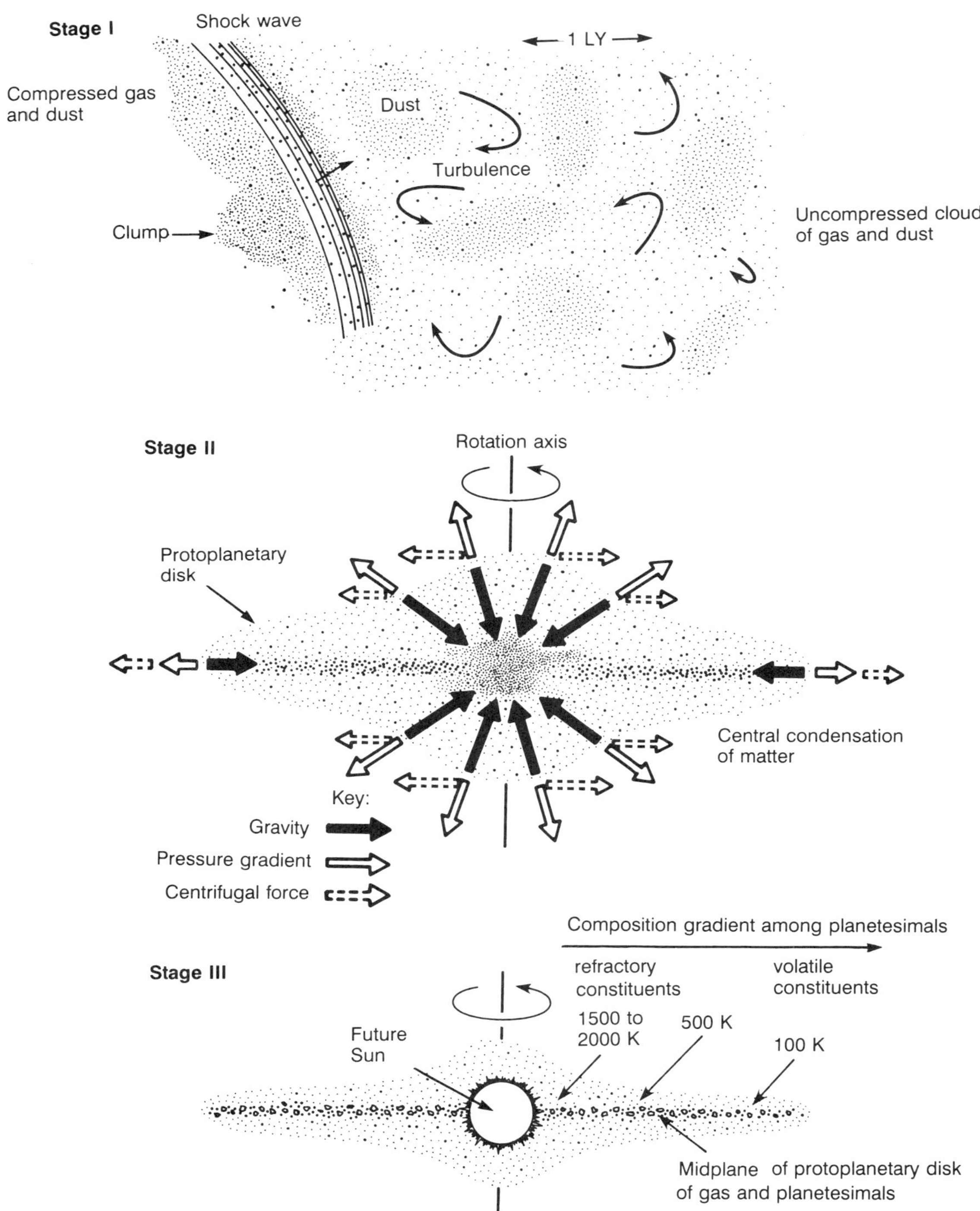

Figure 5.34. Summary of the stages in the birth of the Solar System (see also next page). Details are discussed in the accompanying text and in chapter 3, The Birth of Stars.

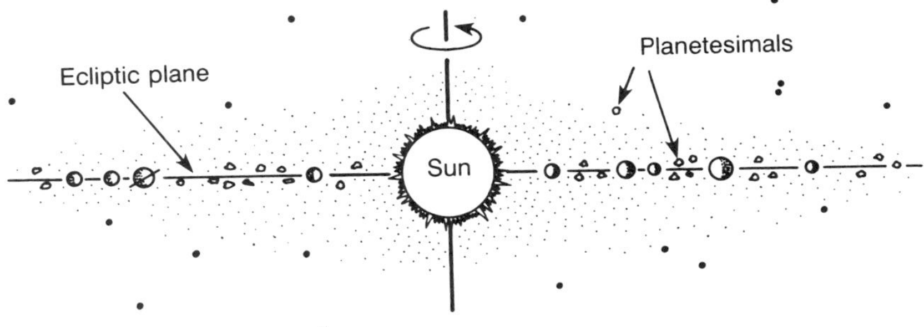

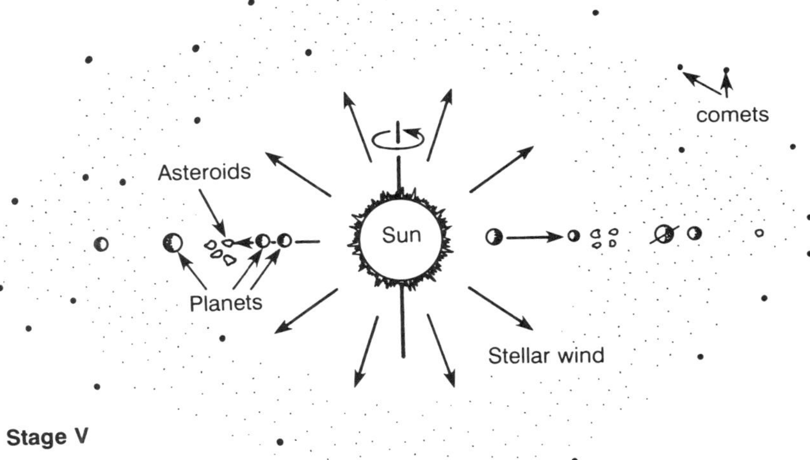

Figure 5.34 (continued)

Stage I

About 4.5 to 4.6 billion years ago, along one of our galaxy's spiral arms, a supernova exploded near an interstellar cloud of gas and dust. As its shock front swept through the cloud, it compressed many of its dense subregions to the point of gravitational collapse. One of the dense clumps thus formed (see chapter 3, The Birth of Stars) contained the raw materials—hydrogen (73.6% by mass), helium (24.8%), and heavier elements (1.6%)—for the birth of the Solar System. Before compression, the clump was roughly 0.1 to 1.0 LY (approximately 10^3 to 10^5 AU) across and it had a mass between 1.5 and 2 solar masses. After compression, its size was reduced by roughly a factor of ten, though there is at present little certainty about these values. The clump rotated slowly; its temperature was about 5–10 K; and most of its material was condensed into molecules and dust grains. Some of the molecules and dust grains had arrived with the shock wave from the supernova and differed in composition from the bulk material of the clump. (For the sake of balance, be aware that some scientists are not convinced that a supernova was responsible for the compression of the cloud to the point of gravitational collapse. They think that this point was reached by the gradual and natural evolution of the cloud, without any external, compressive forces. However, their theory does not readily account for the range of isotopic abundances found in the Solar System today.)

Stage II

As gravity pulled the clump together, it began to rotate faster. Three kinds of forces acted on the clump: gravity, the pressure gradient, and the centrifugal force resulting from rotation. (A fourth possible force—the force from magnetic fields—is disregarded here.) Gravity pulled the matter radially toward the clump's center, the pressure gradient pushed it outward, and the centrifugal force pushed it in directions perpendicularly away from the clump's rotation axis, as shown in figure 5.34. The net result was that the clump slowly contracted into a disk—the *protoplanetary disk*—revolving around a growing central concentration, the future Sun.

Stage III

For some time, the protoplanetary disk and the central concentration probably looked quite similar to the disk and nucleus of galaxies like NGC 3115 (figure 1.9), though, of course, they were very much smaller. Like the disk of a spiral galaxy, the protoplanetary disk rotated differentially, faster close to the central condensation and more slowly further out. As a result, neighboring segments of material that differed slightly in distance from the rotation axis orbited at slightly different speeds and, therefore, rubbed against each other. This created friction (that is, shear forces) and led to local turbulence. Yet another source of turbulence probably was thermal convection due to steep temperature gradients in the protoplanetary disk. The effect of turbulence was that the orbiting matter was slowed down and gradually spiraled inward toward the central concentration. As that concentration of matter grew, it heated up due to the release of gravitational energy and evolved into the Sun.

The disk heated up also because gravitational energy was released in it as matter spiraled through. Furthermore, the disk was heated by the nascent Sun, first by infrared radiation and later by optical and ultraviolet radiation. These heat sources were far stronger close to the Sun than farther out. Consequently, a temperature gradient was established from about 1500–2000 K in the region where Mercury eventually formed to approximately 100 K or less in the disk's outer fringes.

In the course of the growth of the protoplanetary disk, the concentration of its matter increased enormously. This was particularly the case along the disk's midplane, where a number of important and irreversible changes began to take place. The orbiting dust particles were so densely packed there that they collided very frequently with each other and coalesced into growing bodies, first of pebble size and then of boulder size and larger. Quite rapidly, many of the bodies reached diameters of tens to hundreds of kilometers and became the planetesimals, or "little planets."

At first, when the planetesimals were still very small, they had a rather loose and fluffy structure, similar to that of the dust grains. However, as they grew in size, self-gravity compacted them into hard, solid objects. Simultaneously, heat released by gravitational contraction and the decay of unstable isotopes, such as ^{26}Al, caused melting and differentiation.

The growth of the planetesimals was strongly affected by the disk's temperature. In the inner parts of the disk, where temperatures reached 1500–2000 K, the coalescing dust particles lost all but their most refractory constituents. Consequently, the planetesimals that formed there consisted only of silicate rocks and metals, without any volatiles. Toward the middle regions of the disk, corresponding to where Mars and the asteroids eventually formed, the temperature rose to only about 500 K. Hence, the dust particles retained many of their volatiles (such as H_2O, CO, CO_2, H_2S [hydrogen sulfide], N_2 and some organic molecules) and carried them along in the assembly of the planetesimals. Finally, in the outer regions of the disk that were heated the least, virtually all volatiles stayed with the dust particles, including those most easily vaporized, such as CH_4 and NH_3, and became part of the planetesimals. Thus, the temperature gradient of the disk led to a composition gradient among the planetesimals, from those with only refractory constituents to those composed of rocks intermingled with progressively more volatiles at increasing distances from the nascent Sun.

Stage IV

When the growth of the planetesimals first began and for some time thereafter, the gaseous component of the disk (consisting mostly of hydrogen and helium) was still thick and dense, and gas continued to spiral through it toward the Sun. The planetesimals participated only very little in this spiraling motion because they were massive and compact. They plowed right through the gas without being affected much and traveled in nearly circular or elliptical orbits around the Sun. Most of their orbits lay close to the midplane—the future ecliptic plane—of the protoplanetary disk, but some were inclined to various degrees. Large eccentricities and inclinations in the orbits were the result of continuous and ever-varying gravitational pulls that the planetesimals exerted on each other as they orbited the Sun. Planetesimals that were in elliptical and inclined orbits intersected the orbits of other planetesimals and, sooner or later, had near encounters or collisions with them. Some of the collisions were sufficiently energetic to shatter the planetesimals and disperse part or all of the resulting debris. The heavily cratered surfaces of many of the small and intermediate-sized bodies in the Solar System today are evidence of this violence. However, most often the collisions were rather gentle because near-circular and little-inclined orbits predominated. Hence, most collisions led to merger and growth.

The planetesimals that grew most rapidly were those that by chance acquired above-average masses early on and thus were most successful in gravitationally attracting and assimilating other smaller planetesimals. As illustrated by a sequence of computer simulations made by George W. Wetherill of the Carnegie Institution of Washington (shown in figure 5.35), dominant planetesimals established themselves at regular orbital intervals. In the course of time, they swept up nearly all of the other smaller bodies in their orbital neighborhoods and evolved into the terrestrial planets we observe today.

Collisions and accretion of solid planetesimals, as described so far, were the principal mechanisms of planet formation only in the inner part of the Solar System. In the outer parts, where temperatures were lower, planet formation was greatly influenced by the gas component of the protoplanetary disk. In fact, hydrogen and helium became the major constituents of Jupiter and Saturn and, to a smaller degree, became incorporated in Uranus and Neptune as well. At present it is not known whether, in the case of these four planets, solid planetary cores formed first, followed by the gravitational acquisition of envelopes of hydrogen and helium; or gaseous spheres formed first, followed by differentiation and the settling of rocky components toward their centers. Whichever sequence was the actual one, all four giant planets probably were, early in their formations, surrounded by their own protosatellitic disks of gas and dust, with temperature and composition gradients, from which satellites formed.

Stage V

While the planetesimals and planets were forming, the young Sun was also evolving. Within several hundred thousand to a million years of the initial compression of the primordial clump, the Sun became a T Tauri star and experienced the surface eruptions, rapid variations in light output, and mass loss in the form of a strong wind that are typical of that evolutionary stage. The wind blew for millions of years and swept away all of the gas and dust that had not yet coalesced into planets, satellites, or planetesimals, exposing these bodies to interstellar space.

Orbits, Rotations, and Composition Gradients

Some of the specific features of the Solar System can be explained in terms of the formation theory just described, but some cannot. Because the planets were the result of accretion of many small bodies and, in the case of the Jovian planets, of initially widely distributed gaseous material, their final orbits turned out to be averages of the individual orbits of the many small bodies and the gas. These averages added up to the near-circular and little-inclined orbits we observe today. The planets' directions of revolution around the Sun were also the same as the rotational direction of the protoplanetary disk, namely from west to east. The only exceptions were Mercury and Pluto, both of which acquired above-average orbital inclinations and eccentricities. We do not know what caused these results. However, the case of Pluto is not surprising, for it is very small and resembles more an asteroid than a genuine planet.

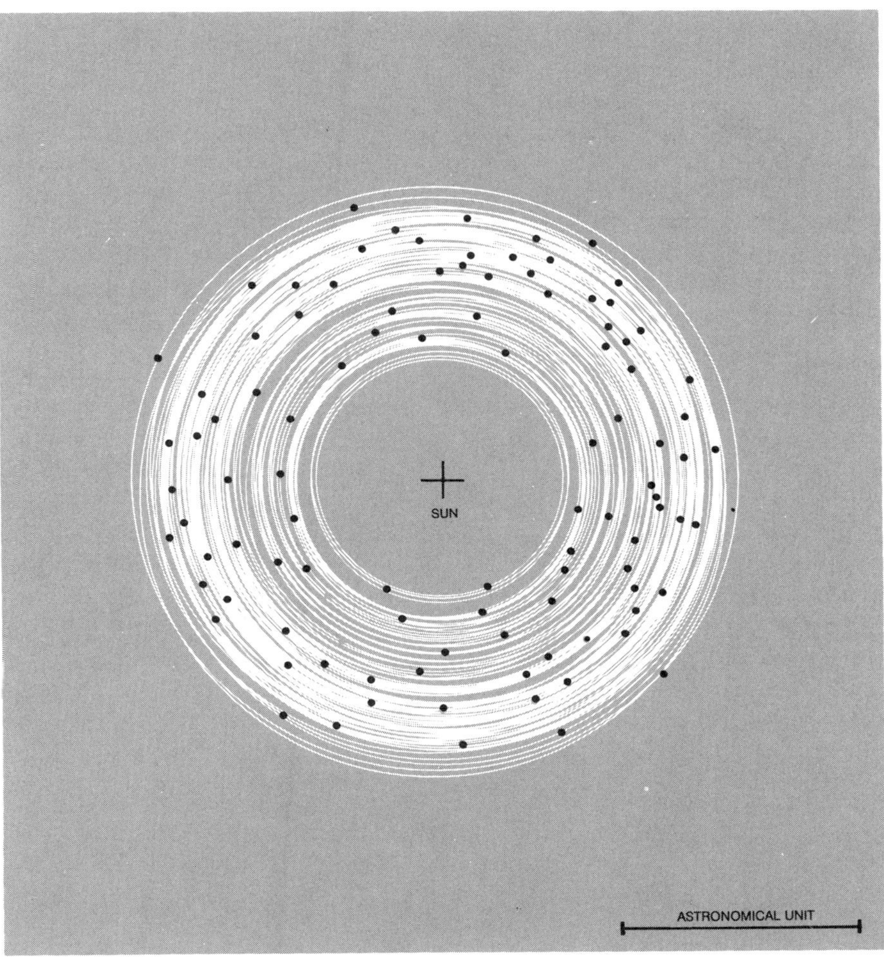

DIAGRAM 1

Figure 5.35. Wetherill's computer experiment simulating the formation of the terrestrial planets of the Solar System (continued on pages 246–247).

Diagram 1. The experiment begins with 100 planetesimals, each containing 2/100 earth masses and revolving around the Sun in random elliptical orbits as shown. Their total mass equals approximately that of Mercury, Venus, Earth, the Moon, and Mars.

Diagram 2. 30 million years later, near encounters and collisions leave only 22 bodies of various masses, as indicated by the sizes of the dots. Their orbits have become more elliptical and the range of their distances from the Sun has broadened.

Diagram 3. After 440 million years, the simulation has produced four fully-formed planets in isolated, nearly circular, and stable orbits. The largest of the planets is the outermost one. It has been built up from 34 of the original planetesimals.

The simulation was carried out in three dimensions, but only the projections of the orbits onto the midplane of the protoplanetary disk are shown. No gas and dust component or the perturbing effects of more distant giant planets were included in the experiment. (From Wetherill 1981, 162–174.)

DIAGRAM 2

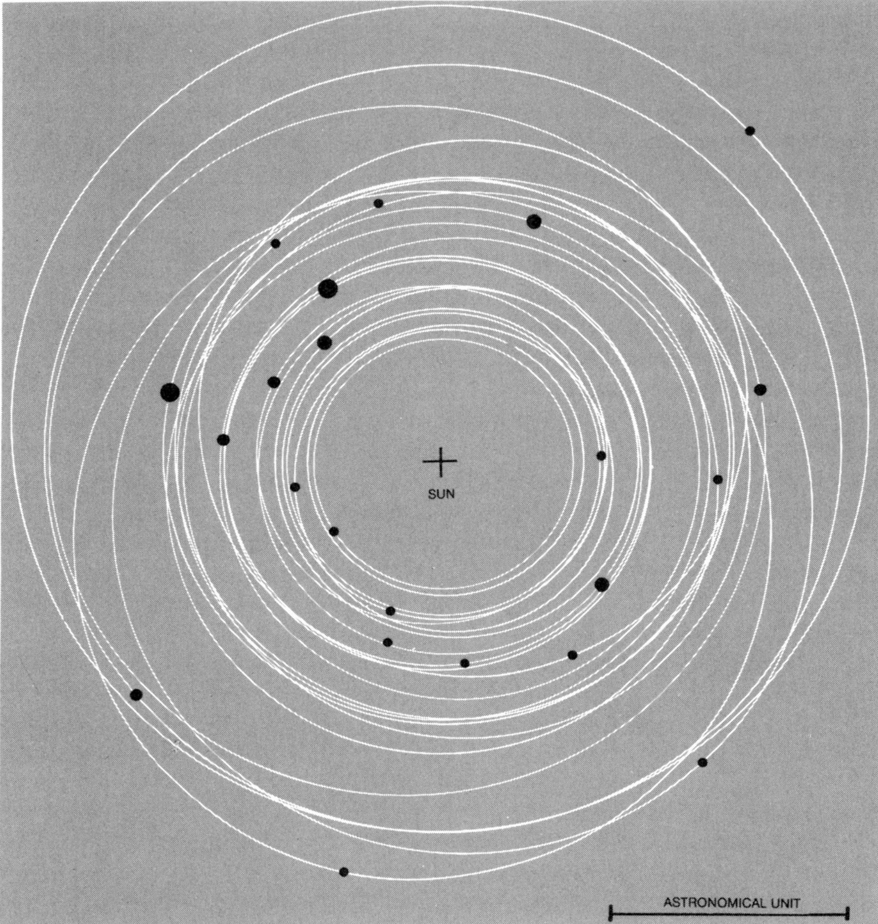

Figure 5.35 ***(continued)***

Most of the collisions between planetesimals were probably not head-on, but off-center or at glancing angles. The same was most likely true of the coalescence of gases in the formation of the outer planets. Thus, all planets ended up with net rotations that, with the exceptions of Venus and Uranus, were also in the direction from west to east. However, it is not clear at present why the planets have the particular rotation rates we observe, or why two rotate from east to west and the rest from west to east.

The development of a composition gradient among the planetesimals combined with the substantial amounts of hydrogen and helium the outer planets managed to acquire, explains the different compositions in the planets today. The Earth, for example, formed in a part of the protoplanetary disk in which heating was so severe that most of the volatiles that originally were present in the dust grains were vaporized and driven off. Consequently, the Earth acquired

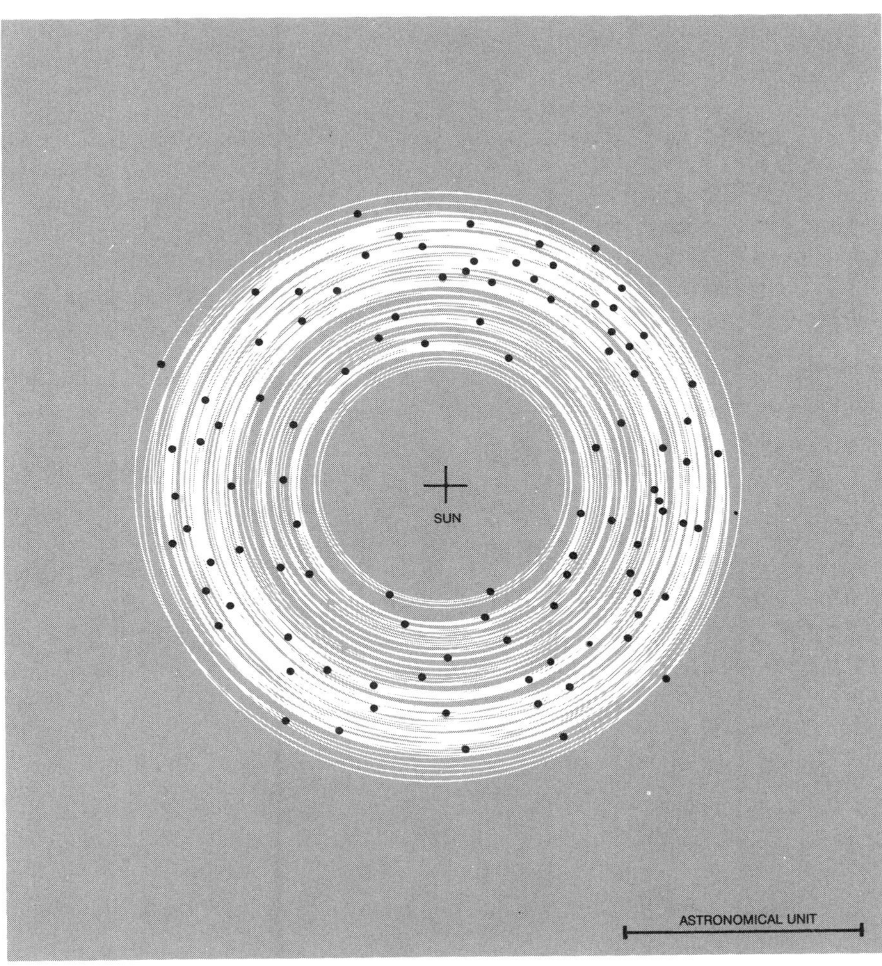

DIAGRAM 1

Figure 5.35. Wetherill's computer experiment simulating the formation of the terrestrial planets of the Solar System (continued on pages 246–247).

Diagram 1. The experiment begins with 100 planetesimals, each containing 2/100 earth masses and revolving around the Sun in random elliptical orbits as shown. Their total mass equals approximately that of Mercury, Venus, Earth, the Moon, and Mars.

Diagram 2. 30 million years later, near encounters and collisions leave only 22 bodies of various masses, as indicated by the sizes of the dots. Their orbits have become more elliptical and the range of their distances from the Sun has broadened.

Diagram 3. After 440 million years, the simulation has produced four fully-formed planets in isolated, nearly circular, and stable orbits. The largest of the planets is the outermost one. It has been built up from 34 of the original planetesimals.

The simulation was carried out in three dimensions, but only the projections of the orbits onto the midplane of the protoplanetary disk are shown. No gas and dust component or the perturbing effects of more distant giant planets were included in the experiment. (From Wetherill 1981, 162–174.)

DIAGRAM 2

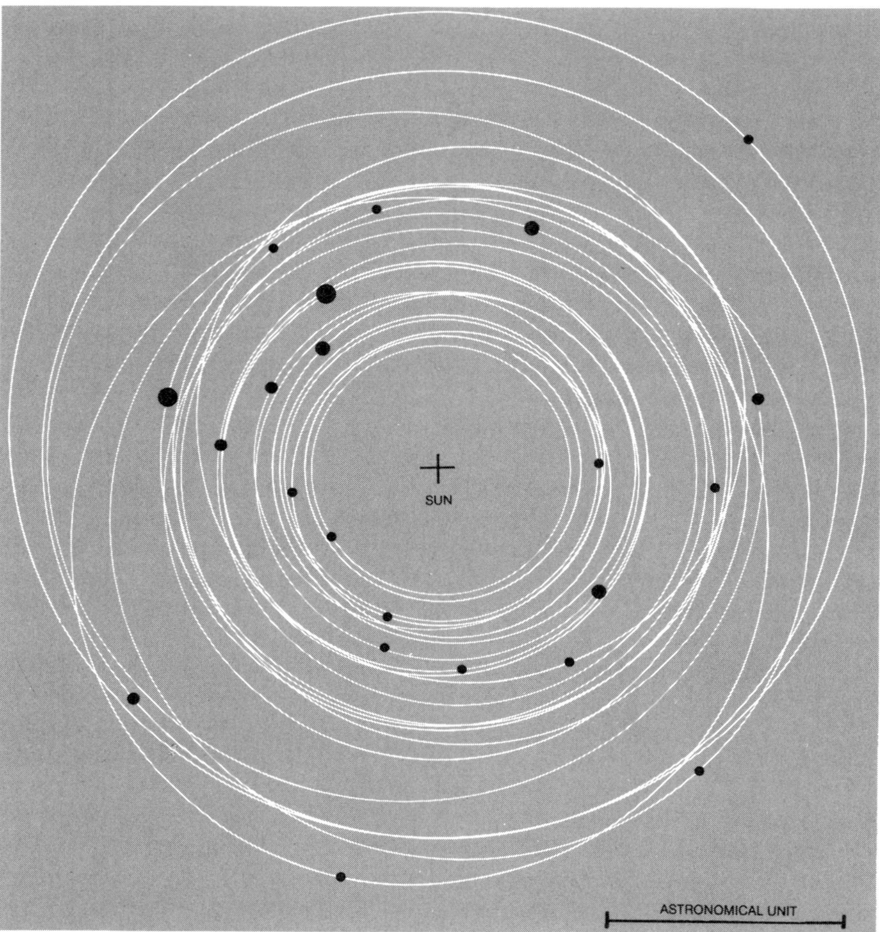

Figure 5.35 **(*continued*)**

Most of the collisions between planetesimals were probably not head-on, but off-center or at glancing angles. The same was most likely true of the coalescence of gases in the formation of the outer planets. Thus, all planets ended up with net rotations that, with the exceptions of Venus and Uranus, were also in the direction from west to east. However, it is not clear at present why the planets have the particular rotation rates we observe, or why two rotate from east to west and the rest from west to east.

The development of a composition gradient among the planetesimals combined with the substantial amounts of hydrogen and helium the outer planets managed to acquire, explains the different compositions in the planets today. The Earth, for example, formed in a part of the protoplanetary disk in which heating was so severe that most of the volatiles that originally were present in the dust grains were vaporized and driven off. Consequently, the Earth acquired

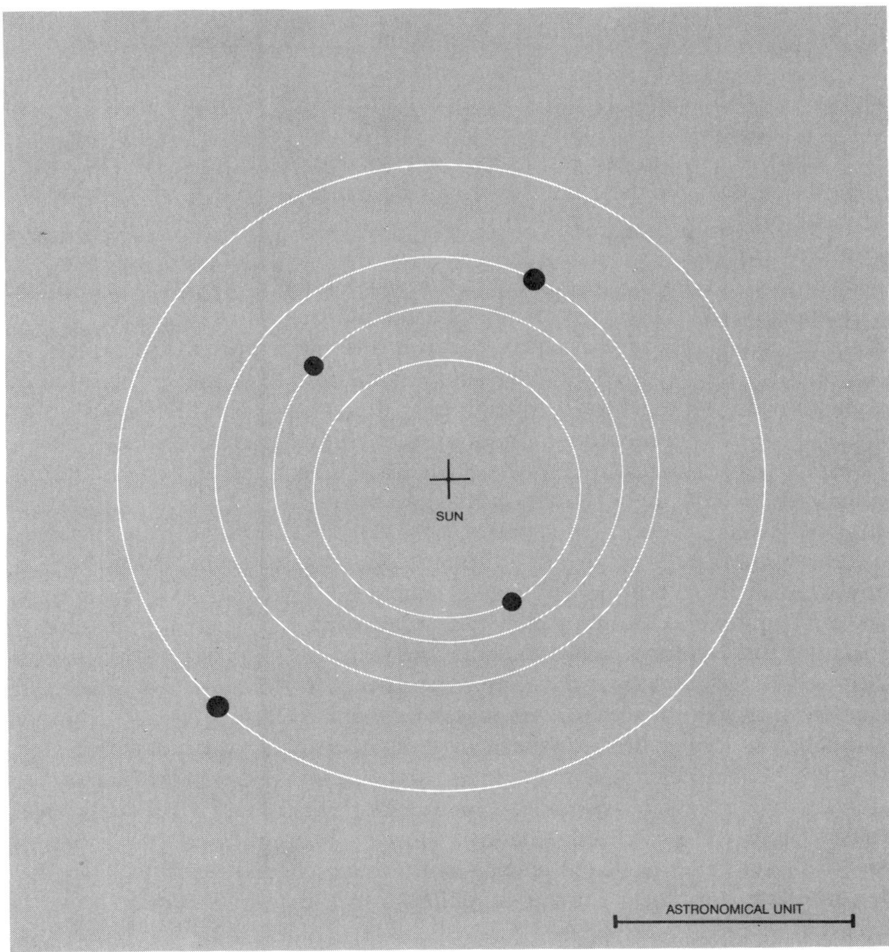

DIAGRAM 3

Figure 5.35 ***(continued)***

very few volatiles, compared to its bulk constituents of silicate rocks and iron–nickel metals. Most of the volatiles it possesses today—such as those constituting its ocean and atmosphere, and the CO_2 that became buried as dolomite and limestone in its crust—were probably brought to it by planetesimals from the vicinities of Mars and the Asteroid Belt. Perturbations of the planetesimals' orbits by near encounters with other planetesimals and nearby Jupiter deflected them inward toward Earth, which then captured them. As Wetherill's computer diagrams show, this kind of scrambling of the planetesimals' orbits was common in the protoplanetary disk. Thus, each planet had a much wider "feeding zone," from which it drew its raw materials, than today's spacing between planetary orbits.

The planet that had the greatest effect on the scrambling and broadening of planetesimal orbits was Jupiter with its large mass. It perturbed the planetesimals in its immediate vicinity so strongly that, for example, at 2.8 AU from the

Sun no planetary mass accumulated where, according to the Bode–Titius law, a planet should revolve. Most of the planetesimals of that region were dispersed to other parts of the protoplanetary disk. The few that survived there are today's asteroids. Likewise, Jupiter's perturbing effect removed many of the planetesimals from the neighborhood of the future planet Mars (at 1.5 AU), which explains why this planet acquired only about one-tenth the mass of Earth. Furthermore, many of the comets in the Oort Cloud, which extends 100,000 AU outward from the Sun, are believed to have been deflected there by Jupiter and the other giant planets.

The deflection of planetesimals from one part of the Solar System to another was responsible for the heavy bombardment that the planets experienced for hundreds of millions of years after they were formed. Only as the supply of planetesimals ran low, about 3.8 billion years ago, did the bombardment diminish. However, even today, the supply—the asteroids and comets—is not fully gone and the bombardment continues, though at a very low rate.

Despite the many features of the Solar System that are explained, at least qualitatively, by the theory just discussed, there remain numerous important and unresolved questions. For instance, how did it happen that a few, relatively tiny, planets swept up nearly all of the material between them? Or, put differently, why are there not more planets, each with a smaller mass? Why are the masses of Jupiter and Saturn roughly one order of magnitude greater than those of Uranus and Neptune, and two orders of magnitude greater than those of Earth and Venus? What was the time scale of planet formation? Was the main assembly phase completed in a hundred million years? Or did it take much more or much less time? Are planetary systems inevitable by-products of star formations?* Are they found only around single stars, or can binary stars have planets, too (perhaps in large orbits around close binaries and in relatively small orbits around widely spaced binaries)? Only by continuing investigations of the Solar and other stellar systems—both observationally and theoretically (in particular, through computer modeling)—can answers be found to such questions.

The origin of the Solar System dates back some 4.5 billion years, when the protosolar nebula contracted under the influence of its own gravitational forces. The greatest condensation of matter occurred at the center of the nebula and gradually evolved into the Sun. Orbiting the central condensation was a disk of gas, dust, and rapidly growing planetesimals. Collisions and coalescence among the planetesimals led to the formation of the small and intermediate-sized

*The first disk of rocky rubble surrounding another star was discovered in 1983 around Vega, with the U.S.–Dutch–British Infrared Astronomy Satellite (IRAS). The disk has a radius of about 80 AU, roughly twice the size of Pluto's orbit. Two more examples of stars circled by grainy or rocky rubble are Fomalhaut and Beta Pictoris, both visible from the Southern Hemisphere. Still other nearby stars are now being examined by the IRAS team for such circumstellar disks. In the case of Beta Pictoris the evidence even suggests, though not conclusively, that some of the rocky rubble may have coalesced into planetesimals or planets. Additional evidence for the existence of a planet or planets beyond the Solar System comes from a wobble in the motion of the nearby star Van Biesbroeck 8. (Similar wobbles reported during the past few decades in the case of several other nearby stars, and also interpreted as being due to planetary companions, have not been confirmed by recent, more sensitive observations.)

planetary bodies, while coalescence of the gas (at distances of Jupiter's orbit and beyond) produced the Jovian planets. All remaining uncondensed gas and dust were driven off into interstellar space by the strong wind that blew from the young Sun as it evolved through its T Tauri stage.

After the bodies of the Solar System were formed, they continued to evolve. In the case of most of them this evolution was strictly physical. For example, the Sun kept burning hydrogen and pouring fourth radiant energy. Venus, Earth, Mars, and most other bodies of intermediate size experienced major geologic changes and some acquired atmospheres. Many of the asteroids collided, some of them with the planets. The member bodies of the Oort Cloud, in the outer fringes of the Solar System, were periodically perturbed by gravitational forces of uncertain origin, and some of them were sent, as comets, in elongated orbits toward the inner part of the system. There, after numerous circuits around the Sun, they either crashed on one of the planets or the Sun, or they broke up and became part of the interplanetary debris.

However, the most remarkable evolution in the Solar System was not physical; it was biological. Within a relatively short time after the system's origin, perhaps even before the major accretion phase of the terrestrial planets had been completed, life emerged on Earth. The initially simple, microscopic organisms gradually became more complex and in the course of some 4 billion years evolved into an enormous variety of species, differing in size, habitats, and adaptive strategies. Most astonishingly, one of the species—*Homo sapiens*—evolved language and culture, the ability to think and to reason, and the urge to embark on a systematic investigation of the Universe near and far. The study of biological evolution on Earth will be the topic of the second part of this book.

> One generation passeth away, and another generation cometh; but the earth abideth forever. The sun also ariseth, and the sun goeth down, and hasteth to the place where he arose. The wind goeth toward the south, and turneth about unto the north; it whirleth about continually, and the wind returneth again according to his circuits. All the rivers run into the sea; yet the sea is not full; unto the place from whence the rivers come, thither they return again.
>
> Ecclesiastes—1:4–7

Exercises

1 Traditionally, the Sun's planetary system has been described as consisting of nine planets, the planets' satellites, asteroids, comets, and interplanetary matter. In the text, I did not follow this tradition, but defined just three families of bodies. Why? Discuss the most important characteristics of each of these families.

2 (*a*) Why are chondrites (ordinary and carbonaceous) said to be more primitive than the achondrites and the stony-iron and iron meteorites? (*b*) How do astronomers explain the origin of these different types of meteorites?

3 Distinguish between the following objects:
(*a*) meteors, meteorites, and meteoroids.
(*b*) asteroids and comets.
(*c*) stars and planets.

4 The radioactive isotope ^{14}C has a halflife of 5730 years. Analysis of skeletal remains unearthed by anthropologists at an ancient North American Indian village indicates that the bones contain only one-third of the ^{14}C originally present (assumed to have been the same as in our bones today). Using figure A, box 5.1, estimate the age of the bones.

5 (*a*) Calculate the average densities of Mercury, Venus, Earth, and Mars, using the planets' masses and radii listed in table 4.1. Compare your answers with the actual densities of these planets given in the table on page 211.
(*b*) Why are the densities you calculated in (*a*) larger than the planets' uncompressed densities? (Note: the average density of a spherical body = $M/\left(\frac{4\pi}{3}R^3\right)$, where M and R are the body's mass and radius, and $\pi = 3.14$.)

6 Describe the physical and orbital characteristics of the Galilean satellites, and discuss in which ways they and Jupiter resemble the Solar System. In which important ways do they differ from the Solar System?

7 Name the planetary bodies of intermediate size that have the most impact craters, the greatest geologic activities, the most water (liquid or ice), and the thickest atmospheres.

8 Why is Earth a geologically very active planetary body, but Mars is not?

9 Why does liquid water cover roughly two-thirds of the Earth's surface, but today is absent on the surfaces of its sister planets, Venus and Mars?

10 What mechanisms contribute to the formation and retention of planetary atmospheres?

11 Why is there no planet at 2.8 AU from the Sun, as predicted by the Bode-Titius law?

12 Water is thought to have been a relatively abundant molecular component of the protosolar disk. That is apparently why Ganymede, Callisto, the comets, and many of the other bodies in the other part of the Solar System contain so much of it. Why did the bodies in the inner part of the Solar System, such as the terrestrial planets, end up with much less water?

13 How does the presence of life affect the Earth's physical environment? (Think, for example, of the composition of the Earth's atmosphere; the existence of coal and oil deposits; the impact of plants, animals, and microorganisms on our climate, makeup of lakes and other bodies of water, and land forms.)

For **Suggestions for Further Reading** and **Additional References** see chapter 4.

We are not alone in the universe, and do not bear alone the whole burden of life and what comes of it. Life is a cosmic event—so far as we know the most complex state of organization that matter has achieved in our cosmos. It has come many times, in many places—places closed off from us by impenetrable distances, probably never to be crossed even with a signal. As men we can attempt to understand it, and even somewhat to control and guide its local manifestations. On this planet that is our home, we have every reason to wish it well. Yet should we fail, all is not lost. Our kind will try again elsewhere.

George Wald (1954)
(*Professor Emeritus*, Harvard University)

Epilogue

"ARE WE ALONE in the Universe, or has life evolved elsewhere also?" Attempting to answer this question will be the final topic of this book. In particular, I shall attempt to estimate how frequently intelligent beings, capable of communicating across interstellar distances, might have evolved on other planets circling other stars. If they have, perhaps we can make contact with them and, thereby, join a larger, cosmic fraternity.

The topic of extraterrestrial life is not a new one. It was already being discussed during the first century B.C. by Titus Lucretius Carus, the Latin poet and philosopher, who had no doubt at all that life, including intelligent life, exists elsewhere in the Universe. Lucretius (see Goldsmith 1980, 4) justified his belief by a simple, yet very persuasive logical argument in his famous poem *De rerum natura* ("On the Nature of the Universe"):

> For since infinite space stretches out on all sides, and atoms of numberless number and incalculable quantity fly about in all directions quickened by eternal movement, it can in no way be considered likely that this is the only heaven and earth created, and all those other atoms there beyond are doing nothing. For this world was created by Nature after atoms had collided spontaneously and at random in a thousand ways, driven together blindly, uselessly, without any results, when at last suddenly the particular ones combined which could become the perpetual starting points of things we know—earth, sea, sky, and the various kinds of living things. Therefore, we must acknowledge that such combinations of other atoms happen elsewhere in the universe to make worlds such as this one, . . . with races of different men and different animals . . .

Today many scientists believe that life, including intelligent life with advanced technology (namely the ability to communicate across interstellar distances), is not unique to Earth. This belief is based on logical arguments that are in principle no different from that of Lucretius. The main difference is that today we express our reasoning in more quantitative terms, mainly because we know much more about the physical structure and content of the Universe as well as about life and its evolution.

However, we must be aware that quantitative estimates of the frequency of extraterrestrial life may be no more right than Lucretius's qualitative statements. They may, in fact, be very misleading because they might delude us into thinking that we know more than we actually do. The reason is that at present Earth is our only example of life's origin and evolution. We know of no other case either on other planetary bodies in our Solar System or elsewhere in the Universe. Therefore, in estimating how widespread life is in the Universe, we must extrapolate from our terrestrial biosphere to other places. Unfortunately, we have no way of knowing how right or wrong such extrapolations are because we have no observational basis to go by. The best we can do is to decide which developments on Earth were essential for the origin and evolution of life, and then to assign low probabilities to those that on our planet required the simultaneous occurrence of rather specialized conditions (for example, geologic and climatic conditions) and proportionately higher probabilities to those that were less constrained. This strategy may give us a reasonably accurate estimate, though there is no guarantee. We simply must accept that, given the limits of our current scientific understanding, this uncertainty cannot be avoided.

Given the present dearth of observational and experimental data concerning extraterrestrial life, we should not expect a consensus among scientists with regard to the probable frequency of such life. In fact, the published numerical estimates range over many orders of magnitude. The numbers I present here do not reflect the judgment of the scientific community as a whole; they merely represent my personal opinion and, possibly, say more about my beliefs and prejudices than about the actual evolution of extraterrestrial life. Because of these uncertainties, readers are as much entitled to their opinions on this subject as I am to mine or the "experts" are to theirs, and I invite them to substitute their own numerical estimates for mine and to draw their own conclusions from the final result.

BASIC ASSUMPTIONS

Before we begin calculating the probable frequency of extraterrestrial intelligent life with advanced technology, we must agree on a number of assumptions:

1. The fundamental unit of life is always the cell, except during the earliest stages of life's origin.
2. Life is never static, but always evolves, and does so as explained by Darwin and Wallace. This requires molecular information carriers and the passage of the information from one generation of organisms to the next, including the continual introduction of changes in the information through mutations and recombinations. The changes in information create diversity, upon which natural selection can act. Very likely, these processes require catalysts to control the rates of chemical reactions. We also assume that only organic chemistry offers such information carriers and catalysts, because we think only carbon bonded to hydrogen, nitrogen, oxygen, phosphorus, sulfur, and numerous other elements provides molecules of sufficient variety, stability, and chemical reactivity to do the job.
3. The chemical reactions of life require not only the ready availability of hydrogen, carbon, nitrogen, oxygen, and so forth, but also liquid water. This is certainly the case on Earth. The condition of liquid water places severe limitations on the environment in which life can arise. It requires a planet circling a star. Furthermore, it requires that the planet has roughly the same mass and composition as Earth, with a solid surface, volcanism, outgassing, and temperature-pressure conditions at which water condenses but does not freeze, at least not everywhere. The limitation on temperature, in turn, requires a stable planetary orbit of low eccentricity and in the right distance range from the central star.
4. The central star must be stable and radiate at a nearly constant rate for at least 3 to 4 billion years (that is, it has to remain on the main sequence for at least that long). Otherwise there probably would be insufficient time for intelligent beings to evolve. We further assume that the star radiates in a wavelength range that peaks neither in the ultraviolet nor in the infrared. Ultraviolet radiation would be too damaging to life, especially land-based life; and infrared radiation might not be sufficiently energetic to induce some of the

chemical reactions crucial during the earliest stages in the evolution of life, before efficient light-gathering (i.e., photosynthetic) molecules have evolved.

5. Intelligent beings capable of developing advanced technology are land-based and have passed through a tree-living phase in their earlier evolution. Only such a phase is likely to produce the large brain and grasping hands, including the ability to execute a precision grip, that appear to be essential for the evolution of such beings. Note that intelligence alone is not sufficient for the development of advanced technology that includes the ability to communicate across interstellar distances. Finely tuned manipulative skills are required as well, to handle the nuts and bolts, to build the transmitters and receivers of electromagnetic signals, and, in general, to assemble the hardware required for such a technology. Of the many millions of species of animals that have arisen on our planet, only among the primates have manipulative appendages evolved, namely our human hands, that are up to the task. For example, such appendages as the trunks of elephants, the eating instruments of insects and other arthropods, or the beaks of birds are not so capable, despite their many unusual and often surprisingly versatile uses. Nor, in my opinion, do these or other nonprimate appendages have much of a potential to evolve to a stage at which they could be used in the development of advanced technology.
6. Finally, let us agree to consider only our Galaxy in estimating the probability of finding extraterrestrial intelligent life. This will simplify the calculations. We assume that the result can be readily extended to other galaxies, provided they contain, besides hydrogen, sufficient amounts of carbon, nitrogen, oxygen, and other heavy elements.

Most scientists probably would agree that these assumptions are quite reasonable, though some might argue that we have set our limits too stringently. Maybe life does not necessarily have to be cellular or be based on organic chemistry, or maybe it can arise in places other than earthlike planets, such as in the upper atmospheric layers of jovian planets or in interstellar clouds. Maybe the manipulative skills required for the development of advanced technology might arise along evolutionary routes quite different from the one we assumed: However, at present we have no scientific basis to believe in such scenarios. Hence, I prefer to take the conservative approach and use the assumptions listed.

METHOD OF ESTIMATING THE PROBABILITIES

Next, let us discuss a method of estimating how frequently we expect that life with advanced technology has arisen on other planets circling other stars in our Galaxy. This method was first suggested in 1965 by Frank D. Drake of Cornell University, one of the pioneers in the search for extraterrestrial intelligence. The method involves two steps. In the first step, the values of several probabilities are estimated:

1. The probability that any given star in our Galaxy has planets suitable for supporting life.

2. The probability that on such planets life arises.
3. The probability that such life evolves into complex multicelled forms.
4. The probability that such evolution leads to intelligent beings with advanced technology.

In the second step, the four probabilities are multiplied together, which gives an estimate of the probability per star in our Galaxy that intelligent beings with advanced technology might arise. Multiplying this result by the number of stars in the Galaxy yields the number of times that technologically advanced civilizations are likely to have arisen in our Galaxy.

CALCULATING THE PROBABILITIES

We are now ready to begin the actual calculations.

Calculation 1: The Probability That Any Given Star in Our Galaxy Has Planets Suitable for Supporting Life

This probability is itself the product of three numbers, namely the fraction of stars in our Galaxy capable of sustaining life in their vicinity, the probability that such stars are circled by planets, and the probability that one or several of such planets are suitable for the origin and evolution of life.

For numerous reasons, not all stars are capable of sustaining life in their vicinity. First, none of the oldest stars in the Galaxy, such as those in globular star clusters, are good candidates. They are made of mostly primordial material (that is, hydrogen and helium) and contain very little carbon, nitrogen, oxygen, silicon, metals, or other elements needed for the making of planets with a solid surface and for the origin and evolution of life. Second, only about fifteen percent of all stars are single. The rest are members of binary or higher order systems, which reduces the probability of finding stable planetary orbits at suitable distances from those stars. Third, all stars more massive than about $1.2\,M_{\odot}$ or less massive than about $0.8\,M_{\odot}$ should be excluded. The more massive stars emit too much UV radiation, which would probably be harmful to life. Besides, such stars become red giants in fewer than 3 to 4 billion years, which allows insufficient time for intelligent life to evolve. Stars less massive than about $0.8\,M_{\odot}$ emit mainly red and IR radiation, which might be insufficiently energetic to induce some of the chemical reactions essential for the origin of life. Furthermore, such low-mass stars may have no continuously habitable zones in their vicinity. This means that planets circling them may not be able to maintain for any great length of time the mild climatic conditions that have characterized our planet. They probably either freeze over or heat up excessively due to runaway greenhouse effects (a fate that may yet befall Earth). Not all scientists agree with this assessment. Nevertheless, even if habitable zones surround low-mass stars, those zones would be quite close to the stars. Planets orbiting there would experience enormous tides, which would slow their rotation rates until they were synchronous with the orbital rates (that is, the durations of the planets' days and years would become equal, as has happened in the case of the Moon with respect to Earth). Such planets would then forever point the same hemispheres toward the central stars. These hemispheres would heat up and probably turn into permanent deserts, while the opposite hemispheres would be in darkness and freeze. If we take these three restrictions into account, it turns out that only *about two percent of all stars in our Galaxy are capable of sustaining life in their vicinity*.

Next let us estimate the probability that a given star is circled by planets. We have little firm observational data to go by in this estimate, for in the case of only two stars other than the Sun, namely Beta Pictoris and Van Biesbroeck 8 (see footnote on p. 248), do we at present have evidence that they might be circled by planets. And even that evidence is not fully established. This lack of firm observational evidence is almost certainly due to the great technical difficulties involved in detecting distant planets, rather than to their nonexistence. Judging from the preponderance of binary and higher order star systems, the fact that even in our Solar System there exist several "planetary" systems (for example, those of Jupiter and Saturn), and the discovery of rocky debris circling Vega, Fomalhaut, and Beta Pictoris, many astronomers think that perhaps all stars between 0.8 and 1.2 $M_{\odot}$, which are potentially capable of sustaining life in their vicinity, are surrounded by planets. Other astronomers are a bit more cautious and would reduce this estimate somewhat. I, too, shall be cautious and assume that *two-thirds of all stars between 0.8 and 1.2* $\mathrm{M}_{\odot}$ *are circled by planets*.

But not all planets are suitable for the origin of life. They need to satisfy the conditions described in the assumptions above. For example, massive planets like the Sun's Jovian gas giants or very small planets like Mercury and Mars will not do. Probably their masses have to lie within plus or minus a few ten percent of the mass of Earth, and their surface layers must contain water and other molecules rich in hydrogen, nitrogen, oxygen, silicon, phosphorus, sulfur, and various other kinds of elements. Only such combinations of mass and composition would give rise to a solid crust, volcanism, outgassing, oceans, and a suitable atmosphere. Furthermore, planets capable of supporting life must lie within the central star's habitable zone, which in the case of the Solar System extends at most from somewhere between 0.8 and 0.9 to between 1.2 and 1.3 astronomical units from the Sun. For example, Venus, which orbits at 0.72 AU from the Sun, developed a runaway greenhouse effect; Mars, which orbits at 1.52 AU, is frozen. Very likely, the planet also must rotate at a moderate rate, similar to Earth. If it rotates much more slowly, one hemisphere would heat up excessively; if it rotates much more rapidly, hurricanelike storms would blow almost incessantly in the lower atmosphere. Thus, the restrictions on a planet to be suitable for life are rather severe and, consequently, the probability that they are satisfied is quite small. How small is difficult to judge, given our current limited knowledge of planetary system formation and evolution. However, it appears that *the probability of a given star being circled by a planet suitable for supporting life cannot be assumed to be greater than about one-tenth*.

Multiplying together these three numbers (2/100, 2/3, and 1/10) gives 1/750. This means that according to my estimate *on the average one star out of every 750 in our Galaxy is surrounded by a planet suitable for supporting life. And because there are approximately 150 billion stars in our Galaxy, there may exist 200 million such planets*.

Calculation 2: The Probability That on Suitable Planets Life Actually Arises

We might be tempted to assume that if a planet is suitable for supporting life, life will inevitably arise. This may well be true. Nevertheless, our understanding of the processes involved in the origin of life is so limited that we cannot be certain of this conclusion. We simply don't know under what range of conditions, or sequences of conditions, life can come into existence. For example, can life originate on a planet that is entirely covered by water (perhaps to depths of several kilometers), without any islands or continents; or are shorelines needed? What if there is much less water than on Earth, collected in a few salty little pools, and the rest of the planet is arid land? (Either of these conditions might arise just as likely as those on Earth, which has two-thirds of its surface covered by water, see p. 229.) Does the planet have to be circled by a massive satellite to cause the daily rising and falling of large (but not too large) tides? Does the origin of life depend on special atmospheric conditions or on particular seasonal variations in its climate? Are, perhaps, unusual clay crystals needed to act as catalysts during prebiotic chemical synthesis? We don't know the answer to any of these questions. Besides, there are probably many other questions of this kind that we don't even know how to ask.

Thus, we cannot simply take it for granted that life will inevitably arise on all planets that are suitable for supporting life once it has arisen. Put differently, we cannot assume that time will always be "the hero of the plot," as George Wald suggested in his discussion of terrestrial prebiotic evolution (Wald 1954, 48). On some planets, time alone may not be able to bring about the origin of life.

Given these uncertainties, but also recognizing that life did arise on at least one planet, namely Earth, I shall assume that the probability of this happening on other similar planets is *one-fifth*. Some biologists would assign a much lower value to this probability, while others might set it closer to one. At present, we have to accept this uncertainty. Multiplying this probability by the previously derived number of suitable planets in our Galaxy yields *40 million. This is the number of planets in our Galaxy on which I expect that life has actually arisen.*

Calculation 3: The Probability That Life, Once It Has Arisen on a Planet, Evolves into Complex Multicelled Forms

Judging from the events as they occurred on Earth, it seems very probable that once life has arisen on a planet competition for the limited resources available will sooner or later lead to the development of photosynthesis and respiration (both anaerobic and aerobic). Of course, aerobic respiration would evolve only after molecular oxygen had begun to accumulate in the planet's atmosphere and oceans, due to its release by aerobic photosynthesis. With the accumulation of molecular oxygen, it seems probable that eukaryotic cells as well as mitotic cell division would evolve. If this happened, meiosis and sexual reproduction could evolve (or some other, comparable mechanism that would shuffle the genetic information of two, or possibly more, parents during reproduction). It is then very likely that complex multicelled life will arise.

As simple as it sounds, this sequence of events is not easily accomplished. It takes time. On Earth it took more than 3 billion years, from roughly 4 billion to somewhat less than 1 billion years ago (see figure E2). Would it also be just a matter of time on other planets, or might there arise insurmountable obstacles on some of them? We don't really know. But it seems unlikely that such obstacles would arise often. As George Gaylord Simpson (1967, 242), the noted American paleontologist and evolutionist, put it, there exists " . . . a tendency for life to expand, to fill in all the available spaces in the livable environments, including those created by the process of that expansion itself." (Simpson meant here life as a whole, not just particular groups or species of organisms.) Expressed differently, natural selection will probably always force life to evolve in directions that allow it to make use of the widest possible resource base, including the resources created by the process of evolution itself. That would probably be an evolution toward eukaryotic cells, as well as toward sexual reproduction and complex multicelled forms. I shall assume that this kind of evolution takes place approximately *three-fourths* of the time.

We must still take into account the fact that only a fraction of the planets in our Galaxy on which life may have arisen are roughly 3 to 4 billion years old (or older), which we assumed is the time required for the evolution of complex multicelled and, possibly, intelligent life. Very approximately this fraction is *two-thirds*. Multiplying together this number, the number three-fourths, and my earlier estimate of the number of planets in our Galaxy on which life may have arisen gives *20 million. This is the number of times I expect that complex multicelled life has evolved in our Galaxy*.

Calculation 4: The Probability That Complex Multicelled Life Evolves into Intelligent Beings with Advanced Technology

Traditionally, most physical scientists who have concerned themselves with the issue of extraterrestrial life have tended to oversimplify the issue of the probability that complex multicelled life would give rise to intelligent beings.* Their reasoning has essentially been as follows: Life always evolves from the simple toward the complex; intelligent beings are the most complex of all; therefore, the probability is high that intelligent beings, including ones capable of developing advanced technology, will evolve in time.

This argument is flawed on two accounts. First, life does not always evolve toward greater complexity. It evolves in directions that allow it to exploit the widest possible resource base, as noted above. If increased complexity in certain species helps to accomplish this, then evolution will progress in this direction;

*Since Frank Drake published his now famous equation (the probability formula we are using), the main users of it have been astronomers and other physical scientists. Relatively few biologists, anthropologists, and social scientists have, even though their expertise is very pertinent to the calculation of the probability that life will originate and evolve on other planets circling other stars.

if it does not, then different evolutionary routes will be followed. For instance, quite often complexity has decreased in the course of evolution, as in the case of certain cave fishes whose eyes have become nonfunctional, primates whose sense of smell has been reduced, cetaceans whose limbs have evolved into tails and flippers (which are anatomically simpler than walking legs), and certain mollusks who have lost their ancestral shells. Furthermore, the complexity of bacteria does not appear to have increased measurably during the past billion or so years, even though these organisms have been evolving and adapting to their changing environments.

Second, implicit in the above argument is the assumption that intelligence is generally an advantage in the struggle for survival and, hence, is selected for in the course of evolution. This is not always the case. For instance, whales and dolphins are the most intelligent animals in the sea, but by number they constitute only a minuscule fraction of all large marine animals. Elephants, too, are very intelligent. Yet today only two species survive of a once much larger order. In contrast, worms, arthropods, and fishes, which are not characterized by great intelligence, have been enormously successful, if we accept number and longevity of phylogenetic lineages as measures of success.

It appears that high intelligence is an advantage in the competition for survival mainly if it is combined with suitable physical traits allowing the animals to make effective use of it. For example, in humans the combination of intelligence together with the precision grip of our hands, our bipedalism (which frees the hands for uses other than locomotion), and the wide range of our vocalization abilities (which is crucial for language development) has made us successful. In fact, no other terrestrial animal, past or present, comes even close to matching us or our *Homo* ancestors (such as *H. habilis*, *H. erectus*, and the Neanderthals) in intelligence and its effective application.

Thus, it seems not justified to assume *a priori* that complex multicelled life will necessarily, or very likely, give rise to intelligent beings capable of developing advanced technology. If we use our terrestrial experiences as a guide (see Assumption 5 above), only evolution on land, including a tree-living phase, is likely to accomplish this.

In order to estimate the probability that complex multicelled life will evolve into intelligent beings with advanced technology, following approximately the route we assumed, let us begin by listing those developments that were central to this evolution on Earth (also see Table E1):

1. Emergence of the chordate phylum during the early diversification of complex multicelled life, with members characterized by an internal skeleton and specialized for a swimming mode of life.
2. Evolution of several distinct lineages among the early chordates, including one—the vertebrates—that allowed some of its members successfully to invade the forest environments that were spreading across the land from the middle of the Paleozoic Era onward.
3. Evolution among the terrestrial vertebrates of a wide range of adaptations to life in their new habitat, including a stronger internal skeleton suitable for quadrupedal locomotion on land, sense organs adapted to life in the air, changes in jaw structure and dentition, warm-bloodedness, a four-chambered

heart, enlargement of the cerebrum, and new reproductive strategies (for example, mammary glands, placenta).

4. Evolution of terrestrial flowering trees and the successful invasion by several groups of vertebrates of the arboreal habitats thus created, among them animals that evolved into the primate order.
5. Evolution among the primates of prehensile hands and feet (including the precision grip in some of them), excellent eye–hand coordination, a wide range of specialized arboreal locomotions (including quadrupedal walking on branches, brachiation, vertical clinging and leaping, and vertical quadrupedal climbing), increased intelligence, and reproductive strategies characterized by the intensive caring for a small number of offspring.
6. Periodic readaptation of some species of primates to terrestrial living, including one group, the australopithecines, that was preadapted for upright, bipedal walking when, some 8 to 5 million years ago, it came out of the trees and onto the ground.
7. Reliance of one group of australopithecines on their intelligence for survival and their evolution into the *Homo* lineage (figure E3). This evolution was characterized by an extraordinarily rapid increase in brain capacity (particularly of the cerebral cortex) and intelligence, and by the development of tool use and language.

The probability that these or roughly similar developments take place on a planet on which complex multicelled life has arisen is very difficult to estimate. To come anywhere close to estimating it realistically, we must recognize that each of the key developments just described depends on the simultaneous occurrence of sometimes rather stringent geologic, climatic, and biologic conditions. For example, life can invade land only if the planet has islands or continents and is not entirely covered by water. Extensive diversification of land animals probably occurs only if the landmasses have sizes as large or nearly as large as our continents (only then is the competition among the species sufficiently intense) and are not just groups of small islands. Flowering trees suitable for arboreal animal life will probably evolve only if the climate is warm and wet. The return of arboreal animals to the ground as upright, bipedal walkers will probably take place only if the forests change to open savannah *and* simultaneously some of the animals have become preadapted to bipedalism (a condition that is not found among any of the nearly 200 species of today's nonhuman primates, at least not to the degree that it existed in the australopithecines). Finally, the evolution of bipedal land animals into humanoids (beings resembling those of the genus *Homo* on Earth) requires suitable ecological conditions in which reliance on intelligence proves to be advantageous and, hence, is selected for (which did not happen in the cases of *A. robustus* and *A. boisei*). Still other factors that might affect the course of evolution are such random and largely unpredictable events as episodes of severe earthquake activities and volcanism, asteroid or comet impacts, fluctuations in climate, mass extinctions, and the chance evolution of organisms capable of impeding or, perhaps, even terminating the sequence of developments outlined above (such as the evolution of bacteria or viruses that could

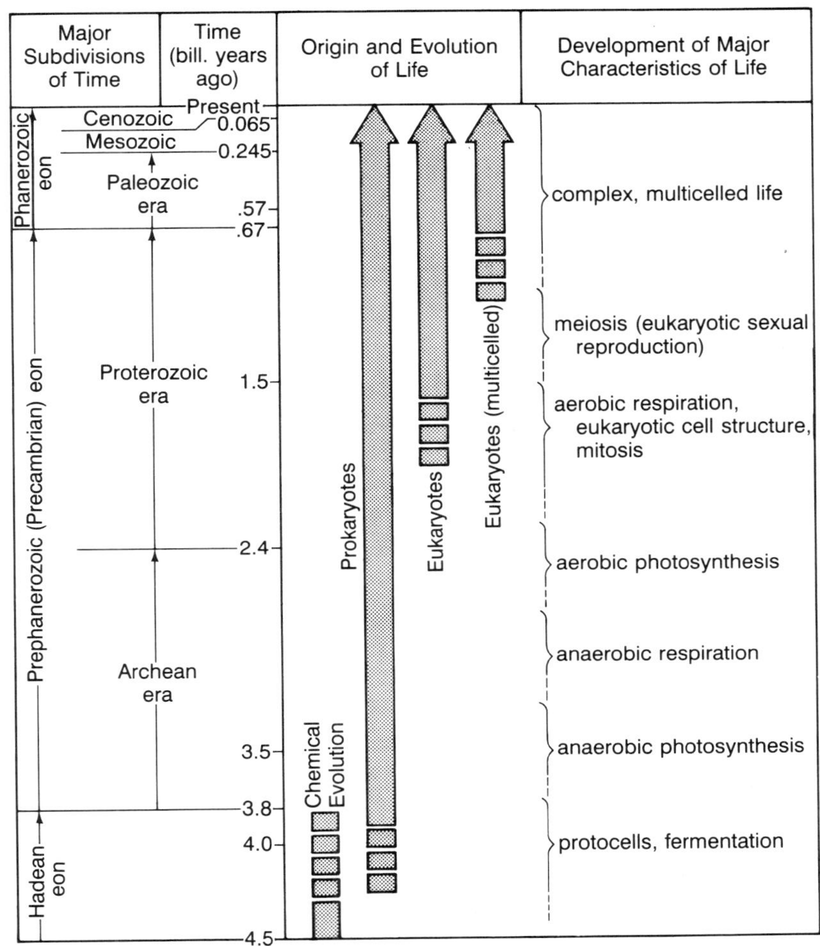

Figure E2. Evolution of prokaryotic and eukaryotic life. (For finer subdivisions of the Phanerozoic eon see table E1.)

cause fatal infections among animal species, or of predators that keep bipedal arboreal animals from moving onto the ground).

Only developments *1*, *3*, and *5* in the above list appear to have a good chance of occurring on a planet on which complex multicelled life has arisen, though even in their cases we cannot be entirely certain. Thus I shall assign the value one-half to their probabilities. Developments *2*, *4*, and *7* seem to have somewhat smaller chances of taking place, mainly because they depend more strongly on suitable geologic, climatic, and biologic conditions. I shall assign the value one-fifth to each of them. Development *6* has the smallest probability of all to be repeated on another planet because it depends most strongly on the simultaneous occurrence of rather specific geologic, climatic, and biologic conditions. Therefore, I shall assign to it a probability of 1/100.

Multiplying these seven numbers together gives 1/100,000 which is my estimate of the probability that complex multicelled life evolves into humanoids possessing language and tool-making abilities. This is the most uncertain of the numbers derived. Some might think it is too small and, perhaps, should be closer to 1/10,000 or even larger. But most paleontologists and primatologists

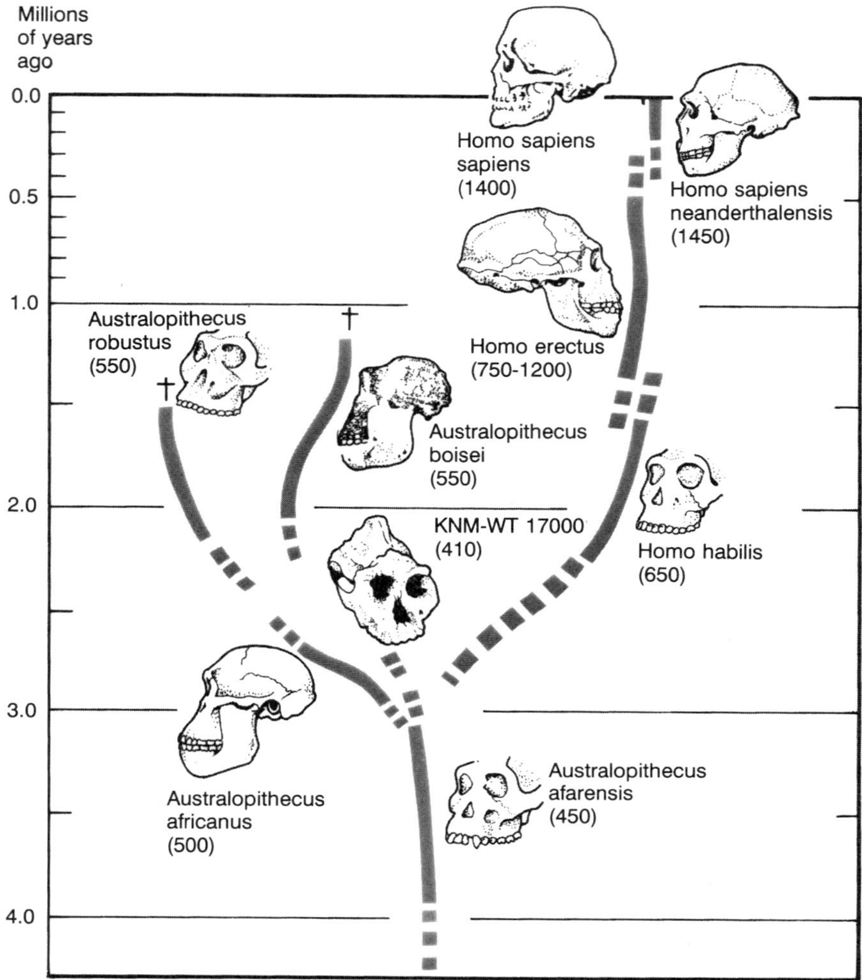

Figure E.3. The evolution of the hominids during the past 4 million years. The dashes signify uncertainties in our understanding of the phylogenetic relationships. Estimates of average brain sizes in cubic centimeters are given in parentheses.

would probably agree that at best it is an upper limit and that a value ten or a hundred times smaller is more realistic. For lack of any observational evidence to guide us, let us work with the value I have derived, acknowledging, however, the great range of uncertainty inherent in it.

To complete this calculation, we still need to multiply this number by the probability that humanoids will, in fact, develop advanced technology. At present we have little understanding of the range of conditions under which this kind of evolution will occur. Surely, geologic, climatic, biologic, as well as historic conditions affect it. Perhaps, the best we can do is to assume that conditions that made possible the origin of humanoids would probably also favor the subsequent evolution toward civilizations with advanced technology. Accepting this assumption as likely, I shall assign the value one-half to the probability that such an evolution takes place. Multiplying this number by 1/100,000 gives *1/200,000*, which is *my estimate of the probability that complex multicelled life evolves into beings with advanced technology*. Finally, multiplying this probability by 20 million, which is my earlier estimate of the number of times complex multi-

Table E.1. The Geologic Time Scale of the Phanerozoic Eon

Era	Period	Epoch	Duration of epochs (10^6 years)	Duration of periods (10^6 years)	Millions of years ago (from start of period)	
Cenozoic (65 million years duration)	Quaternary	Recent	0.01	2	2	Rapid evolution of hominids (human beings), rise of civilizations
		Pleistocene	2.0			
	Tertiary	Pliocene	3	63	65	Widespread diversification of mammals (beginning of Age of Mammals), birds, insects, and flowering plants
		Miocene	20			
		Oligocene	13			
		Eocene	17			
		Paleocene	10			
		Major mass extinction, including the disappearance of the dinosaurs				
Mesozoic (160 million years duration)	Cretaceous			80	145	Major diversification of modern bony fishes; first placental mammals; end of the Age of Reptiles; second great diversification of insects; first flowering plants
	Jurassic			65	210	Flourishing of dinosaurs; first birds
	Triassic			35	245	First dinosaurs and mammals; beginnings of Age of Reptiles; first coniferlike plants
		The most severe mass extinction in Earth history				
Paleozoic (345 million years duration)	Permian			40	285	Major diversification of reptiles; decline of amphibians; last of the trilobites and placoderms
	Carboniferous[a]			75	360	Major diversification of amphibians (Age of Amphibians); first reptiles; first great diversification of insects; great coal-forming forests of fernlike plants
	Devonian			50	410	Major diversification of fishes (Age of Fishes); invasion of land by vertebrates; first amphibians; first insects and seed plants
	Silurian			30	440	First fishes with jaws (acanthodians); invasion of land by plants and by arthropods
	Ordovician			65	505	Origin and expansion of many invertebrate classes

Note: All ages are approximate.

[a]In North America, the Carboniferous period is often divided into two separate periods: the Mississippian period (40 million years duration), corresponding to the Lower (earlier) Carboniferous, and the Pennsylvanian period (35 million years duration), corresponding to the Upper Carboniferous.

continued

Table E.1. continued

Era	Period	Epoch	Duration of epochs (10^6 years)	Duration of periods (10^6 years)	Millions of years ago (from start of period)	
	Cambrian			65	570	First widespread appearance of organisms with hard skeletons; final stages of the major diversification begun during the Ediacarian period; first vertebrates (jawless fishes)
		Transition from soft-bodied organisms to organisms with hard skeletons				
—	Ediacarian			100	670	Earliest known fossil evidence of soft-bodied, multicelled marine organisms; rapid evolution of the major body plans of the animal phyla

celled life has evolved in our Galaxy, yields the number *100. This is how many times, according to my estimate, civilizations with advanced technology are likely to have evolved in our Galaxy.*

What does my final result mean? It is the product of probability estimates obtained from data belonging to the fields of astronomy, geology, biochemistry, microbiology, paleontology, primatology, anthropology, and history. All of the estimates are based on extrapolations from our terrestrial environment and experiences to other places and contain progressively greater uncertainties as we come closer to *Homo*-like beings. Clearly, little trust can be placed in the final result, as I warned at the outset. Certainly answers of 10,000 or 1 are just as defensible as mine. Some physical scientists even think that an answer in the millions is not out of the question, but biologists tend to be more conservative in this regard. Most of those who have concerned themselves with this topic believe that an answer of 100, like mine, constitutes at best an upper limit of the number of times intelligent life with advanced technology is likely to have evolved in our or other similar galaxies, and that very probably a value smaller by several orders of magnitude lies closer to the truth.

The biologists' view is generally based not on multiplying together a string of probabilities, but on the realization that evolution takes place in the context of ecosystems, all of whose countless components—both biological and physical—are evolving and continually affecting each other. To a large extent this evolution is a statistical process and, hence, at least in detail is neither predictable nor repeatable . Should life start over again on Earth, with identical conditions, the probability that any particular species alive today would evolve again is essentially zero, and that surely includes *Homo sapiens*. This is why George Gaylord Simpson (1964, 267) wrote, "This essential nonrepeatability of evolution on earth obviously has a decisive bearing on the chances that it has been repeated or closely paralleled on any other planet. The assumption, so freely made by astronomers, physicists, and some biochemists, that once life gets started anywhere, humanoids will eventually and inevitably appear is plainly false."

However, neither Simpson nor other responsible biologists deny categorically that intelligent beings, including ones capable of developing advanced technology, could not have arisen elsewhere. Such beings might be quite different from us and have evolved along quite different routes than we assumed. This sentiment of caution has been clearly expressed by Francisco J. Ayala and Theodosius Dobzhansky (Gould 1983, 65): "Granting that the possibility of obtaining a man-like creature is vanishingly small even given an astronomical number of attempts . . . there is still some small possibility that another intelligent species has arisen, one that is capable of achieving a technological civilization."

Keeping in mind the range of opinions that exist on this subject and the inherent uncertainties, and also remaining true to my own rather conservative tendency on such matters, I would interpret the numbers derived above as follows:

> Life is not uncommon in the Universe. It probably has arisen tens of millions of times—perhaps even hundreds of millions of times—in our Galaxy alone. In many cases, life probably managed to evolve into complex multicelled forms. In some cases, evolution probably continued and produced ecosystems comparable to ours in complexity and biological diversity. But probably only in the rarest of instances did evolution lead to intelligent beings with culture and advanced technology. It may have happened a hundred times in our Galaxy, though this is likely too optimistic an estimate. If we also keep in mind that, like all biological species, beings with intelligence probably endure for only a limited stretch of time (particularly if they have the technology to destroy themselves), it appears conceivable that we are the only civilization with advanced technology presently existing in our Galaxy. However, just as likely, given the billions of galaxies that populate the observable part of the Universe, we are not entirely alone. Surely our kind has evolved, and survived, elsewhere also.

If this interpretation is correct, then the probability that we will make radio contact with another civilization, separated from us by interstellar or perhaps intergalactic distances, is rather slim indeed. Still, if we have learned anything at all from the study of science, it is to expect the unexpected. Maybe by pure chance there presently exist intelligent beings on another planet circling a nearby star, sending out messages just in case someone is listening. And because we have the ability to listen, without undue expense compared to many of our other technological undertakings, it surely is worth the effort. In fact, radio astronomers in the United States and abroad have been doing this since 1960, when Frank Drake first pointed the 26-meter radio telescope of the U. S. National Radio Observatory in Green Bank, West Virginia, at two nearby sunlike stars, Epsilon Eridani and Tau Ceti, and listened for signals bearing the signature of another civilization. But, to date none of these attempts has been successful.

Should we some time in the future detect such signals, it may well turn out to be "the most cataclysmic event in our entire intellectual history," as Stephen Jay Gould (1983, 65) so aptly put it. We would then know for certain that we are not alone. This might give us hope for our own future, for chances are good that any extraterrestrial civilization with whom we establish communication is technologically more advanced than we. And if they were able to survive the dangers that seem to accompany the early years of technological development, so can we. We would also then know that "we do not bear alone the whole burden of life and what comes of it" (Wald 1954, 53). Perhaps this thought might allow us to take ourselves less seriously and to be less self-centered. It might even motivate us to put more of our talents and intelligence into solving some of the social, ecological, and political problems we have brought upon ourselves.

But even if we should fail to make contact, benefits would still be reaped from a search for extraterrestrial civilizations. Like all large scientific enterprises, such a search would probably be of international scope and, thereby, contribute to bringing nations of different political ideologies closer together. Furthermore, in its broadest sense such a search is a very interdisciplinary endeavor and would promote the lowering of some of the barriers that have arisen among physical, biological, and social scientists. Finally, like all other great scientific undertakings of the past, a search for extraterrestrial civilizations is bound to yield unanticipated discoveries and deepen our understanding of the Universe at large as well as of ourselves and our terrestrial environment. This kind of understanding can only help to enrich our lives now and in the future.

Definitions of biochemical, biological, and anthropological terms used in the Epilogue

Australopithecines. Members of an extinct genus that lived approximately 4 million to nearly 1 million years ago in Africa and from which the genus *Homo* evolved. The known members of the australopithecines include *A. afarensis*, *A. africanus*, *A. robustus*, and *A. boisei*.

Biological evolution according to Darwin and Wallace. Biological evolution refers to the process by which the genetic information of populations of organisms and the phenotypic expression of this information changes with time. According to Darwin and Wallace, evolutionary change is the result of natural selection and takes place in two steps, one random and the other specific:

1. *Creation of genetic variation (random process).* From generation to generation the genetic information of a population is altered by mutations and by recombinations during sexual reproduction, creating a large degree of genetic variation among the members of the population.
2. *Selection dictated by the requirements of the physical and biological environment (specific process).* Because of genetic variation, the members of a population are not all equally competent in surviving and reproducing under the prevailing conditions of their environment. Some are more likely than others to pass their particular genes on to the next generation.

The second step acts like a filter on the reservoir of genetic variation created by the first step, rejecting individuals with inadequate competence in the struggle for survival and passing those best adapted to the current conditions. The physical and biological conditions of the environment change with time. Therefore, the selection criteria change also. The result is biological evolution.

Chordates. A phylum of the animal kingdom, whose members possess, at least during some stage of their development, a notochord, a dorsal nerve chord, and pharyngeal gill slits. One subphylum of the chordates is the vertebrates, which include the fishes, amphibians, reptiles, birds, and mammals. In adult vertebrates, the notochord, nerve chord, and gill slits have evolved into an internal skeleton, central nervous system, and (in the case of most of them) jaws and teeth.

Eukaryotic cell. A cell possessing a true nucleus, which is a membrane-bounded body containing the cell's genetic material (*i.e.*, the chromosomes). The cells of all animals, plants, fungi, and protists are eukaryotes.

Fermentation. Cellular metabolism by which energy is derived from the biochemical breakdown of complex organic molecules (*e.g.*, glucose) into simpler organic molecules (*e.g.*, lactic acid, ethanol). Yeast cells and many kinds of bacteria use fermentation. The earliest cellular organisms on Earth are believed to have been fermenters.

Genotype. An organism's full set of genes.

Hominids. Bipedal primates, including the australopithecines and members of the genus *Homo*. The hominids are believed to have evolved from arboreal (*i.e.*, tree-living) ancestors between 8 and 5 million years ago.

Homo. The genus that includes us (*H. sapiens*) and our immediate extinct ancestors (*H. habilis* and *H. erectus*).

Meiosis. A process of eukaryotic cell division, producing reproductive cells (*e.g.*, ova and sperm). During meiosis, each reproductive cell receives *half* as many chromosomes as are present in the parent cell. Meiosis is an essential component of sexual reproduction.

Mitosis. A process of eukaryotic cell division, in which the nucleus divides and a *full* complement of the cell's genetic material (*i.e.*, the chromosomes) is passed on to each offspring cell. Mitosis is the cell division by which our bodies grow and maintain themselves.

Organic chemistry. The chemistry involving carbon *and* hydrogen, though other elements may be present also. Because of carbon's ability to readily form single, double, and triple bonds, organic molecules assume a greater variety of different forms and are able to participate in complex, self-regulating chemical reactions. That is why organic chemistry is the chemistry of life. (All chemistry not involving carbon *and* hydrogen is called inorganic chemistry.)

Phenotype. The physical or visible expression of an organism's genotype.

Photosynthesis. The synthesis of organic compounds through the use of energy obtained from light. Examples of photosynthetic organisms are the green plants, some protists (*e.g.*, algae), and many of the bacteria (*e.g.*, blue-green bacteria).

Phylogeny. The evolutionary history of organisms or groups of organisms.

Precision grip. The grip by which we pick up and handle small or delicate objects, such as when picking up a penny or writing with a fountain pen. The precision grip is achieved by moving the thumb towards the fingers. Old World monkeys, apes, and humans possess the precision grip, but it is most highly developed in our species. (The precision grip is to be contrasted to the power grip, which we use when picking up a shovel or an ax.)

Prokaryotic cell. A cell that has no true nucleus and no nuclear membrane. Bacteria are prokaryotes. The first cellular organisms on Earth were prokaryotes.

Respiration. Cellular metabolism by which energy is derived from the biochemical reactions between organic (*e.g.*, glucose) and inorganic (*e.g.*, O_2, $SO_4^{=}$, NO_3^{-}, $CO_3^{=}$) molecules. If O_2 is involved, respiration is aerobic; otherwise it is anaerobic. All eukaryotes and many prokaryotes are capable of respiration.

Vertebrates. See definition of *Chordates*.

Suggestions for Further Reading

Ball, J. A. 1980. "Extraterrestrial Intelligence: Where Is Everybody?" *American Scientist*, November–December, 656–63.

Barrow, J. D., and **F. J. Tipler**. 1985. *The Anthropic Cosmological Principle*. Fair Lawn, N.J.: Oxford University Press. The anthropic cosmological principle, based in Protagoras's assertion that "man is the measure of all things," has been interpreted more recently to mean that life of any sort is impossible unless the basic laws of nature are exactly as they are. Barrow and Tipler discuss four versions of the anthropic cosmological principle. This gets them deeply into the underlying philosophical principles of quantum mechanics and cosmology, the question of the existence of a Creator, and the problem of extraterrestrial life, whose existence they deny. The book is controversial, but worth reading for anyone with patience and a philosophical bent. (Also see "The Anthropic Principle" by G. Gale, referenced in chapter 1).

Billingham, J., ed. 1981. *Life in the Universe*. Washington, D.C.: Scientific and Technical Information Branch, NASA. A collection of reports presented at a 1979 conference held at the NASA Ames Research Center (Moffett Field, California) on the broad question of the evolution of life on Earth and possibly elsewhere, and on the possibility of detecting the existence of extraterrestrial life.

Bracewell, R. N. 1975. *The Galactic Club: Intelligent Life in Outer Space*. San Francisco: W. H. Freeman and Co. An excitingly written little book about the possible existence of intelligent civilizations beyond the Solar System and our joining this "galactic club."

Cameron, A. G. W., ed. 1963. *Interstellar Communication: A Collection of Reprints and Original Contributions*. New York: W. A. Benjamin. Though published more than 20 years ago, many of the 32 articles in this collection—all concerned directly or indirectly with the question, "How many technologically advanced societies exist in our galaxy and how can we communicate with them?"—are still timely and worth reading.

Feinberg, G., and **R. Shapiro**. 1980. *Life Beyond Earth: The Intelligent Earthling's Guide to Life in the Universe*. New York: William Morrow and Co. The nature of life, its possible forms, and its likely distribution in the Universe.

Goldsmith, D., comp. 1980. *The Quest for Extraterrestrial Life: A Book of Readings*. Mill Valley, Calif.: University Science Books. A collection of 58 articles by authors both modern and from the past about the possibility of the existence of extraterrestrial life.

Goldsmith, D., and **T. Owen**. 1980. *The Search for Life in the Universe*. Menlo Park, Calif.: Benjamin/Cummings Publishing Co.

Gould, S. J. 1983. "The Wisdom of Casey Stengel." *Discover*, March 1983, 62–65. A critique of probability estimates of the existence of extraterrestrial life, focusing on the repeatability of "humanoids" elsewhere and on the much more general issue of the repeatability of "intelligence."

Milne, D., et al., eds. 1985. *The Evolution of Complex and Higher Organisms.* Washington, D.C.: Scientific and Technical Information Branch, NASA. A report prepared by the participants of workshops held in 1981 and 1982 at the NASA Ames Research Center (Moffett Field, California) on the likelihood that extraterrestrial events played a role in the development of complex life on Earth and on whether knowledge of such events could contribute to a better understanding of the nature and distribution of complex extraterrestrial life.

Morrison, P., **J. Billingham**, and **J. Wolfe**, eds. 1977. *The Search for Extraterrestrial Intelligence (SETI)*. Washington, D.C.: Scientific and Technical Information Branch, NASA. A report from a series of workshops sponsored by the NASA Ames Research Center (Moffett Field, California). For advanced students.

O'Neill, G. K. 1976. *The High Frontier: Human Colonies in Space*. New York: William Morrow and Co.

Papagiannis, M. D., ed. 1980. *Strategies for the Search for Life in the Universe*. Dordrecht, Holland: D. Reidel Publishing Co. A collection of reports presented during the International Astronomical Union General Assembly (Montreal, Canada, 1979). For the advanced student.

——— ed. 1985. *The Search for Extraterrestrial Life: Recent Developments*. Dordrecht, Holland: D. Reidel Publishing Co. A collection of reports presented at the 112th symposium of the International Astronomical Union (Boston University, 1984). Topics include the history of the search for extraterrestrial life, current search efforts, extraterrestrial organic matter, universal aspects of biological evolution, and space colonization.

Sagan, C., and **F. Drake**. 1975. "The Search For Extraterrestrial Intelligence." *Scientific American*, May, 80–89.

Shapley, H. 1964. *Of Stars and Men: The Human Response to an Expanding Universe*, revised edition. Boston: Beacon Press. A beautiful little book, in an easy, popular style, by one of the great astronomers of the twentieth century. Though many of the facts presented are now dated, the book still lives up to its intended purpose: "To present some . . . ideas, new and old, bearing on the position of mankind in the universe of physics and sensation; [and to write] a tentative obituary . . . of anthropocentrism in our description of the universe."

Shklovskii, I. S., and **C. Sagan**. 1966. *Intelligent Life in the Universe*. San Francisco: Holden-Day. A classic on the topic of life on Earth and the possibility of its existence beyond, by two eminent scientists, one Russian and the other American. Though much of the information in this book has been superseded by new discoveries, the spirit of adventure in which it was written has not.

Additional References

Lucretius (1st century B.C.). 1980. *De rerum natura*, trans. Mary-Kay Gamel Orlandi. In D. Goldsmith, ed., *The Quest for Extraterrestrial Life: A Book of Readings*, 4. Mill Valley, Calif.: University Science Books.

Simpson, G. G. 1967. *The Meaning of Evolution: A Study of the History of Life and of Its Significance for Man*, rev. ed. New Haven and London: Yale University Press.

——. 1964. *This View of Life: The World of an Evolutionist*. New York: Harcourt, Brace & World.

Wald, G. 1954. "The Origin of Life." *Scientific American*, August, 44–53.

Author Index*

*Numbers in italics refer to figures and tables; numbers followed by *n* refer to footnotes.

Subject Index*

*Numbers in italics refer to figures and tables; numbers followed by *e* refer to exercises; numbers followed by *n* refer to footnotes; numbers followed by *r* (*author's name*) refer to Suggestions for Further Reading.